Peter Sprent

Mathematics Without Fears

o o o o o o o o o o o o o o o o o o o o o o o o o o o o o o o o o o o o o o o o o o o o o o o o o o o o o o o o

First published 2009

ISBN 978-1-4092-5670-0

To minimize storage and distribution costs so as to ensure availability to as wide a readership as possible at a reasonable price, this book is available only on a print-on-demand basis either directly from the publishers lulu.com, or by ordering through a bookstore. Purchases may be made from the lulu.com website

www.lulu.com

o o o o o o o o o o o o o o o o o o o o o o o o o o o o o o o o o o o o o o o o o o o o o o o o o o o o o o o o

Preface

The last few decades have been eventful for mathematics and for mathematicians. The subject, and how it is taught, has changed dramatically. Computers carry out in hours, minutes or even seconds, calculations once impossible because they would have taken several lifetimes. New technologies have stimulated fresh approaches.

Mathematics has long had a role in the natural and physical sciences, engineering, technology and economics. Applications are now common also in medicine, agriculture, psychology and the social sciences, in industry, commerce and business — even in areas as diverse as the law, forensic science, the arts, and the entertainment industry.

A growing need for mathematical expertise is allied to a decline in the number of people with the required skills. This inbalance is of increasing concern to governments and to public and private sector employers.

The decline in numbers studying mathematics to University entrance level and beyond is often attributed to the notion that mathematics is inherently difficult and requires special aptitudes. This has only a grain of truth. Fear of mathematics more often has its roots in misunderstandings about the nature of the subject.

Computers have taken drudgery out of tedious arithmetic calculations, reducing the need for certain more or less mechanical mental skills. They have not removed a need for mathematical competence. An understanding of the basic properties of numbers helps us avoid the nonsenses that stem from a computer's inability to think. Errors in data entry, for example, often go unnoticed, yet many of these may be detected by a glance at the output if one has a good 'feel' for numbers. This comes only with understanding.

Major changes in curricula and in teaching methods reflect the new skills needed to apply mathematics in other disciplines. Yet, despite the many new developments, the total time now devoted to mathematics in schools is often less than it was when a classic overview of the subject by Richard Courant and Herbert Robbins, *What is Mathematics?* — a favourite of countless mathematicians — first appeared in 1941.

Like Courant and Robbins, I assume that most readers will have a basic knowledge of mathematics as met in secondary school or college. My overview is less sophisticated than that by Courant and Robbins, but I have introduced topics such as *operations research* and *chaos* that were nonexistent, or virtually so, 60 years ago.

The goal is to help senior secondary school pupils and first-year University or College students who are either contemplating a career in Mathematics, or who need it as a tool in other disciplines.

I have much sympathy for those who need to study the subject beyond the basic level, but are apprehensive about doing so. The problems they face are not always well understood. For such people I want to show that the subject has a lot to offer, even though worthwhile achievement calls for hard work. For those approaching further studies of mathematics with enthusiasm, I try to provide a helpful insight free from the many pressures implicit in a crowded curriculum.

Mathematics (and mathematicians) can sometimes be irritating, seeming to be over-pedantic in demands for rigour. It can be infuriating when intuition leads one astray. The other side of the coin is the elation one feels when intuition guides one along a fruitful path, the satisfaction when a rigorous argument produces a fascinating, or unexpected, result.

A key aim has been to show links between various strands of the subject. These links are often hidden at the elementary level because each branch is presented in a more-or-less watertight compartment. Fortunately, this tendency is disappearing in the better modern syllabuses.

I hope also that teachers of mathematics, from primary school to college or university undergraduate level, will find the book a source of ideas for presenting a subject that has brought me considerable pleasure — and only occasional heartaches — over many years.

Peter Sprent
Wormit, Fife
January, 2009

How to Read Mathematics

To get best value from a book on mathematics — both information and pleasure-wise — needs a different approach from that for reading a novel, short story, biography, account of an historical event, or even textbooks on other aspects of science.

Opinions differ about how best to read mathematics. There is no quick and easy way. Some people like to read slowly and carefully, making sure they understand each sentence, each paragraph, each formula, each proof, before they continue. This approach appeals mainly to those with considerable experience in, or a great enthusiasm for, the subject. Most prefer to start with a relatively quick — but not too casual — perusal of the text, perhaps taking a section or chapter at a time, to get a general impression of what the writer is trying to convey.

Such a scan must not be too cursory. The aim should be to glean as much information as possible without worrying unduly about details that seem incomprehensible at this stage. The danger is that if there are too many things that seem incomprehensible, everything fuses into one big muddle. If that happens it is wise to put the book aside, or move to something new, and come back to the troublesome section or chapter later.

With the 'quick perusal' approach one must later return for a more systematic appraisal. Well-understood material need only be skimmed through at this second reading. More time, often a lot more, must be devoted to mastering things that were difficult or obscure at first reading.

Whatever approach one takes, essential adjuncts at all stages are pencil, paper and a pocket calculator or computer. Mathematical writing abounds in phrases like 'It only requires simple algebra to show that this reduces to … ', or 'It is easy to see that this is equivalent to …', or 'The key steps in the proof are …'. In these situations the reader must do some work to master the argument. This is where the pencil and paper come in.

Reading mathematics has analogies to using an instruction manual for assembling a do-it-yourself furniture kit. Some things in the manual only become clear if we put the components together step-by-step as we read. In the same way, a statement like 'It only requires simple algebra to show that this reduces to ...' can only be fully appreciated if one does that simple algebra — again with pencil and paper. Here 'pencil and paper' are the analogues of the tools needed to assemble the furniture.

If the approach of trying to master everything at one reading is followed, this will almost certainly call for pencil, paper and calculator as one proceeds.

Those who take the quick perusal path, followed later by a more detailed reading, should, at the first stage, make a note of where there is a need to do pencil and paper, or calculator, work at the next reading. They should also list anything that seems obscure, such as the proof of a theorem. This provides a reminder to look at this more thoroughly next time. Use pencil and paper — and calculator or computer where appropriate — freely at a second reading.

There will inevitably be some sections of this book that are of more interest and greater relevance to some readers than to others. A comparatively quick — but reasonably careful — reading will draw attention to many interesting links between topics. It is important to recognise these if one is to get a real appreciation of — and true satisfaction from — mathematics.

Depending upon one's training and interests there may be, indeed almost certainly will be, some topics that are already familiar to you. Others may not appeal to you. At a detailed reading you should concentrate on things that are both new and interesting. At some later date a topic that appeared to be of little interest may take on new significance. It may turn out to be relevant to work you are doing in another discipline. Then is the time to return to study that topic in more detail.

This book makes extensive use of cross references, but most chapters are designed to be readable on their own. How easy that will be inevitably depends upon how familiar one is with basic mathematical concepts.

Contents

1

Mathematics in Perspective

1.1 Mathematical Diversity

- Do you like mathematics?
- Or does it frighten you?
- Do you want to become a professional mathematician?
- Are mathematical skills needed in the disicipline or profession you work in or hope to enter?
- Do you want to know more about mathematics because experts say we need more people with mathematical skills?

If you answered *yes* to one or more of these questions this book is designed to help you.

It is not a textbook, but a book to be picked up and read at leisure. If a chapter bores you after reading the first couple of pages leave it alone and go to the next chapter. Come back to the boring one later. It may be more interesting second time round. Some chapters are about topics not usually dealt with in school or preliminary undergraduate mathematics. In others I look behind the scenes at elementary notions — even some met in primary school — to show how concepts and rules have developed.

If you are apprenesive about, or bored by, matheatics take heart. It may not be your fault. The way it is taught and examined varies from excellent to very bad. Sloppy or badly explained use, or even abuse, of mathematical ideas and terms in everyday life is another irritant. Here the prime offenders are politicians, journalists and advertisers.

Sloppy mathematics, often designed to deceive, is common in advertising. Sales promotions for utility services and for financial products are prime offenders. It is ethical — but often not helpful — to design advertising to show a product in the best possible light, providing no demonstrably false claims are made. Information given is often incomplete, or uses jargon that is hard to understand.

Promoters of investments, savings schemes, mortgages or credit cards, take a delight in highlighting interest rates in a way that favours their particular product. The quoted rates may all be legitimate, but care is needed when comparing those for competing products, because there are many different ways of expressing interest rates. These include

- Annual percentage rate (APR)
- Annual equivalent rate (AER)
- Gross rate
- Net rate
- Monthly rate

Most of us have seen them all, but definitions of each may differ between users. Even when each is clearly defined, it takes some mathematical skill to compare, say, a *monthly interest rate* with an *annual equivalent rate.*

A bank where I have some of my savings advertises (in large print) an interest rate of 5 per cent per annum. In small print appears the information that this is an 'AER', and in very small print, that it is 'equivalent to a gross rate of 4.89 per cent per annum'. It does not tell us that 'AER' is an abbreviation for *annual equivalent rate*, nor does it explain why this corresponds to a 4.89 per cent 'gross rate', or indeed what is meant by a gross rate. Confusing, isn't it?

I leave you in suspense until Sect. 4.4, where I explain that if we interpret the percentages correctly, there is no contradiction.

Ploys used by competing utility companies when setting out complex tariffs for commodities such as gas and electricity, or services like mobile phones or internet connections, make it virtually impossible for potential customers to decide which deal is best without, among other things, a good grasp of basic mathematics.

Advertisers are adept at showing products in the best possible light. A bank that increases the interest it pays on savings from 2 percent to 3 per cent per annum may, almost legitimately but somewhat confusingly, claim it is 'giving you 50 per cent more interest'. I say confusingly because at first reading you might think you are going to get a total of 52 per cent interest!

If the bank reduces the interest rate from 3 per cent to 2 per cent per annum it is unlikely to draw your attention to the fact that it is giving you 33.33 per cent less interest. It is more likely to tell you that it is 'reducing the rate by only 1 percentage point'.

Comparing rates and percentages is just a small part of mathematics. The scope of the subject is made clear in *Mathematics Unlimited – 2001 and Beyond* [18], edited by Björn Engquist and Wilfried Schmid. In it some 90 authors, each an expert, writes about his or her own field. Some write from an elementary viewpoint; others in a way that will only be fully comprehensible to those who have completed, or are in the late stages of, an undergraduate course in Mathematics.

The titles of the articles shows the diversity of present-day mathematics. Many reflect the impact of computers as well as the ever increasing demands for mathematical skills and techniques in disciplines as diverse as engineering, space travel, medicine, information technology, economics, business practice, and even law and entertainment.

Articles of a different nature tell of developments important primarily to mathematicians working on specific topics. Some, but not all, have applications in other disciplines.

Some titles involving applications are:

Computational Mechanics: Where is it Going?
A Basis for a New Relationship Between Mathematics and Society
Communicating Mathematics Across the Web
Computational Neuroscience: More Math is Needed to Understand the Human Brain
Developments in Insurance Mathematics
Modeling the Cardiovascular System: A Mathematical Challenge
Math in the Entertainment Industry

Titles of articles more specifically of interest to mathematicians include *From Finite Sets to Feynman Diagrams*, *Whither Applied Nonlinear Dynamics* and *Primitive Automorphic Forms*, titles that may mean little to other than professional mathematicians.

Today the subject is so vast that hardly any mathematician, however eminent, has anything approaching a complete understanding of the subject. Most are expert in only one, two, or perhaps three branches, sometimes with only an elementary grasp of a handful of other topics.

This does not mean the subject is a patchwork of loosely connected, or even disconnected, strands. There are many connecting threads. Some delicate and intellectually appealing in their simplicity; others difficult to unravel.

Realizing that connecting strands were often not recognised motivated two eminent mathematicians, Richard Courant and Herbert Robbins, to write in 1941 the book mentioned in the preface *What is Mathematics?* [12, 13]. This masterpiece was greeted enthusiastically by teachers and students worldwide. It is one of the most influential books about general mathematics ever written. The authors took some 500 pages to answer the question posed in their title. To do so they used only what their sub-title aptly called *An Elementary Approach to Ideas and Methods*. This classic is still in print, with only one chapter added in a 1996 revision to cover some later developments. A feature of the book is its strong emphasis on the connections between diverse strands that make up the whole.

1.2 Mathematics in Action

In this book we look at the fascinating mix of, on the one hand, purely intellectual challenges, and on the other, developments triggered by practical demands. The driving force for the latter has been an ever-increasing need to understand the quantitiative elements of the natural and social sciences, medicine, engineering, technology, business and commerce, and other facets of increasingly complex societies. We shall see that the intellectual and practical strands often interlock.

Mathematics abounds in dilemmas. Things that seem obvious may prove on closer inspection not to be true. Or when they are true, it may be difficult to prove them so. Failure to recognise a key point may mean an argument that looks sound leads to an obviously false conclusion. Here is a simple example. Given the equation

$$35x - 65 = 28x - 52 \tag{1.1}$$

factorising each side we get

$$5(7x - 13) = 4(7x - 13)\,. \tag{1.2}$$

Dividing each side of (1.2) by the same divisor, $7x - 13$, we reach the nonsense conclusion that $5 = 4$.

What has gone wrong? The rule that allows us to divide each side of an equation by the same divisor has an exception. That divisor must not be zero. 'Thou shalt not divide by 0' is a golden rule of mathematics.

If you are not convinced we have divided by zero, try solving (1.1) in a more orthodox way. First subtract $28x$ from, then add 52 to, each side. This gives $7x - 13 = 0$. That is why division by $7x - 13$ is not allowed. Mathematicians do not allow division by 0 just to be difficult. They forbid it because, as the above example shows, doing so may lead to a nonsense.

Mathematicians' dreams may be quickly shattered by new developments. It is little more than 50 years since the first 'electronic' computers appeared. Only major universities or large corporations could afford these expensive mainframe giants. Despite their physical dimensions, most completely filling one or more large rooms, their computing capability was a small fraction of that of a modern PC or laptop. Nevertheless, their invention inspired mathematicians to believe that it would soon be possible to make accurate weather forecasts weeks, or even months, in advance.

Atmospheric conditions affecting weather had already been modelled accurately by large systems of what are called *differential equations*. To use these models for long range forecasting all that appeared to be needed was an expanded grid of weather stations to collect data to feed into these equations, and more powerful number-crunching computers to solve the equations in hours or even minutes rather than the weeks or months needed for manual calculations. It seemed only a matter of time until we could make accurate forecasts of the weather in London, New York or Sydney days, weeks, or months ahead.

Long before the needed number crunchers arrived, computers that we today would regard as dwarfs helped shatter that dream. During the 1960s and 1970s several workers noted unexpected, and somewhat startling, traits when they tried to solve even relatively simple systems of equations. They had stumbled upon a new mathematical phenomenon, now given the name *chaos.* This showed that the systems of equations needed to make long term forecasts were incapable of doing so, because their own solutions were as unstable as the weather they were trying to predict.

1.3 Mathematics and Society

Looking at historical developments helps us understand the structure of mathematics and explain some interlocking threads. The main driving forces have been

- human curiosity;
- meeting the practical needs of ever more complex societies.

Mathematics was born when primitive man needed to count. By a quirk of fate there still existed some 200 years ago a closed society where mathematics had made no progress beyond what was known to our early ancestors.

In 1803 British colonists settled in the island of Tasmania, then called Van Diemen's Land. They came face to face with a unique race — the Tasmanian aborigine. Inhumane treatment of the aborigines by the colonists, coupled with inadequacies of an immune system that could not cope with diseases brought to the island by white colonists, were recipes for disaster. In 1876 the last full-blooded Tasmanian aborigine still resident in her homeland died.

Tasmanian aborigines used a language with no recognisable connection to any spoken elsewhere. Their mathematical vocabulary was limited to equivalents for our *one, two, plenty.*

Rising sea-levels at the end of an ice-age cut Tasmania off from mainland Australia some 10 000 years ago. The aboriginal life-style probably changed little from that time to the arrival of white colonists. Its simplicity meant there was no pressure to develop sophisticated number systems like those needed for commerce, building, navigation, science, technology, and the other activities of more developed communities.

The history of mathematics shows that its development from a basic counting tool to the powerful science it is today has been strongly motivated by practical needs. The simple *one, two, plenty* of the Tasmanian aborigines sufficed because it met all their needs. They used primitive tools. They only needed a count to tally their possessions, the spoils of the hunt, or the haul when they dived for shellfish. Quantitative expressions like 'two spears', 'one wallaby', 'two oysters', 'one wife', 'plenty children' sufficed to describe the simple, but fairly stable, status quo in Tasmania's precolonization days. Indeed, the estimated population of some 5000 at the time white settlers arrived — this in a land roughly the area of Ireland — even suggests that 'plenty children' may have been a rarity.

Only if they had had a more advanced society, would they have needed a more advanced counting system. There is, incidentally, slender evidence from aboriginal cave drawings that one or two more sophisticated tribes may have progressed to a system equivalent to our '1, 2, 3, 4, *plenty*'.

1.4 Short Steps and Giant Leaps

The interplay between developments stimulated by practical needs, and initiatives inspired by human curiosity, is lost in antiquity. It was well advanced by the heyday of Greek civilization.

Centuries may elapse before a curiosity-driven discovery is found to have a practical application. We learn at school that a prime number is one divisible only by itself and by unity. A proof, first given by Euclid, that there are infinitely many prime numbers may seem no more than an interesting curiosity, one of no practical importance. But not so, for one implication of Euclid's proof is that primes exist with 50 or even 100 million or more digits. Today, more than 2000 years after Euclid gave his proof, the existence of such large primes is the key to many internet security systems to protect sensitive information such as credit card details when we buy on the internet. We say more about this in Sect. 3.6.

On the other hand, proposing a symbol '0' for *nothing*, or *zero*, can hardly be deemed a serious intellectual challenge, yet it instantly revolutionized the way we do arithmetic. A symbol to represent nothing is a simple, but not a trivial, idea. The Romans never thought of it. The price they paid was that they were lumbered with a number system that made addition difficult, and multiplication almost impossible, unless one counted on ones fingers or used some primitive calculating machine such as the counting frame or *abacus*. A Roman soldier who marched XIII mille passus (the Roman mile of one thousand paces) each day, would have been hard pressed without an abacus to work out that in IX days he would have marched CXVII mille passus, while in X days he would have covered CXXX mille passus.

Mathematics has reached its present position by an evolutionary process — a mixture of short steps and giant leaps. Some short steps, like inventing a symbol '0' for nothing, have had consequences that those who took them may never have envisaged. Some giant leaps have revolutionized the way we think.

1.5 Who Needs Mathematics?

Intending professional mathematicians must study the subject in depth. There is also, as already indicated, an increasing number of disciplines where, like it or not, a core knowledge of mathematics is at least desirable, and more often essential.

For an engineer that core is substantial. If you are headed for the world of finance it is a different, somewhat smaller, but still quite appreciable, core. The mathematics of risk assessment — the key element of actuarial science — is vital to the insurance industry. Statistical methods ensure the validity of medical research findings. At the clinical level they may be used to indicate the likelihood of incorrect diagnoses.

My earlier reference to problems stemming from mathematical half-truths in advertising, suggests that a better understanding of mathematics may help us all in managing our financial affairs — finding the best home for our savings, minimizing the amount we pay for our gas, electricity, telephone services, choosing the credit card that gives us the best deal.

I shall often use text boxes like this:

Only 200 years ago
there was a civilization
whose entire counting system
consisted of three words

I think of these as *surprise packets* or *attention grabbers*. The content of some is more surprising than that of others. They sometimes draw attention to things that are interesting or important, but not obvious or well-known. The example above is just interesting, referring to something earlier in this chapter. Others are important, many are both. The content of each box is elaborated upon in nearby text.

At times I may irritate because I seem pedantic. At others I may appear to trust too much to intuition. Sometimes I point out that my arguments are incomplete. This may be because completion calls for a deeper knowledge of mathematics than I assume. In other cases it is to avoid tedious technical detail that is hardly a source of intellectual satisfaction. I take this approach because I aim to drive away any fears you may have, and to add to your enjoyment of a suject that has given me great pleasure.

1.6 Further Reading

Details of books mentioned in this section, and of books or papers referred to in later chapters, are given in the *References* at the end of this book.

I have already mentioned *What is Mathematics?* [12, 13] and *Mathematics Unlimited* [18]. The former is an excellent picture of mathematics as it was some 60 years ago. Most of it is still topical. The latter gives an overview of present day (or almost present day) mathematics. It is ideal for selective reading. I particularly like the interviews with eminent mathematicians. These give their views, usually without technical detail, about how mathematics is being used at present, and how they think the subject should develop. At the back of the book there are thumbnail biographies of the many illustrious contributors. Reading these gives a picture of the diverse facets of modern mathematics. Another broad survey, aimed rather more at the general reader, is *Excursions into Mathematics* by A. Beck, M.N. Bleicher and D.W Crowe [8].

There are many books that deal at varying levels with aspects of the subject that appeal both to laymen and students of mathematics, and do this without getting lost in technicalities — an approach that stimulates one to follow up a topic in depth. Two enjoyable short books designed to alleviate many fears are those by W Timothy Gowers [28] and by David Acheson [1]. Both survey a range of key topics in less than 200 pages. M.W. Liebeck [33] gives a slightly more advanced treatment of many topics, especially some in pure mathematics.

Two books by Robert B Banks [6, 7] assume some familiarity with undergraduate mathematics. Each highlights relatively straightforward applications in real-world problems, largely but not exclusively, associated with the physical sciences. The titles, *Towing Icebergs, Falling Dominoes and other Adventures in Applied Mathematics* and *Slicing Pizzas, Racing Turtles, and Further Adventures in Applied Mathematics*, indicate the spirit of the author's approach.

There are many popular, yet serious, books about mathematics written by first rate mathematicians adept at explaining often complicated notions in easily understood terms. This is something many of their peers find difficult. Keith Devlin [14, 15, 16, 17] and Ian Stewart [43, 44] are two such writers. Each has written several books in this category, as well as more comprehensive works. Devlin's titles include *Mathematics: The New Golden Age*, *All the Math That's fit to Print*, *Life by the Numbers* and *The Maths Gene*, a study of what makes mathematicians tick. Ian Stewart's books include *Does God Play Dice?*, a good introduction to chaos, and *From Here to Infinity.* Brian Everitt's [19] *Chance Rules* is an entertaining introduction to mathematical ideas relevant in statistics.

There are many books written at a less sophisticated level. These are usually aimed more specifically at the layman. Martin Gardner [23, 24] specializes in the unusual and the puzzling aspects of mathematics. Typical Gardner titles are *Mathematical Circus* and *A Gardner's Workout.*

Another classic that first appeared in 1945, and is available in a 2004 reprint, is G. Polya's [38] *How to Solve It.* It is a mine of information about how to tackle problems.

If you are interested in the history of mathematics, and famous names in all its branches, a classic reference work is E.T. Bell's [9] *Men of Mathematics.* It should be in any major mathematics library. Published originally in 1937, a 1986 paperback reprint was still available at the time of writing.

There has recently been a near epidemic of books giving biographical accounts of eminent mathematicians of the 19th and 20th centuries. There are also numerous books covering the historical development of particular branches, or of the steps in solving famous problems such as *Fermat's last theorem*, or the *four colour problem.* Some of these are referenced in later chapters.

A basic dictionary of mathematics is useful when one meets an unfamiliar term. The wide-ranging *Penguin Dictionary of Mathematics* edited by R.D. Nelson [37], or the *Oxford Concise Dictionary of Mathematics* edited by Christopher Clapham [10], are suitable choices. There are literally thousands of important, and widely used, terms in mathematics that are not mentioned in this book because any meaningful explanation would be too long. Sometimes it would also be impossible to indicate why they are important without introducing further new ideas.

New books on popular mathematics appear regularly. A leading London academic bookshop devotes some 7 metres of shelf space to books on popular and recreational mathematics. Keep an eye open in your local high-street bookstore, at your internet book supplier's web site, or on academic, educational and technical publishers' web sites. If you have access to a scientific or mathematics library, look too at the shelves housing books on general mathematics, or on the history of mathematics.

Happy reading!

2

The Counting House

2.1 The Natural Numbers

In the rest of this book, feel free to work rapidly through material with which you are already familiar, and with which you feel at ease. What that material will be depends very much on one's position on the mathematics learning curve. Take a little longer when you find yourself in unfamiliar territory.

> **Advice from a mathematician**
>
> ***Begin at the beginning and go on until you come to the end and then stop***

This quote is from Lewis Carroll's *Alice in Wonderland.* 'Lewis Carroll' was the pseudonym of Oxford University mathematics lecturer, Charles Lutwidge Dodgson (1832–1898).

The first part of this advice is fine. I follow it and start an account of numbers spread over several chapters, with a discussion of the aptly named *natural numbers* 1, 2, 3,. . . ,. The dots here mean that, unlike the Tasmanian aborigines, we continue writing them in increasing order for as long as we please. These numbers are the foundation stones of mathematics. Their properties range from fascinating to frustrating, from the immediately obvious, to the yet to be proved.

I am less happy about the second part of Dodgson's advice. Mathematics may have no end. It is growing at the moment faster than ever before.

Our complete number system, ranging from integers through rational and irrational numbers to real and complex numbers, has evolved over thousands of years. Practical needs have motivated many developments. Others have been curiosity driven.

Early in our schooldays we learn the rules for operating with numbers whether they be positive or negative, whole numbers (integers) or fractions, rational or irrational. Bear with me while we look again at things most of us regard as routine. I do so because the logic behind some routines may not be evident at first meeting.

Here is an example. We know that $(-3)+(-5)=-8$, and that $(-3)\times(-5)=15$. Why does adding two negative numbers give another negative number, while multiplying the same two negative numbers gives a positive number? It is easy to see the logic behind the addition rule. It is harder to see why the product of two negative numbers should be positive. 'Because teacher said so' is not very convincing. I give a better reason in Sect. 2.6.

For fractions the multiplication rule is simple. For example,

$$\frac{2}{3}\times\frac{5}{7}=\frac{2\times 5}{3\times 7}=\frac{10}{21},$$

i.e., the numerator of the product is the product of the numerator terms, and the denominator is the product of the denominator terms.

For addition there is not an analogous rule with the $\times$ sign replaced by a $+$ sign. We know

$$\frac{2}{3}+\frac{5}{7}\neq\frac{2+5}{3+7},$$

where the symbol $\neq$ means 'is not equal to'. Instead, we have the more complicated

$$\frac{2}{3}+\frac{5}{7}=\frac{2\times 7+3\times 5}{3\times 7}=\frac{29}{21}.$$

To see why the correct rules are logical we need to look at how the number system has developed form the simple counts of $1, 2, 3, \ldots$. We know these natural numbers also as *whole numbers* or *positive integers.*

Can you define a natural number? Don't worry if you are lost for words. Even famous mathematicians have shirked the issue, some claiming them to be products of divine inspiration.

Despite the difficulty of putting it into words — and that is frustrating — most of us understand what these numbers mean. We recognise the symbols used as a valuable shorthand. Thus, 1763 is a short way of writing *one thousand seven hundred and sixty three.* The symbols are international, but the word forms are language dependent. Five, cinque, cinq, finf, fünf all represent the number 5.

We are also familiar with, and often frustrated by, the less elegant, and less useful, Roman shorthand, e.g., MDCCLXIII for 1763. The Greeks had an even less user-friendly system. It was closely linked to their alphabet, and the ordering of the letters in it.

The counting system now used worldwide is the *Arabic* or *Hindu-Arabic* system, so called because of its origin. In it, any natural number, however large, is expressed uniquely by appropriate positioning of the ten digits 0, 1, 2, 3, 4, 5, 6, 7, 8, 9. There are more spartan systems that use as few as two symbols, but we shall see later that such parsimony may be a false economy.

The Roman shorthand, as mentioned in Sect. 1.4, makes arithmetic difficult, and requires new symbols for each of what we now write as 1, 5, 10, 50, 100, 500, 1000, 5000, ...

The Romans didn't realize that nothing was important

This somewhat ambiguous boxed comment means that in Julius Caesar's day the then greatest civilization did not realise the value of a symbol like 0 for *zero* or *nothing.*

The Roman and Arabic number systems do have some common features. There are comparable symbols (I and 1) for the number one. In each system ten, and multiples of ten, are important, being given letter symbols X, C, M by the Romans, but the simpler 10, 100, 1000 in Arabic notation. The Romans also had intermediate equivalents of our 5, 50, 500 using the symbols V, L, D.

We were born with built in counting machines

The prominence of ten, and multiples of ten, in number systems has its origins in the earliest counting machines, our hands — each with 5 digits or fingers — 10 digits in all. In the Arabic system the relative positioning of only ten symbols, representing digits, lets us designate large numbers easily. The Roman system also invokes positioning rules, but these are more complicated than ours. This makes addition harder, and multiplication almost impossible without some calculating aid. Try multiplying CVI by XXI without first translating these numbers into Arabic notation. It is not immediately obvious that the answer is MMCCXXVI.

2.2 Alternative Groupings

The basic role of 10 in the Arabic and other counting systems is well established, but several 'bases' other than 10 are widely used. In the USA, and also in the United Kingdom, though now to a lesser extent than it once was, groups of 12 are important, This base had origins in using the ten fingers plus two hands as counting tools to form groups. There are 12 inches in a foot. Eggs, or small cakes, are often sold by the dozen. In pre-decimal British currency there were 12 pence in a shilling.

Groups of 12 appealed at an early stage in several civilizations, perhaps partly because of the convenience when sharing between 2, 3, 4 or 6 recipients. We can give equal shares of a dozen eggs to 2, 3, 4, 6 or 12 people without breaking any, but not to 3, 4, 6 or 12 people if we start with 10 eggs.

A base 20 persists in the word *score*, again inspired by a natural human counting machine — 10 fingers and 10 toes. There is a reminder of this in the French *quatre-vingt* (four twenties), equivalent to our eighty (eight tens).

In recording time, and in trigonometry, the base 60 often appears [60 seconds = 1 minute, 60 minutes = 1 hour (1 degree in trigonometry)]. Use of the base 60 can be traced back to Babylonian astronomers. Sixty also has the appeal that it is exactly divisible by no less than 10 smaller numbers, 2, 3, 4, 5, 6, 10, 12, 15, 20 and 30, as well as by 1 and by 60.

A parsimonious base for counting is groups of two. This base has a key role in electrical or electronic systems such as computers. At a basic level there are two options — either a current flows in, or a pulse passes through, a component (represented by 1) or it does not (represented by 0). This situation inspired the binary system of counting. This uses only the digits 0, 1 to represent any positive integer — but more about that in Sect. 3.2.

2.3 A Price to Pay

A Paradox

Computers — even pocket calculators — free us from the arithmetic drudgery of adding long columns of figures, or the tedium of doing long multiplication or long division, or worse, manually calculating a square root. The down side is that they open up new ways to produce nonsense if one does not appreciate how numbers work. That this is so, may only become clear after you have read a large part of this book, but that much quoted computer maxim

Garbage in Garbage out

is all too true. Understanding how the number system works helps us spot the garbage.

I am not deriding computers. Their advent has probably had a more profound impact on mathematics than any other single external factor. That too will become ever clearer as we proceed.

2.4 Abstract by Nature

I mentioned the difficulty of defining a number. Tautological phrases like ‘the number *one* describes the property of *oneness*’ don’t help. What the number *one* does is to describe a unique property, irrespective of any concrete object to which it applies. One apple is not the same as one pear, or one man, but each shares a property we designate by a count of *one*.

A pair of ones is assigned the number *two* to describe the associated abstract concept — two men, two pears, or two rifles. Two implies more than one. More precisely, items having a count of *two* may be separated into single items, each with a count of one. This is the genesis both of *counting* and of *addition*. *Two* is the sum of two *ones* (addition).

Most of us learnt this, and more, about counting when we were only a few years old. Yet, what I described in the last paragraph, was about as far as the Tasmanian aborigines, and probably our own primitive ancestors, got with arithmetic in some 10 000 years. The concept of 'plenty', for counts exceeding two, carries the mathematical notion of order implicit in the relationship *one* is less than *two* is less than *plenty*.

Some things we think are obvious about numbers are not. In Sect. 3.2 we shall see that 1763 and 3343 are different ways of writing the same number. Is that obvious?

The 'groups of ten' idea, fundamental to the Arabic notation, is closely associated with the idea of counting on our fingers. To add 7 and 5 by counting on fingers, when we have added '3' of the '5' to '7' we think of this as '10' with '2' left over. We write this as '12'. This idea of forming and moving groups, here groups of 10, in a 'positioning' notation resembles the reasoning used historically with a counting frame or abacus. There, positioning of beads on wires determined the size of the group each bead represented.

Early forms of counting frames, predating our decimal system, were used by the Chinese and the Babylonians. Later versions were widely used for commercial calculations throughout Europe and Asia up to the Middle Ages.

2.5 Keeping Within the System

Starting with natural numbers, addition and multiplication simply produce other natural numbers.

Addition and multiplication follow rules that we tend to regard as trivial — but they are not quite. Although we add and multiply more or less mechanically, care is needed, especially when using a pocket calculator.

What does $7 + 5 \times 3$ mean? Multiplying 5 by 3 gives 15, then adding 7 gives 22. However, if we add 7 to 5 to get 12, and now multiply by 3, we get 36. It matters whether we do the addition or the multiplication first. We learn at school that multiplication takes precedence over addition.

If we want to multiply the sum of 7 and 5 by 3 (a perfectly valid aspiration) we write $(7 + 5) \times 3$, and apply the rule that operations within brackets are carried out first. Thus $(7 + 5) \times 3 = 12 \times 3 = 36$. Alternatively, we may first multiply 7 by 3 to get 21, then 5 by 3 to get 15, then adding $21 + 15$ gives 36. That is, $(7 + 5) \times 3 = 7 \times 3 + 5 \times 3$, where, in the latter form, we only need to invoke the rule that multiplication takes precedence over addition. Calculations using pocket calculators often come unstuck because people ignore these 'order of operation' rules.

Pocket calculators are dumb. They don't tell you when you are not obeying the rules.

We generalize from particular numbers like 3, 5, 7 by replacing them with symbols to represent any natural number.

Let a, b, c each represent a natural number or positive integer. This may seem a trivial development, but only because we are used to it. It is the basis of *algebra.*

With this notation the basic rules for addition and multiplication of integers are

$$a + b = b + a \tag{2.1}$$

$$a \times b = b \times a \tag{2.2}$$

$$a + (b + c) = (a + b) + c \tag{2.3}$$

$$a \times (b \times c) = (a \times b) \times c \tag{2.4}$$

$$a \times (b + c) = a \times b + a \times c \tag{2.5}$$

The rules have names. Rule (2.1) is the *commutative law of addition*, (2.2) is the *commutative law of multiplication*, (2.3) is the *associative law of addition*, (2.4) is the *associative law of multiplication* and (2.5) is the *distributive law.*

You will know that it is common practice when symbols like a, b denote 'any' integer, to abbreviate $a \times b$ to ab. Obviously we cannot apply this for specific numbers such as 23, because this is equivalent to $20 + 3$, not 2×3, so perhaps the idea is not so clever after all, but it is a convention we learn to live with.

You may want to jump ahead here and point out that rules (2.1)–(2.5) still apply if a, b, c are fractions. Be patient, because it is useful to look at the logic behind that extension after we consider subtraction and division when applied to natural numbers.

A simple extension of the natural numbers is to include zero, (denoted by '0') for the count of no objects. Zero has the properties $a + 0 = a$, and $a \times 0 = 0$ for any natural number a. Rules (2.1)–(2.5) hold for this slightly extended set of natural numbers if we define addition and multiplication by 0 that way.

Just as multiplication is a useful shorthand for addition of the same number a fixed number of times, there is a shorthand for multiplication of the same number n times, e.g., $a \times a \times \ldots \times a$, where the dots indicate that there are a total of n numbers or 'factors' each a. The shorthand for this is a^n, e.g.,

$$a \times a = a^2,\ a \times a \times a = a^3 \text{ and } a \times a \times b \times b \times b = a^2 \times b^3 .$$

The latter is commonly abbreviated to a^2b^3. In particular $100 = 10\times 10 = 10^2$, $1000 = 10 \times 10 \times 10 = 10^3$, etc.

What if we want to multiply a by itself 0 times? If we write this a^0, what does that mean? We define a^0 to equal 1. This makes certain rules for operating with indices (the n in a^n is called an *index*) work properly. These rules are found in most books on elementary algebra. A brief summary of them is given in the Appendix in Sect. A.2. One that is easily seen to hold is that $a^m \times a^n = a^{m+n}$.

2.6 Subtraction

Developing societies soon discovered that the positive integers did not meet all their needs. If a baker has 50 loaves of bread available, and a customer buys 40, the baker has 10 remaining. Mathematically, this is expressed as $50 - 40 = 10$. If the customer had wanted 60 loaves, then either he or she would have to go elsewhere for 10 of these, or the baker would have to find 10 loaves from another source. We need a new type of number to cover this situation. Today, we happily express this as $50 - 60 = -10$.

These new numbers representing a *deficit* are familiar to us as *negative integers*, written as $-1, -2, -3, \ldots$.

It makes sense to define addition of negative numbers as a totality in the way we do for a positive count, i.e., $(-2) + (-3) = (-5)$ represents a total 'deficit' of 5. The rule for adding a deficit and a surplus is also simple, and again intuitively reasonable.

Multiplication of a negative number by a natural (positive) number is treated as a special case of addition. Adding the same negative number four times, e.g., $(-2)+(-2)+(-2)+(-2) = (-8)$, may be written $(-2) \times 4 = (-8)$. I mentioned in Sect. 2.1 that it is less obvious what $(-2) \times (-4)$ means, even if we happily accept that the answer is 8.

The explanation is that it allows the rules (2.1)–(2.5) to hold for an extended number system consisting of the positive and negative integers and zero. We refer to these collectively as the *integers*, sometimes conveniently subdivided into positive and negative integers. Positive integers correspond to natural numbers. If we want to consider zero and the natural numbers we speak of the *non-negative integers*.

It is easy to verify that (2.1)–(2.5) hold for addition and multiplication in the extended system, providing we define the product of two negative integers to be $(-a) \times (-b) = a \times b$. This is needed to make the distributive law hold.

If a and b are both natural numbers, it is easy to see that $b + (-b) = 0$. Therefore

$$(-a) \times [b + (-b)] = (-a) \times 0 = 0 .$$

If the distributive law holds this requires

$$(-a) \times [b + (-b)] = (-a) \times b + (-a) \times (-b) = 0 .$$

Since $(-a) \times b = -ab$, the last equality is only true if we define $(-a) \times (-b)$ to equal $a \times b = ab$. This justifies defining the product of two negative integers to equal the product of the corresponding natural numbers.

Writing negative integers in brackets, e.g., $(-a)$, highlights the fact that the negative sign and the letter a together designate a negative number, which has the magnitude of the natural number a. There is a logical distinction between the minus sign used in this way and its use as an operational sign meaning subtraction as an arithmetic operation. This does not worry us in practice, because subtraction of $(-b)$ from a is equivalent to adding b to a; i.e. $a - (-b) = a + b$. Effectively, we are saying $-(-b) = (-1) \times (-b) = +b$. The minus sign outside the bracket used to indicate subtraction, behaves in this context as though it were equivalent to a multiplier (-1) preceded by an operator $+$. This sounds complicated, but it is how the notation works in practice.

Mathematicians sometimes describe subtraction as the *inverse* of addition. In algebraic terms, if a and b are any two natural numbers, addition defines a natural number, s, their sum, written symbolically

$$a + b = s .$$

The number s answers the question 'What is the sum of a and b?' Subtraction addresses the problem where we are given two numbers, a and b, and are asked to find a number d such that $a + d = b$. This number d, the difference between b and a, is determined by

$$d = b - a .$$

Unlike addition and multiplication, subtraction is not commutative, i.e., $b - a \neq a - b$ in general. Can you think of special cases where $b - a$ equals $a - b$?

2.7 Division

Sharing 12 items equally between 4 people is easy — give 3 to each. Mathematically, knowing that $3 \times 4 = 12$ leads to an inverse process written as $12 \div 4 = 3$, or more commonly as $12/4 = 3$. We are still in the natural number system.

To share 12 items equally among 5 people, we could start by giving each person two items. We then have two left over. If we split these two remaining items each into 5 smaller items of equal size, we have 10 of these smaller items.

Giving each of the 5 people 2 of these ensures equal shares. Remembering that 5 of these smaller parts is the same size as one original item, 2 smaller parts represent a proportion '2 among 5' which we write as 2/5. Since 2 among 5, is the same proportion as 4 among 10, we may also write this as 4/10, or in decimal notation as 0.4. This is a concrete illustration of how we arrive at the abstract notion that $12 \div 5 = 2.4$.

Introducing fractions to deal with the remainder left over in division is a mathematical extension needed when we start with natural numbers a, b, but cannot find a natural number q such that $aq = b$.

Division of b by a may also be considered as an operation giving rise to a relationship between natural numbers a, b, q^* and r of the form $b = a \times q^* + r$. This relationship is not unique. For example, if $b = 12$ and $a = 5$, the relationship is satisfied if either $q^* = 1$ and $r = 7$ or if $q^* = 2$ and $r = 2$. Further, if we allow r to be a negative integer, the relationship still holds for these values of a, b if $q^* = 3$ and $r = -3$. The relationship is uniquely determined if we restrict r to values between 0 and $a - 1$. We then call r the *remainder* when b is divided by a.

2.8 Prime Numbers

Negative integers and rational fractions evolved to meet practical needs. A curiosity driven approach leads to numbers even more fundamental than the natural numbers. These are the *prime numbers.*

For any natural number t, say, there is only a limited choice of natural numbers $a, b, c \ldots$ such that their combined product is t. If $t = 5$, the only natural numbers giving that product are 1 and 5. If $t = 12$, we may express 12 as the products 1×12, 2×6, 3×4 or $3 \times 2 \times 2$. The commutative law for multiplication lets us change the order in which each number enters the product. The number of product groups does not always increase as t increases. For example, only 13×1 (or 1×13) equals 13.

A number like 5 or 13 that is expressible as a product of only 1 and itself is called a *prime number.* Any natural number that is not a prime is said to be *composite.*

It is easy — though tedious — to verify that the prime numbers less than 100 are 2, 3, 5, 7, 11, 13, 17, 19, 23, 29, 31, 37, 41, 43, 47, 53, 59, 61, 67, 71, 73, 79, 83, 89, 97.

At first prime numbers look like a potential mathematical nightmare. They are lacking in, though not devoid of, pattern. Knowing that the largest prime number less than 100 is 97, is no help in telling us that the next prime number is 101. Again, knowing this does not help us predict that the next is 103.

Unpredictable does not mean unimportant

Don't let this unpredictability delude you into believing that prime numbers are of little practical interest. They are more basic than the natural numbers. This is because all natural numbers are either prime, or the product of two or more prime numbers. What is even more important is that this decomposition, or factoring, into primes is unique – a result referred to as *the fundamental theorem of arithmetic.* This may be stated formally as

Every integer greater than one can be factored into primes in only one way.

Because of the commutative law of multiplication, we make the proviso that reordering terms in the product gives the same factorization. Thus 12 may be factored as $2 \times 2 \times 3$, or equivalently as $3 \times 2 \times 2$ or $2 \times 3 \times 2$. Most of us think decomposition into primes is obvious, because we have done it nearly all our life.

You may feel that the fundamental theorem of arithmetic is obvious, but it is not trivial. I do not give the proof, for although it is reasonably simple it needs some care. Euclid, better known in elementary mathematics as the father of geometry, gave one proof, but there are alternative proofs. One is given in [12, 13] (pp. 23–24), as well as by others.

We can show that the theorem holds for all numbers less than 100 by checking that it holds for each such number separately. It would be extremely tedious to demonstrate it this way for all numbers less than 1000, and virtually impossible to do so for all numbers less than a million. To establish the universal truth requires logical arguments to cover every conceivable case.

In arithmetic, we often factorize integers to simplify computation by allowing what is usually called *cancelling out.*

Asked to divide 810 by 54 a pocket calculator, if available, quickly gives the answer 15 — providing we hit the right keys. We could check the result by long division (tedious), or by factorizing 810 and 54. It is easy to verify that

$$\frac{810}{54} = \frac{3 \times 3 \times 3 \times 3 \times 5 \times 2}{3 \times 3 \times 3 \times 2} ,$$

and 'cancelling' factors common to numerator and denominator gives

$$\frac{810}{54} = (3 \times 5)/1 = 15 .$$

This works only because factorization to primes is both possible and unique.

No power of 2 is divisible by 3

True or False?

If you didn't know the answer to this boxed poser before, it should now be obvious. If you are stumped I give it at the end of the chapter.

Although primes do not have a pattern that helps us predict higher value primes when we know smaller ones, they are not patternless. Apart from 2, they are all odd, since any even number greater than 2 is divisible by 2, and is therefore composite. Looking at the primes less than 100, we see a tendency for them to occur in pairs that differ by 2, e.g., (3, 5) (5, 7) (11, 13) (17, 19) (29, 31) (41, 43) (59, 61) (71, 73). These do seem to be further apart as the integers become larger. Mathematicians have for long wondered whether or not such pairs are limited in number. For example, can such pairs be found only among integers not greater than 1 000 000 000 000 000? The answer is that they occur among larger numbers, and this has led to the conjecture that there are infinitely many such pairs.

I show later in several contexts that we must be careful about using terms like *infinitely many.* The concept of infinity is surrounded by a multitude of difficulties if approached without caution.

In 1742, Christian Goldbach (1690–1764), in correspondence with Leonhard Euler (1707–1783), conjectured that every even number greater than 2 could be expressed as the sum of exactly two prime numbers. Nobody has ever proved or disproved this. To disprove it we would only need to find just one even number for which it is not true.

Little progress was made towards a proof until 1931 when a young Russian mathematician showed that every positive integer could be expressed as the sum of not more than 300 000 primes. It was a step in the right direction, but hardly strong supporting evidence for Goldbach's conjecture!

Progress since has been relatively rapid. In 1934 the 300 000 was dramatically reduced to 4 providing the number, n, was even and sufficiently large. By 1966 it had been established, but still with restriction to an unspecified sufficiently large n, that all even integers were the sum of a prime, and a product of at most $c = 2$ primes. We still wait for the crucial step to $c = 1$ and the removal of any reference to sufficiently large n before the Goldbach conjecture is proved true.

Computer searches had established by 1965 that the conjecture is true for all even numbers up to 100 000 000 ($= 10^8$). By 1989 this had increased to 2×10^{10}, by 2003 to 6×10^{16}, and by 2005 to 2×10^{17} , and that number is still creeping up, a reflection of the rapid increase in the number-crunching ability of modern computers.

To learn more about this conjecture I recommend [13] (Chap. 9, Sect. 2). Entering 'Goldbach conjecture' into an internet search engine also produces a mass of information of varying reliability.

It has been suggested that one reason the Goldbach conjecture is hard to verify is that primeness is essentially to do with multiplication, whereas the Goldbach conjecture is about addition. What do you think? I find that argument unconvincing.

The conjecture has spawned numerous 'related' hypotheses. One such is that every integer greater than 17 is the sum of three *distinct* primes. It is easily verified that $18 = 2 + 5 + 11$, $19 = 3 + 5 + 11$, $20 = 2 + 7 + 11$,

$21 = 3 + 5 + 13$, $22 = 2 + 7 + 13$, giving a hopeful start. It is relatively easy to see that the restriction to numbers greater than 17 is required. It is impossible to find three different primes that add to 17. You should not find it hard to verify that the only integers less than 18, that are the sum of 3 distinct primes are $10 = 2 + 3 + 5$, $12 = 2 + 3 + 7$, $14 = 2 + 5 + 7$, $15 = 3 + 5 + 7$.

Attempts have been made to find formulae that always produce prime numbers. Several produce a restricted number of primes, but all are doomed to break down at some stage. One that produces primes for all values of n less than 41 is $n^2 - n + 41$. When $n = 41$ it gives

$$41^2 - 41 + 41 = 41^2 = 41 \times 41 = 1681 ,$$

which is the square of a prime, and therefore not prime. The formula $n^2 - 79n + 1601$ yields primes for $n < 80$, but when $n = 80$ it also takes the value $1681 = 41^2$. Seeking a simple formula that will give only primes, let alone one that will generate all primes, is a futile task.

Euclid gave us more than geometry

Primes look to occur less frequently among larger natural numbers, suggesting the total number of primes is perhaps finite. An elegant proof that this is not so was given by Euclid who, as already mentioned, also proved the fundamental theorem of arithmetic.

Curiosity driven developments have provided us with several basically different techniques for proving things. The one Euclid used here has three steps:

- Assume what we want to prove is not true.
- Show this leads to a logical contradiction.
- This implies what we want to prove must be true.

This method is referred to as an *indirect proof*, or *proof by contradiction*, or *reductio ad absurdum* (reduction to absurdity).

Before we look at Euclid's proof, here is a helpful notation. It may seem obvious, but it has only been widely used in mathematics in the last couple of centuries.

So far we have used single letters such as a, b, c to denote unspecified natural numbers. If we want to refer to many numbers, we soon exhaust the 26 letters in the alphabet. A more versatile notation, called an *alphanumeric* notation, is to represent all unspecified natural numbers by a single letter associated with a numeric subscript. This subscript is different for each unspecified number. Thus, we may express, say, 5 numbers a, b, c, d, e as a_1, a_2, a_3, a_4, a_5. The advantage is clear if we want to refer to, say, 1000 such numbers. We designate these as

$$a_1, a_2, a_3, \ldots, a_{1000} ,$$

where the dots imply that all numbers with a subscript between 4 and 999 are also included. An alternative shorthand is

$$\{a_i\};\ 1 \leq i \leq 1000 .$$

In words, we refer to this as 'the set of numbers a_i where i takes all integer values between 1 and 1000 inclusive'. If we want to specify an unlimited set of natural numbers we may write

$$a_1,\ a_2,\ a_3,\ \ldots ,$$

or perhaps more clearly for those not familiar with the notation,

$$a_1, a_2, a_3, \ldots\ ad\ inf ,$$

where *ad inf* is an abbreviation for *ad infinitum.* The choice of the letter a is arbitrary. It is common to use letters near the start of the alphabet for fixed sets of numbers. If we want to refer to specific kinds of numbers, such as prime numbers, we might use the letter p with numerical subscripts, writing, for example, n primes as

$$p_1,\ p_2,\ p_3,\ \ldots,\ p_n\ .$$

We have used the word *set* above with its everyday meaning. In the context here this does not conflict with a specialised use of the term in mathematics that we discuss briefly in Sect. 4.2, and in more detail in Sect. 16.2.

2.9 Euclid's Proof of the Infnity of Primes

Using the alphanumeric notation makes Euclid's proof easier to follow. He began by assuming that there are only a finite number of primes, i.e., that what he wanted to prove was not true. He did not need to specify that finite number. It could be 1000, 1 000 000, 2 519 711, or some other number. All he required was that such a number existed. We denote it by N. These primes may be written $p_1,\ p_2,\ p_3,\ \ldots,\ p_N$.

To show this leads to a contradiction, Euclid used a clever trick. He introduced a number, A, that was equal to one more that the product of the N primes. That is

$$A = p_1\, p_2\, p_3\ \ldots\ p_N + 1.$$

It is easy to see that A is greater than any of the p_i. The assumption that there are no primes other than the p_i implies A is composite. This is not so.

Again, it is easy to see why. The product $p_1\, p_2\, p_3\, \ldots\, p_N$ is divisible by any p_i. This means that if A is divided by any p_i the remainder must be 1. Thus A is not divisible by any prime p_i. This implies A must be prime. But A cannot be both prime and composite.

Thus, the assumption that there are only a finite number of primes leads to a logical contradiction, because A is a prime not in the original set $p_1, p_2, p_3, \ldots, p_N$. Therefore, the assumption that there are only a finite number of primes cannot be correct. This implies that there are infinitely many primes.

With hindsight this proof looks deceptively simple. It is in the choice of A that Euclid showed a flash of inspiration. Or a stroke of genius?

Euclid's proof does not tell us whether any particular large number is a prime. Various algorithms have been developed to tackle this problem, many of these being computationally slow. In 2002 Manindra Agrawal, a mathematician at the Indian Institute of Technology at Kanpur, working with two of his students Neeraj Kayal and Nitin Saxena, developed an algorithm that pointed the way to more efficient determination of whether any given integer is prime or composite.

2.10 Fermat's Last Theorem

Studies of natural numbers have led to many conjectures, some interesting or fascinating rather than important. Obviously, all positive integers other than 1 can be expressed as the sum of two smaller integers in the form $a + b = c$. As c increases so does the possible choices for a and b.

If n, a, b, c are all positive integers, $a + b = c$ is the special case where $n = 1$ of a more general relationship, $a^n + b^n = c^n$.

Every school pupil studying geometry meets the relationship $a^2 + b^2 = c^2$. This is an algebraic equivalent to the famous *Pythagoras's theorem* which states that if x, y, z are the length of the sides of a right-angled triangle, where z is the length of the hypotenuse (the side opposite the right-angle), then $x^2 + y^2 = z^2$. This geometric result — that in any right-angled triangle the area of the square on the hypotenuse is equal to the sum of the area of the squares on the other two sides — takes us beyond the range of the positive integers. It is still true when x, y, z are not integers, providing they are associated with side lengths of a right-angled triangle.

Pythagoras's theorem is well known to be true for certain sets of integers. The classic example is $x = 3$, $y = 4$, $z = 5$, for then

$$3^2 + 4^2 = 9 + 16 = 25 = 5^2.$$

The triplet of integers (5, 12, 13) also has this property, since $5^2 + 12^2 = 13^2$. There is an unlimited mumber of triplets with this property. If we multiply each element in a triplet such as (3, 4, 5) by *any* integer r we get another triplet satisfying the condition. For example, setting $r = 5$ gives the triplet (15, 20, 25), and

$$15^2 + 20^2 = 225 + 400 = 625 = 25^2.$$

Moving from $n = 1$ to $n = 2$ has dramatically reduced the number of triplets of integers satisfying the relationship $a^n + b^n = c^n$, though there are still 'infinitely many'.

Infinity is a trap for the unwary

When $n = 2$ it is easier to find triplets for which the relationship does not hold than it is to find triplets for which it does. Thus, there are also infinitely many triplets for which it does not hold, a hint that infinity is not a simple concept.

What happens if we seek integers satisfying the relationship $a^3 + b^3 = c^3$? Try to find a set for yourself, but don't waste too much time on your search. We may suspect these will be fewer than for the case $n = 2$, yet feel intuitively that there should be some sets.

Intuition is a good servant but a bad master

Here intuition lets us down. In many places in this book I urge you to remember the warning in the above box. The study of integer solutions to the equation $a^n + b^n = c^n$ is the subject of *Fermat's last theorem.* Pierre de Fermat (1601-1665) claimed to have proved that there are no triplets of positive integers (a, b, c) satisfying the relationship $a^n + b^n = c^n$ for any positive integer $n \geq 3$. If Fermat had a proof he never wrote it down. Despite many attempts by practically every mathematician of note since then, it was only in the 1990s that a proof was found by Andrew Wiles, using very advanced mathematics. The quest for what had almost become a holy grail of mathematics is described in popular accounts by Aczel [2] and by Singh [41].

Fermat's last theorem is an example of a puzzling — and in that sense interesting theorem — that has, so far as anyone knows, no practical importance. This is a rash assertion! In Sect. 3.6 we meet an example of a mathematical concept that is basic to protecting information transmitted on the internet, yet prior to 1973 it was little more than an interesting — or maybe just a frustrating — curiosity.

There is at least one other conjecture by Fermat, although he made no claim to have proved it, that has since been shown to be incorrect. He conjectured that all numbers of the form

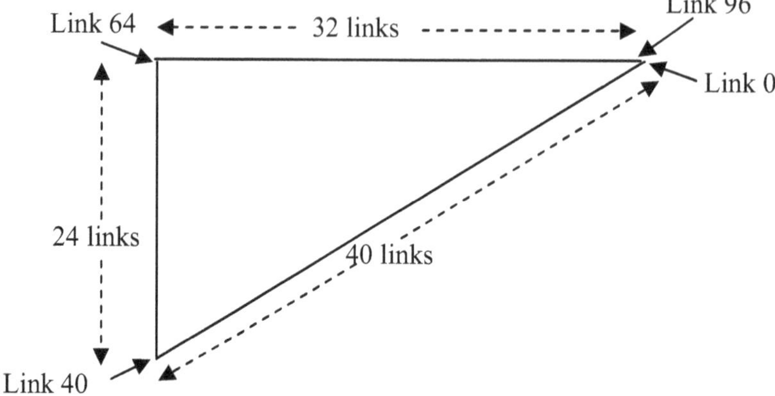

Fig. 2.1. Using a surveyor's chain to determine a right angle in the field

$$2^{2^n} + 1 ,$$

where n is a non-negative integer, are prime numbers. This is true when $n \leq 4$, but for $n = 5$ we find $2^{32} + 1 = 4\,294\,967\,297 = 641 \times 6\,700\,417$, and so it is not prime.

In the absence of a proof that Fermat claimed to have for his last theorem, a doubt persisted until the appearance of the proof by Wiles. That proof used concepts that were not available in Fermat's lifetime. Whether Fermat had a simpler valid proof that he did not write down remains a matter for conjecture.

It is intriguing that for natural numbers the relationship $a^n + b^n = c^n$ is trivial to the point of being uninteresting when $n = 1$. Yet it provides a link between numbers and geometrical shape when $n = 2$, but for all greater integer values of n it has no solution in terms of positive integers.

The 3, 4, 5 side lengths for a right angle triangle based on the Pythagorean relationship was probably put to practical use by the Egyptians and other early civilizations to align walls, or foundations, at right-angles when building.

As recently as 40 years ago the surveyor's chain – a set of 100 linked rods, each rod of the same length, with a total length of 66 feet – was still used sometimes by engineers and others to lay out a right-angle on a piece of ground. Horticulturalists laying out an orchard could use a chain to align trees on a rectangular grid.

I worked in the 1960s at a horticultural research station where we did this. Once the direction of one side of the grid was decided upon, we used a chain to determine a direction at right-angles, making use of a triangle with sides 24, 32, 40 links (in the ratio 3, 4, 5). We did this by putting pegs at links 40 and 64 after having stretching the section between these links tightly in the chosen direction, thus marking off in that direction a length $64 - 40 = 24$ links. One person then lifted the unpegged section from 0 to 40 at link 0, and another person lifted the section pegged at 64, and located link 96. The two then walked with the unpegged sections until they met with the sections fully stretched. The direction of the section from links 64 to 96 (length 32) then gave the direction of the line at right angles to the original. Can you see why? Fig. 2.1 may help.

Choosing lengths of 24, 32, 40 rather than, say, 9, 12, 15, made use of most of the chain (i.e., all but 4 links). This could be expected to reduce the error below that likely with comparatively short lengths when using so primitive a device.

2.11 Loose Ends

The final section in many chapters is headed *Loose Ends.* These contain material that is either interesting in its own right, or shows how concepts in that chapter fit in a broader mathematical canvas. References in some of these sections point to sources where those who want to can find out more about a particular topic.

I have in a few cases, without going into technicalities, mentioned not-always-obvious applications in other disciplines. In some of these the mathematics involved may be simple, but it is used in a non-elementary way.

Other number systems. Its use even today on memorials, etc., keeps the archaic Roman number system alive. As mentioned in Sect. 2.1 the Greeks had a number system based on the ordered letters of their alphabet, e.g., $\alpha = 1,\ \beta = 2,\ \gamma = 3,\ \delta = 4,\ \epsilon = 5$, and so on. Had they continued this way they would have run out of symbols by the time they got to the end of the alphabet, with counts only to the mid-twenties. They got round that problem much as the Romans did. They stopped the individual letter allocation at 10, and provided letter symbols for 50, 100, 500, 1000 etc.

The Egyptians had a system in some ways superior to that of the Romans. It was based on using different symbols for each power of 10. The symbol for *one* was | and that for *ten* was ∩. The number 74 was written using repetitions of the appropriate symbols, i.e., the tens symbol repeated 7 times and the unit symbol 4 times. The positioning rules used by the Egyptians for larger numbers were somewhat complicated.

A common feature of many early number systems is the use of something equivalent to a notch or slash for a unit, with repetition for small counts. This is indicated by the following equivalents to our 1, 2, 3:

Egyptian	|	||	|||
Indian	−	=	≡
Roman	I	II	III

With primitive writing or drawing tools, a slash or stroke was easy to make. This doubtless explains its use as the basis for several early counting systems.

A positional system with the base ten was first used in India during the sixth century AD, although the use of the symbol 0 for zero appeared later. It was not introduced into Europe until about the twelfth century, probably via trade links with Arabia. The use of 0 was further stimulated by a translation about that time of the book *Algebra*, written by an Arabian mathematician al-Khwarizmi who lived about 800 AD.

A possible origin of the signs + and − was their use on tea chests in European warehouses to indicate whether these were above or below an agreed standard weight.

Fermat again. We saw in Sect. 2.10 that Fermat made at least one conjecture that later proved to be wrong. However, he established many important results. One sometimes referred to as *Fermat's little theorem* states that if p is a prime number, and a is a number not divisible by p, then the remainder when a^{p-1} is divided by p is 1. An immediate consequence of this theorem is that if n is any odd number other than 1, and the remainder when 2^{n-1} is divided by n is not equal to 1, then n must be composite. For example, if $n = 9$, then $2^8/9 = 256/9$ gives remainder 4. So 9 is, as is in any case obvious, composite. If $n = 13$ then $2^{12}/9 = 4096/9$ which gives remainder 1. This does not prove that 13 is prime (though of course it is) because although the condition is *necessary* for a number to be prime, it is not *sufficient.*

There are some composite n for which the remainder is 1. One such is $n = 341$. This is composite since $341 = 11 \times 31$, yet $2^{340}/341$ gives a remainder 1. You will be hard put to show this by direct division and your pocket calculator is unlikely to help. We look at an easy way to establish this in Sect. 3.3.

In mathematics the terms *necessary condition* and *sufficient condition* occur frequently. We often have to decide whether an assertion satisfies both, only one, or neither, of these conditions. That all three angles of a plane triangle are equal, is both a necessary and a sufficient condition for the triangle to be equilateral. Fermat's little theorem gives, as we have just seen, a necessary but not a sufficient condition for an odd natural number to be prime.

My poser. Finally, the answer to the poser in the box in Sect. 2.8. Both 2 and 3 are prime numbers, and 2^n is the product of n primes all 2. Therefore none of these is divisible by 3. Hence no power of 2 is divisible by 3.

3

Some Properties of Numbers

3.1 A Look at Structure

In Sect. 2.4 I claimed that 1763 and 3343 were different ways of writing the same number. To understand why we look more closely at the structure of numbers.

The Roman numeral MCCXIII (1213) has an additive structure in the sense that it equals

$$\mathrm{M} + \mathrm{C} + \mathrm{C} + \mathrm{X} + \mathrm{I} + \mathrm{I} + \mathrm{I}\,.$$

Arabic numbers have a different additive structure. We know that 1213 does not equal $1 + 2 + 1 + 3$. The meaning attached to each digit depends on its position in the number. The leading 1 in 1213 means 1000, the 2 means 200, the second 1 means 10, and the 3 means 3 units. That is,

$$1213 = 1000 + 200 + 10 + 3\,.$$

It is convenient to write the base units associated with each position, i.e., 1, 10, 100, 1000, 10000, ... as $1 = 10^0$, $10 = 10^1$, $100 = 10 \times 10 = 10^2$, $1000 = 10 \times 10 \times 10 = 10^3$, $10000 = 10 \times 10 \times 10 \times 10 = 10^4$ and so on, where the superscript — called the *index* or *power* — indicates the position of a digit, counting from 0 as the position of the unit digit. In this notation

$$1213 = 1 \times 10^3 + 2 \times 10^2 + 1 \times 10^1 + 3 \times 10^0\,.$$

Any number involving the power s, say, as the highest power of 10 may be written

$$a_s\, a_{s-1}\, a_{s-2}\, \ldots\, a_2\, a_1\, a_0\,.$$

Here a_s is a digit between 1 and 9, and the remaining subscripted a are digits between 0 and 9 inclusive. This is a shorthand for

$$a_s \times 10^s + a_{s-1} \times 10^{s-1} + a_{s-2} \times 10^{s-2} + \ldots + a_2 \times 10^2 + a_1 \times 10 + a_0. \quad (3.1)$$

More fully, we may write $a_1 \times 10^1$ and $a_0 \times 10^0$ for the last two terms, but this is usually an unnecessary elaboration.

The key advantage of Arabic over Roman numerals is that addition and multiplication only require rules for operations with the digits 0 to 9. These rules are the same irrespective of the digit position. For addition, when we say $723 + 89 = 812$, algebraically what is happening is that we follow the rules of addition and multiplication in the form

$$(7 \times 10^2 + 2 \times 10 + 3) + (8 \times 10 + 9).$$

We regroup in powers of 10, starting with the units (corresponding to 10^0). Then, using the associative, commutative and distributive laws and rule for precedence of operations given in Sect.2.5, we get

$$(3 + 9) + [(2 + 8) \times 10] + (7 \times 10^2)\ .$$

The addition of 3 + 9 gives one group of 10 and 2 left over. We express this sum as 10 + 2. The 10 may be incorporated with the 2 + 8 groups of 10 already present, giving in all (1 + 2 + 8) groups of 10. This is just expressing formally what, in the language of school arithmetic, is meant by 10 + 2 = 12 when we describe it as '2 and carry 1'. Here 'carry 1' means take it across to the next (here the tens) group. If we now add the 1 + 2 + 8 = 11 in the tens column we have 10 groups each of 10 units (equivalent to 1 group of 100 units), and 1 group of 10 left over (in all, the '1 and carry 1' of school arithmetic). We record the 1 in the tens column, and add 1 to the 7 already in the hundreds column, giving 8 in that column. So finally 723 + 89 = 812.

The key point is that the effect of adding digits is the same in any group or column. The consequences of that simplicity are far from trivial. This can be summarized in a simple addition table for digits like Table 3.1. The number at the intersection of the ith row and jth column of the table is the sum of the number at the top of that column, and the first entry in that row, i.e., of the column and row numbers. To add natural numbers we need only memorize this table, and the rule that we carry over to the next column a '1' corresponding to each group of 10. Remember also that adding zero leaves the sum unaltered.

Most of us carry the simple addition table in our heads without conscious thought. Multiplication needs a table like Table 3.2. It is sometimes a chore of school arithmetic to learn such a table, with the extra almost trivial, but practically important information, that any digit multiplied by zero is zero. It has been argued that, because one can routinely use a pocket calculator to multiply, there is no need to memorize Table 3.2. Those who disagree — I am one — point out that many mistakes made when using calculators can easily

Table 3.1. Addition table for numbers to the base 10

Col	1	2	3	4	5	6	7	8	9
Row									
1	2	3	4	5	6	7	8	9	10
2	3	4	5	6	7	8	9	10	11
3	4	5	6	7	8	9	10	11	12
4	5	6	7	8	9	10	11	12	13
5	6	7	8	9	10	11	12	13	14
6	7	8	9	10	11	12	13	14	15
7	8	9	10	11	12	13	14	15	16
8	9	10	11	12	13	14	15	16	17
9	10	11	12	13	14	15	16	17	18

Table 3.2. Multiplication table for numbers to the base 10

Col	1	2	3	4	5	6	7	8	9
Row									
1	1	2	3	4	5	6	7	8	9
2	2	4	6	8	10	12	14	16	18
3	3	6	9	12	15	18	21	24	27
4	4	8	12	16	20	24	28	32	36
5	5	10	15	20	25	30	35	40	45
6	6	12	18	24	30	36	42	48	54
7	7	14	21	28	35	42	49	56	63
8	8	16	24	32	40	48	56	64	72
9	9	18	27	36	45	54	63	72	81

be detected if the user has a good 'feel' for numbers. Having the multiplication table at one's finger tips is essential for developing that 'feel'.

Table 3.2 may be applied to multiplication of digits in any position, with the usual rules for carry over. Multiplication of 236 by 37 is in essence

$$(2 \times 10^2 + 3 \times 10 + 6) \times (3 \times 10 + 7).$$

Multiplication is based primarily on the distributive law. We first multiply each term in the first bracket by 7 (the unit digit in the second bracket), with the usual rules for carry over once we get a product composed of two or more than two digits. Thus

$$(2 \times 10^2 + 3 \times 10 + 6) \times 7 = 14 \times 10^2 + 21 \times 10 + 42 = 1400 + 210 + 42 = 1652.$$

We next multiply $(2 \times 10^2 + 3 \times 10 + 6)$ by (3×10), which equals

$$(200 + 30 + 6) \times 30 = 200 \times 30 + 30 \times 30 + 6 \times 30 = 6000 + 900 + 180 = 7080.$$

Finally, we add 1652 + 7080 = 8732. You should verify that the usual way of writing long multiplication leads to this result, and see how the steps are effectively carried out, e.g.,

$$\begin{array}{r} 2\,3\,6 \\ 3\,7 \\ \hline 1\,6\,5\,2 \\ 7\,0\,8 \\ \hline 8\,7\,3\,2 \end{array}$$

Some people prefer to multiply by the leading digit first, effectively reversing the order of the two steps above.

Addition and multiplication use only the associative, commutative and distributive laws, together with the multiplication and addition tables for digits (including the rules that multiplication by zero give zero and addition of 0 to any digit leaves it unaltered).

The above is a formal account of what most of us (or a pocket calculator) do more or less mechanically. Here is something that is less well known. If we divide 8732 by 10 the quotient is 873 and the remainder 2. If we now divide 873 by 10, the quotient is 87 and the remainder 3. Dividing 87 by 10, the quotient is 8 and the remainder 7. Finally, dividing 8 by 10, the quotient is 0 and the remainder 8. The successive remainders 2, 3, 7, 8 are the digits in 8732 in reverse order.

This is no fluke. If we divide any natural number by 10 until we get a quotient 0, the remainders are the digits in reverse order. Can you see why? Considering successive division by 10 of a number written in the form (3.1) helps one see this. I have no idea who first noted this property; it is almost certainly a curiosity driven finding. In the next section we see that an extension of it meets a practical need.

3.2 Changing the Base

What about my claim in Section 2.4 that 3343 is another way of writing 1763? Are you surprised? I cheated a little. I did not say that for each the rules for addition and multiplication are different. In 3343 we use sets of 8, not sets of 10, as a basic group. That is, 3343 is a shorthand for

$$3 \times 8^3 + 3 \times 8^2 + 4 \times 8 + 3\,.$$

By the rules of arithmetic

$$\begin{aligned} 3 \times 8^3 + 3 \times 8^2 + 4 \times 8 + 3 &= 3 \times 512 + 3 \times 64 + 4 \times 8 + 3 \\ = 1536 + 192 + 32 + 3 = 1763 &= 1 \times 10^3 + 7 \times 10^2 + 6 \times 10 + 3. \end{aligned}$$

Table 3.3. Addition table for numbers to the base 8

Col	1	2	3	4	5	6	7
Row							
1	2	3	4	5	6	7	10
2	3	4	5	6	7	10	11
3	4	5	6	7	10	11	12
4	5	6	7	10	11	12	13
5	6	7	10	11	12	13	14
6	7	10	11	12	13	14	15
7	10	11	12	13	14	15	16

Table 3.4. Multiplication table for numbers to the base 8

Col	1	2	3	4	5	6	7
Row							
1	1	2	3	4	5	6	7
2	2	4	6	10	12	14	16
3	3	6	11	14	17	22	25
4	4	10	14	20	24	30	34
5	5	12	17	24	31	36	43
6	6	14	22	30	36	44	52
7	7	16	25	34	43	52	61

Had we been born without thumbs

Had we chosen 8, rather than 10, as our base for counting — as we might have done had we been born without thumbs — we would need only 8 different digits 0, 1, 2, 3, 4, 5, 6, 7. These have the same meaning as they have in a decimal (base 10) count. With base 8, we discard the digits 8 and 9. Instead of 8 we write 10 ($= 1 \times 8 + 0$). Instead of 9 we write 11 ($= 1 \times 8 + 1$). That is, 9 in decimal (base 10) notation corresponds to 1 group of 8 and 1 remainder when we change the base to 8. Base 8 is called the *octal* base. The decimal number 10 becomes 12 (base 8), implying 1 group of 8, and 2 left over.

Multiplication and addition in this system are based on Tables 3.3 and 3.4 for digits in any position. As with the base 10, any digit added to zero is unaltered, and multiplication by zero gives zero.

To multiply 17 by 42, both numbers expressed using base 8, we proceed as in the decimal system, but use Tables 3.3 and 3.4. First multiply 17 by 2, and from these tables $7 \times 2 = 16$ (recorded as 6 and carry 1). Also $1 \times 2 = 2$ and, adding the 1 that was carried over, we get $17 \times 2 = 36$. We next multiply 17 by 4 and you should verify by using Table 3.4 that this gives 74. Remember, the 4 is in the 'group of 8' column, so this means the complete multiplication is 36 + 740 = 776. We may verify that 17 (base 8) is equal to 15 (base 10), that 42 (base 8) is equal to 34 (base 10) and that 776 (base 8) is equivalent to 510 (base 10), and that 15 × 34 = 510 (base 10).

Now divide 510 (base 10) successively by 8 (base 10); 510 divided by 8 gives a quotient 63 and remainder 6. Then division of 63 by 8 gives quotient 7 and remainder 7. Finally, division of 7 by 8 gives quotient zero and remainder 7. The remainders 6, 7, 7 are the digits of 776 in reverse order. This result generalizes, giving an easy way to convert any number (base 10) to the corresponding number (base 8). Try this to verify that 1512 (base 10) = 2750 (base 8). In what other way might we check this?

The base 8, or *octal system*, is used in computer science to represent bytes of information. One byte is divided into 8 bits. The importance of powers, or multiples, of 8 in computing shows itself in familiar expressions like 32, 64, 128 mb (megobytes) of RAM. Less obviously, the so-called kilobyte is not, as one might expect, 1000 bytes. It is actually 1024 ($= 2 \times 8^3$) bytes.

We mentioned in Sect. 2.2 that the base 2 of the binary system is also important in computing. It reflects the on/off situation relevant to whether or not a pulse or current flows in some component.

Symbolically, in the binary system we need only two digits corresponding to 0 and 1. Any number is written in the form

$$b_s\, b_{s-1}\, b_{s-2}\, \ldots\, b_2\, b_1\, b_0$$

where $b_s = 1$, and all other b_i ($i = 0, 1, 2, \ldots, s-1$) are either 0 or 1. This is shorthand for

$$b_s \times 2^s + b_{s-1} \times 2^{s-1} + b_{s-2} \times 2^{s-2} + \ldots + b_2 \times 2^2 + b_1 \times 2 + b_0 .$$

In binary notation 1001110 is the decimal number

$$1 \times 2^6 + 0 \times 2^5 + 0 \times 2^4 + 1 \times 2^3 + 1 \times 2^2 + 1 \times 2 + 0 ,$$

which, in decimal notation is $64 + 0 + 0 + 8 + 4 + 2 + 0 = 78$.

Tables 3.5 and 3.6 are the addition and multiplication tables for the binary digits 0, 1.

10 000 000 000 need not be a large number

The simple addition and multiplication tables suggest the binary system may be ideal for everyday use. The snag is that large numbers involve long strings of ones and zeros. For example, 1025 (base 10) becomes 10 000 000 001 and 1 048 608 (base 10) becomes a 21 digit 100 000 000 000 000 100 000. You

Table 3.5. Addition table for numbers to the base 2

Col	0	1
Row		
0	0	1
1	1	10

Table 3.6. Multiplication table for numbers to the base 2

Col	0	1
Row		
0	0	0
1	0	1

should check that the decimal numbers 0 to 20 in the binary system are those in Table 3.7.

We have seen that we can convert a number base 10 (decimal) to base 8 (octal) by finding the octal digits in reverse order as remainders on successive division by 8. What happens if we replace 8 by 2 and dividing successively by 2? For example, if we divide 19 successively by 2 we get quotient 9 and remainder 1, then quotient 4 and remainder 1, quotient 2 and remainder 0, quotient 1 and remainder 0, and finally quotient 0 and remainder 1. The successive remainders 1, 1, 0, 0, 1, when reversed, give the binary number 10011 for the decimal-based 19. The procedure works for other cases.

Binary multiplication and addition are easy using Tables 3.5 and 3.6. The carry-over rule is that each time a group of 2 occurs in any column we carry over a 1 to the digit column on the left. The only possible remainders in

Table 3.7. Binary equivalent of numbers 1 to 20

Decimal	Binary	Decimal	Binary
0	0	11	1011
1	1	12	1100
2	10	13	1101
3	11	14	1110
4	100	15	1111
5	101	16	10000
6	110	17	10001
7	111	18	10010
8	1000	19	10011
9	1001	20	10100
10	1010		

any column are 0 or 1. Check for yourself that the following addition and multiplication are done correctly.

$$\begin{array}{r} 1\,1\,0\,1 \\ 1\,1\,1 \\ 1\,0\,0\,1 \\ 1 \\ 1\,1 \\ \hline 1\,0\,0\,0\,0\,1 \end{array}$$

$$\begin{array}{r} 1\,1\,0\,0\,1 \\ 1\,0\,1 \\ \hline 1\,1\,0\,0\,1 \\ 1\,1\,0\,0\,1\,0 \\ \hline 1\,1\,1\,1\,1\,0\,1 \end{array}$$

Counting systems using base 12 have an appeal evident in the dozen and the number of inches in a foot. As suggested in Sect.2.2, this popularity may be partly because of the exact divisibility of 12 by 2, 3, 4 and 6. Using this base, known as the *duodecimal* system, has two minor disadvantages compared to our familiar decimal system. We need two additional symbols for digits corresponding to the decimal-based 10 and 11. The tables for addition and multiplication are a little more complicated.

Symbols often used for counts of 10 and 11 in the duodecimal system are those for 2, 3 written upside down, a nightmarish trap for copy editors, printers and proof readers unfamiliar with the system. They often attribute this usage to a typesetting error and order the inverted digits to be set the right way up! Until recently it also required ingenuity to get such symbols into a word processor document, although that can be done with most modern processors. I do not discuss the use of this base further, but you may like to form an addition and multiplication tables, and verify, for example, that $9 \times 7 = 53$ (base 12). Does intuition suggest an easy way to find the duodecimal number equivalent to the decimal number 962 or 1452? Check whether it works.

Another base used in computing is 16, giving the *hexadecimal system*. You may meet it if you have to devise codes to print special symbols in, for example, a word-processing program. It requires 6 additional digits to represent the decimal numbers 10 to 15. A common practice is to use either upper or lower case letters a, b, c, d, e, f for these.

3.3 Time and the Notion of Congruence

Asked on a Tuesday morning what day of the week it will be tomorrow, our immediate answer is 'Wednesday'. There might be a longer pause if asked what day of the week it will be 288 days hence. Is it obvious that the answer again is Wednesday? The key is to remember there are 7 days in a week. Dividing 288 by 7 gives a quotient 41 and remainder 1. This implies that 288 days represents the passing of 41 weeks and 1 day. If today is a Tuesday, then after exactly 41 weeks we are back to a Tuesday. One day later (represented by the remainder 1) brings us to Wednesday. Similarly, if we want to know the day of the week 5403 days hence, we again divide by 7. The remainder is 6. If it is Tuesday today, then 5403 days hence it will be a Monday.

The link between a current day of the week and the day any specified number of days ahead is characterised by the remainder when that number of days ahead is divided by 7. If today is a Tuesday the key is:

Remainder on division by 7	*Day of week*
0	Tuesday
1	Wednesday
2	Thursday
3	Friday
4	Saturday
5	Sunday
6	Monday

Considering remainders when dividing by a fixed divisor has many uses. In particular, the concept of sets of integers, a, that give the same positive reminder, r, on division by a specified divisor d when a is positive, and a remainder $s = r - d$ when a is negative is important. It is called *congruence.* Because of the relationship between r and s we regard each as 'equivalent' (or the same) remainder.

The notion of *congruence* was introduced in 1801 by one of the great mathematicians of all time, Carl Friedrich Gauss (1777–1855).

By definition

- Two numbers that give the same remainder on division by d are said to be *congruent modulo (d).*

We indicate that integers a, b are congruent modulo (d), i.e., that they have the same remainder on division by d, by the notation

$$a \equiv b \pmod{d} ,$$

spoken as 'a and b are congruent mod d'.

Thus, for example, $8 \equiv 23 \pmod{3}$ since both give remainder 2 on division by 3. Similarly, $-1 \equiv 15 \pmod{8}$ since both give remainder 7 on division by 8. This follows because the relationship $s = r - d$ in this case implies $-1 = 7 - 8$.

Two numbers that are not congruent (mod d) are said to be incongruent (mod d). This is indicated by replacing the symbol $\equiv$ by $\not\equiv$. Thus, $1 \not\equiv 15$ (mod 6), since the remainders on division by 6 are respectively 1 and 3.

Don't believe everything a car salesman tells you

Congruence is important in many everyday situations. An odometer that records the total distance a car has travelled usually returns to zero 1 km after completion of 99 999 km (in the USA or UK, 1 mile after 99 999 miles). This means that if an odometer reads 22 542, all we know is that the car has travelled either 22 542, 122 542, 222 542, or perhaps even 322 542 km (or miles).

It is not hard to confirm that the following statements are equivalent

- (1) $a \equiv b$ (mod d).
- (2) $a = b + nd$ for some integer (positive or negative) n.
- (3) $a - b$ (or $b - a$) is exactly divisible by d (i.e., gives an integer quotient without remainder).

Statement (2) implies that congruent numbers differ by integer multiples of d and (3) follows immediately from (2). For example, it is easy to see that $7 \equiv 187$ (mod 9). Then (2) takes the form $7 = 187 + (-20) \times 9$, and (3) holds since $7 - 187 = -180$ which is divisible by 9.

Gauss showed that numbers which are congruent share, within that context, many properties of ordinary numbers. The following property is a key to the importance of congruence:

- Congruencies *with respect to the same modulus* can be added, subtracted and multiplied in the way we operate with integers.

For example, $7 \equiv 29$ (mod 11) and $2 \equiv 13$ (mod 11), and it is easily verified that $(7+2) \equiv (29+13)$ (mod 11), both giving remainder 9 on division by 11. Also $(7-2) \equiv (29-13)$ (mod 11), both giving remainder 5, while $7\times 2 \equiv 29\times 13$ (mod 11), both giving remainder 3.

In general, if p, q, r, s are positive or negative integers and $p \equiv q$ (mod d) and $r \equiv s$ (mod d), we have $p \pm r \equiv q \pm s$ (mod d) and $pr \equiv qs$ (mod d).

We can use these properties to prove the assertion in Sect. 2.11 that $2^{340}/341$ gives remainder 1. First, we note that $2^{10}/341 = 1024/341$ and this gives quotient 3 and remainder 1 (i.e. $3 \times 341 + 1 = 1024$). Recalling that $a^m \times a^n = a^{m+n}$ (Appendix, Sect. A.2), we may write $2^{20} = 2^{10} \times 2^{10}$. This, on division by 341 gives remainder $1 \times 1 = 1$. Similarly, $2^{40} = 2^{20} \times 2^{20}$ also gives remainder 1. Extending this argument it follows that 2^{80}, 2^{160}, 2^{320} also give remainder 1 on division by 341. Finally, $2^{340} = 2^{320} \times 2^{20}$, and since we have established that both 2^{320} and 2^{20} give remainder 1 on division by 341, so also by the multiplication rule for congruence (mod 341) does 2^{340} give remainder 1.

3.4 Congruence in Practice

I was taught at school, without any explanation, that if the sum of the digits in an integer were divisible by 3 (or by 9), then that number was divisible by 3 (or by 9). I was also taught more complicated rules for detecting divisibility by 7 or by 11.

The relevant rules are easily derived using congruences. We illustrate the approach for division by 9, 11 or 7.

If we divide 10 by 9 we see $10 \equiv 1 \pmod 9$. By the multiplication rule for congruences it follows that $10^r \equiv 1 \pmod 9$ for any positive integer r. Writing an integer in the form (3.1), i.e.,

$$n = a_s \times 10^s + a_{s-1} \times 10^{s-1} + a_{s-2} \times 10^{s-2} + \ldots + a_2 \times 10^2 + a_1 \times 10 + a_0$$

and setting

$$t = a_s + a_{s-1} + a_{s-2} + \ldots + a_2 + a_1 + a_0,$$

i.e., t is the sum of the digits, then

$$\begin{aligned} n - t = a_s(10^s - 1) + a_{s-1}(10^{s-1} - 1) + a_{s-2}(10^{s-2} - 1) + \ldots \\ + a_2(10^2 - 1) + a_1(10 - 1) + a_0(10^0 - 1)\,. \end{aligned}$$

By the subtraction rule it follows that all $10^r - 1 \equiv 0 \pmod 9$ for non-negative integers r. Repeated application of the addition rule for congruences gives $(n - t) \equiv 0 \pmod 9$. This implies that if t is divisible by 9, then both n and t are divisible by 9.

For divisibility by 11, we note that $10 \equiv -1 \pmod{11}$, and by the multiplication rule $10^2 \equiv (-1) \times (-1) \equiv 1 \pmod{11}$. Applying the same rule again $10^3 \equiv (-1)^3 \equiv -1 \pmod{11}$. More generally, $10^r \equiv 1 \pmod{11}$ if r is even and $10^r \equiv -1 \pmod{11}$ if r is odd. If we again write

$$n = a_s \times 10^s + a_{s-1} \times 10^{s-1} + a_{s-2} \times 10^{s-2} + \ldots + a_2 \times 10^2 + a_1 \times 10 + a_0$$

but now set

$$t = a_0 - a_1 + a_2 - a_3 + \ldots + (-1)^s a_s,$$

then

$$n - t = a_1 \times 11 + a_2 \times (10^2 - 1) + a_3 \times (10^3 + 1) + \ldots + a_s \times [10^s - (-1)^s].$$

Each of 11, $10^2 - 1$, $10^3 + 1$, ..., $10^s - (-1)^s$ is congruent to zero (mod 11). It follows that $n - t$ is also congruent to zero(mod 11). This implies that if t is divisible by 11, then n and t are both divisible by 11. So a number is divisible by 11 if the difference between the sums of the odd and even digits is divisible by 11. Thus 93 247 187 is divisible by 11 since

$$(7+1+4+3)-(8+7+2+9)=15-26=-11\ ,$$

and -11 is divisible by 11.

To find a rule for division by 7, we note that $10 \equiv 3 \pmod 7$, $10^2 \equiv 3\times 3 \equiv 2 \pmod 7$, $10^3 \equiv 3\times 2 \equiv 6 \equiv -1 \pmod 7$, $10^4 \equiv 3\times(-1) \equiv -3 \pmod 7$, and continuing this way, that $10^5 \equiv -2 \pmod 7$, $10^6 \equiv 1 \pmod 7$, $10^7 \equiv 3 \pmod 7$. The pattern of remainders now repeats this cycle. Following similar arguments to those used above for division by 9 and 11, it is easy to show that n is divisible by 7 if

$$t = a_0 + 3a_1 + 2a_2 - a_3 - 3a_4 - 2a_5 + a_6 + 3a_7 + 2a_8 + \ldots$$

is divisible by 7. Applying this rule in practice is about as much work as carrying out the actual division. It is even more work than using a pocket calculator to do the job!

3.5 Checking Multiplication

Extending the check for divisibility by 9 provides a way to detect some, though not all, types of errors in long multiplications. The method is called *casting out the nines*. It works this way. By long multiplication, or using a pocket calculator, we find

$$12\ 589 \times 74\ 307 = 935\ 450\ 823\ .$$

The check starts by adding the successive digits first in 12 589. Each time that sum equals or exceeds 9 we subtract 9 from that sum and continue adding that remainder to any following digits, i.e. $1+2+5+8=16$, $16-9=7$ and $7+9=16$, and $16-9$ gives a remainder 7. We note this remainder. We now perform a similar casting out of nines from 74 307. Proceeding in the same way gives $7+4=11$, $11-9=2$, $2+3+0+7=12$, $12-9=3$. We showed above that if the sum of the digits in a number is divisible by 9, then that number is divisible by 9. It is easy to extend that argument to show that the final digit left after casting out the nines is the remainder when that number is divided by 9. In this example this implies $12\ 589 \equiv 7 \pmod 9$ and $74\ 307 \equiv 3 \pmod 9$. By the multiplication rules for congruence, this implies that $12\ 589 \times 74\ 307 \equiv 7\times 3 \pmod 9$, i.e., that each should give the same remainder on division by 9. Since 7×3 leaves remainder 3 on division by 9, it follows that if $12\ 589 \times 74\ 307$ equals 935 450 823, then the latter should leave remainder 3 on division by 9. This may be checked by casting out the nines. We get

$$9 - 9 = 0,\ 0 + 3 + 5 + 4 = 12,\ 12 - 9 = 3,\ 3 + 5 + 0 + 8 = 16,$$
$$16 - 9 = 7,\ 7 + 2 = 9,\ 9 - 9 = 0,\ 0 + 3 = 3,$$

thus establishing the remainder is 3.

This does not prove the result is correct, but if the remainders are not equal the result is certainly wrong. Common mistakes in long multiplication usually lead to failure of the rule for casting out nines.

It is easy to check that when we sum consecutive digits, once a sum exceeds 9 we can always get the remainder on division by 9 simply by adding the digits in that sum. For example, 8 + 7 = 15 gives a remainder 6 on division by 9, and 6 is simply the sum of the digits, i.e. 1 + 5, in 15. Thus, in practice, we may condense the steps outlined above.

Testing a multiplication by casting out nines is less important when using a pocket calculator. Here, the most likely source of error is incorrect data entry. This is a human frailty best detected by repeated calculation. With mechanical, or electronic, computing devices it is best to do a repeated computation with some changes in order. For example, to check the product of 79 513 and 24 917 we might enter the numbers in that order, and on the check enter them in reverse order (i.e., enter 24 917 first). If one enters the numbers in the same order each time, then the way the human brain tackles repetitive or boring tasks makes it likely that an earlier mistake is repeated. For instance, if one misreads 79513 as 79531, it is easy to make the same mistake again if the computation is done in the same order. Psychologists probably have an explanation for this aberration in our mental processes.

3.6 Prime Numbers and Internet Security

Secret codes were used for military purposes in Roman times. Some they invoked may have been as simple as those still used by children to pass surreptitious messages they would prefer parents or teachers not to understand. Basic codes are unlikely to trouble an astute adult for more than a minute or two. See how long it takes you to decode the message

AQW JCXG ETCEMGF OA EQFG

I have used a simple character shift code. Each letter in the original message is replaced by a letter occurring r places later in the alphabet, where r is an integer between 1 and 25. For example, if $r = 3$, A goes to D, E goes to H and so on. Letters near the end of the alphabet may go to letters near the beginning, e.g., for $r = 4$, Y goes to C.

To decode the above message we need to find the value of r. There are only 25 possibilities, so it should not take you long to find one that gives an intelligible message. The deciphered version is given at the end of this section.

A more sophisticated letter swapping code is one where the keys to both encoding and deciphering are the same, but where the encryption has no regular pattern. A possible key under this system would be

$$\begin{array}{cccccccccccccccccccccccccc} A & B & C & D & E & F & G & H & I & J & K & L & M & N & O & P & Q & R & S & T & U & V & W & X & Y & Z \\ P & M & B & A & G & Q & Z & W & C & N & Y & U & O & K & L & V & D & X & F & H & R & E & I & S & J & T \end{array}$$

Both the person who sends the coded message, and the recipient, are assumed to possess the key. Without this key a short message like the one at the start of this section would be harder to unravel. For longer messages this stronger code would not deter a skilled code-breaker. Certain letters, and letter combinations, are more common than others in a given language, and there are patterns in word structure.

In English, a single letter word is almost certainly 'A' or 'I', though it just might be 'O' or perhaps some person's initial. Nearly all words contain a vowel (a, e, i, o, u, or sometimes y has this role). Letters like 'X' and 'Z' occur much less often than letters like 'T' or 'S'. The letter 'X' often appears as the second letter of a word with the preceding letter 'E', e.g., *exception*, *expert*, *extend*, etc. 'Q' is nearly always followed by 'U'. The letter combinations 'ING' or 'TION' at the end of a word are common. These, and other features of a language, make it relatively easy to crack an unpatterned alphabet-based code for all but very short messages.

The main features of any coding system for transmitting information are analogous to a security system, where the sender of information, valuables, or whatever has a key to lock a secure box. After it is locked, the box is sent to a receiver who has a key to open it. In the examples above the same key (the alphabet based code) is used to lock and unlock the box (i.e., to code and decode the message). Security depends upon unauthorized users not having access to the key, or to their lacking the ability to pick the lock. As we have indicated, lock-picking is relatively easy with alphabet-based codes. Also, unless users and receivers carefully guard the common key, this may fall into the hands of unauthorized parties.

The next advance in coding was to devise systems where senders and receivers had different keys. Such systems are especially useful where many senders want to send different coded messages to one receiver. This is the situation in internet purchasing, where many customers need to send details such as credit card information to a trader. Customers want to be confident that such information will not fall easily into the hands of unauthorized people such as computer hackers.

An electronic combination safe

In one such system a customer feeds information into the trader's internet site using a simple key which is common to all senders. This is often called a *public key*. The information is put into a self-locking box using screen instructions. These are encrypted by the locking key.

There is an analogy with a combination safe which may be clicked shut. If it is well built, the only way to open the safe short of using high explosive, is to know the 'combination' of numbers or letters that must be dialled to open it. This is often called a *private key*. The security of the system depends upon authorized users keeping the private key secret. If there are many possible combinations it is extremely unlikely that anyone will hit upon the correct combination by chance, unless they can guess it because some obvious code like 123456 is used.

Modern computing developments have, in theory at any rate, sped up the potential for scanning all possible codes among only a few thousand, or even a few million. So, where confidential information is transmitted electronically, it does not suffice to have a key to decode a message that is one of only a few million possible keys.

It only takes a hacker less than a second to try all six digit numbers, and if one of them is the key it will disclose the encrypted information. To overcome this problem we need a public key to lock the box that is easy to use by many potential customers. The private key to unlock the information must be as hacker-proof as possible.

Prime number keys

In Sects.2.8 and 2.9 we looked at decomposition of an integer into products of primes. The uniqueness of this decomposition is embodied in the fundamental theorem of arithmetic.

Modern computers can find prime numbers with 50 or more digits. Having found two such numbers, we form their product, consisting of 100 or so digits. That is relatively easy to do. Given this composite number of some 100 digits, which factors into just 2 primes, it is much harder to determine those prime numbers. Even the most powerful computers may find this task beyond them.

Therefore, if the prime decomposition is kept a secret by the originator of the locking system, others will be unable to find the key to open the box.

This is the basic dea behind a widely used system for encrypting data in e-commerce. There are many technical considerations in its implementation, but it is a good illustration of the practical application of a mathematical concept — the uniqueness of the decomposition into primes — that may have seemed little more than an intellectual curiosity to its discoverers.

It was only in the 1970s that mathematicians first hit upon the idea of using prime number decompositions for code keys. It was some ten years later that the technicalities to make it effective were sorted out. The basic idea was stumbled upon in 1973 by Clifford Cocks, a mathematician working at a high-security British intelligence post. He was not allowed to publish his proposals because of security implications, so the practical development was left to others.

Several years later three scientists M Helman, R. Merckle and W. Diffie from Stanford University published similar theories to Cocks. They had arrived at these ndependently. Shortly afterwards, three other Americans at MIT — R. Rivest, A. Shamir and L. Aldeman — took this work further. They presented the RSA algorithm (RSA being their surname initials). This algorithm is now owned by RSA Data Security Inc., a major player in the data encryption field for financial transactions.

The RSA method is only one of an expanding number of encoding systems. A readable introduction to this, and several other approaches, is given by Koblitz [31]. There you will find an explanation of the delightfully named hash function. This is used in coding to make possible what are called digital signatures.

Finally, as promised, my coded message at the start of this section deciphers as

YOU HAVE CRACKED MY CODE

the shift number being $r = 2$, i.e. A goes to C, etc.

4

A Trip to Infinity

4.1 The Need for Care

Here are two arguments. Only one can be correct.

> I walk at 4 kilometres per hour (kph). My younger friend Jock walks at 5 kph. We both start at the same time to walk to a shop 5 km from a point A. Jock starts from A, but I start from a point B, which is 1 km nearer to the shop (i.e., at 4 km from the shop). Since Jock has a 5 km walk and his speed is 5 kph, it takes him one hour to get to the shop. Because I walk at 4 kph and have only 4 km to walk, and we both start at the same time, we arrive at the shop together.
>
> Look at the walk another way. I start at B, one kilometre ahead of Jock. By the time he gets to B, I have moved to a point C, which is still ahead of him. By the time he gets to C, I have moved to a point D, and by the time he gets to D I have moved ahead to a point E. This goes on indefinitely through staging points A, B, C, D, E, F, … . Whenever Jock reaches the point I was at when he passed the previous one, I have moved ahead to the next one, so he never catches up with me.

Plausibility v. Commonsense

Both arguments are plausible, but commonsense and experience tells us that the first is right. Jock catches me, and if we continue walking he will overtake me.

The second argument is a modern version of a paradox usually referred to, for fairly obvious reasons, as that of *Achilles and the tortoise*. It is one of several paradoxes attributed to the Greek philosopher Zeno of Elea, who lived in the 5th century BC. Zeno was not a fool. He knew his argument that Achilles would never catch the tortoise, like my argument that my faster walking friend Jock would never catch me, was false. He formulated the paradox because he wanted to see the fallacy exposed.

Infinity – a multi-headed serpent?

Exposure is easy using techniques based on concepts that have been refined only during the last three to four hundred years. Their development altered dramatically both the scope of, and our attitudes towards, mathematics. Many advances in modern mathematics call for a proper understanding of the notion of *infinity* and the often related, but much wider, concept of a *limit.* So far I have been coy about the meaning of infinity, often using the term in a way that might be interpreted as 'and so on without end' when referring to an unending string of numbers.

4.2 Sets

To view infinity from another angle, it is useful to know what mathematicians mean by a *set*, a concept to be describe more formally in Chap. 16.

A shorthand for the aggregate of natural numbers between 1 and n inclusive is

$$1, 2, 3, \ldots, n\ .$$

This 'collection' of numbers is an example of a *finite set.* A set is a collection of items determined by rules which enable us to specify uniquely whether any given item is, or is not, a member of that set. The items need not be numbers. For the above set of positive integers, if we fix n to be 100, then 77 is a member of the set, but 7.25, 102 and 6749 are not. The number of elements in a finite set is called the *size* of the set. Two sets of the same size are often described as *equivalent.* Here equivalent does not mean the individual items (called the *elements*) in the sets are the same. If we have a set consisting of seven bananas and another of seven oranges, these are equivalent. Each is also equivalent to the set of integers 1 to 7 inclusive.

There is no such thing as a largest natural number, for however large n may be there is always a still larger integer $n + 1$. A shorthand for the set of natural numbers (positive integers) without end is

$$1, 2, 3, \ldots$$

which is sometimes made clearer by writing, as I did in Sect.2.8,

$$1, 2, 3, \ldots ad\,inf.$$

where *ad inf* is an abbreviation for *ad infinitum.* Sometimes one sees *ad inf* replaced by the symbol ∞ (called *infinity*). This is unfortunate in this context, because it gives the impression that ∞ is just another natural number, This is not so, for ∞ does not obey the ordinary rules of arithmetic for natural numbers.

Infinity + infinity = ?

For example, for any natural number n we have $n+n = 2n$, but a statement $\infty + \infty = 2\infty$ is not meaningful. There is no number that is twice as large as infinity! In the next chapter we shall see that the properties of infinity depend on what type of infinity we are talking about, for there is more than one kind.

Sets consisting of numbers have an important role in mathematics. They may consist of positive or negative integers, or rational numbers, or other kinds of numbers that we meet later.

The notion of finite sets extends to 'infinite' sets with an unlimited number of elements, such as the set of all integers. Care is needed to define equivalent sets in this case.

The 'equal size' property for equivalence of finite sets implies that we can pair members of each set in such a way that none are left unpaired in either set, and in each set every element is paired with only one element from the other set. Thus the set consisting of the natural numbers 1 to 10, and the set of the first ten prime numbers 2, 3, 5, 7, 11, 13, 17, 19, 23, 29 are the same size, because we can match them up in this way. One such matching is

$$\begin{array}{cccccccccc} 1 & 2 & 3 & 4 & 5 & 6 & 7 & 8 & 9 & 10 \\ \downarrow & \downarrow & \downarrow & \downarrow & \downarrow & \downarrow & \downarrow & \downarrow & \downarrow & \downarrow \\ 2 & 3 & 5 & 7 & 11 & 13 & 17 & 19 & 23 & 29 \end{array}$$

This is only one of many possible matchings. Another valid one is

$$\begin{array}{cccccccccc} 1 & 2 & 3 & 4 & 5 & 6 & 7 & 8 & 9 & 10 \\ \downarrow & \downarrow & \downarrow & \downarrow & \downarrow & \downarrow & \downarrow & \downarrow & \downarrow & \downarrow \\ 5 & 13 & 11 & 19 & 2 & 3 & 29 & 23 & 7 & 17 \end{array}$$

This possibility of matching extends to certain infinite sets, but we shall see in Sect. 5.4, not to all infinite sets. Here is a case where it works. Consider the set of all positive integers and the set of all even integers. One way we can uniquely match all the elements in the two sets is

$$\begin{array}{ccccccccccccccccc} 1 & 2 & 3 & 4 & . & . & . & 99 & . & . & . & n & . & . & . \\ \downarrow & \downarrow & \downarrow & \downarrow & \downarrow & \downarrow & \downarrow & \downarrow & \downarrow & \downarrow & \downarrow & \downarrow & \downarrow & \downarrow & \downarrow \\ 2 & 4 & 6 & 8 & . & . & . & 198 & . & . & . & 2n & . & . & . \end{array}$$

This equivalence is valid, since no matter how large n is we can find another integer $n+1$, and this can be paired with the (necessarily even) integer $2(n+1)$.

When part may equal whole

This result runs counter to intuition. Intuition tells us there are as many odd integers as there are even integers, and therefore that the total number of even integers must be only half the total number of integers. This is true if we fix n, say $n = 20$. Then the number of even numbers less than or equal to 20 is 10, i.e., 2, 4, 6, 8, 10, 12, 14, 16, 18, 20, whereas there are 20 natural numbers less than or equal to 20. This is not relevant to the situation considered above, because the set of integers less than or equal to 20 has 20 elements, whereas the set of even intengers less than or equal to 20 has only 10 elements, so the two sets are not equivalent. For finite set equivalence, we might consider the set with elements 1, 2, 3, 4, 5, 6, 7, 8, 9, 10 and the set with elements 2, 4, 6, 8, 10, 12, 14, 16 18, 20. This generalizes to a pairing of the natural numbers 1 to n with the even numbers 2 to $2n$.

The finite concept that the part is less than the whole may no longer hold when we move to infinite sets. This, perhaps surprising, situation serves well to remind us that in mathematics intuition may be a good servant but a bad master — a warning I have already given and repeat more than any other in this book.

Zeno's paradox of Achilles and the tortoise also illustrates the dangers of intuition.

4.3 Limits

If the elements of a set are placed in an order determined by some rule this ordering forms a sequence. In cases we consider the elements are numbers, or algebraic symbols representing numbers.

For a finite sequence of n elements a useful general notation is to denote the rth element by a_r where $r = 1, 2, \dots, n$. In set terminology the elements of a sequence of size n, have an equivalence to the set of the first n natural numbers.

It is a little irritating to newcomers to the subject, but sometimes mathematicians find it convenient to let r take integer values between 0 and $n - 1$ (instead of between 1 and n) when specifying a sequence with n elements.

If we allow n to become indefinitely large, we call the associated sequence an *infinite sequence*. What happens to a_n when n becomes indefinitely large — usually referred to as 'n tending to infinity' — is frequently of interest.

What is often of more interest than the sequence itself, is the sum of all terms in either a finite, or infinite, sequence. Such sums form a *series*.

To find the sum of the terms of a finite sequence, we apply the rules of addition. Finding the sum of the terms of an infinite sequence is a matter of considering what happens to the corresponding finite sum as n becomes larger and larger. This again is described as 'letting n tend to infinity', which is often written 'when $n \to \infty$'. The symbol $\to$ is mathematical shorthand for 'tends to'. Such studies lead to the notion of a *limit*.

Sometimes it is easy to see what happens to the sum, S_n, of the first n elements of a finite sequence as n becomes larger and larger. For example, if that finite sequence is the first n natural numbers, then as n increases their sum becomes larger and larger. If the terms of a sequence become progressively smaller as n increases, it may be less obvious what happens to their sum as $n \to \infty$.

The successive sums in a series $S_1, S_2, \ldots, S_n$ themselves form a sequence.

For our first example of a finite sequence we set $a_r = 1/10^{r-1}$, where r takes all integer values between 1 and some fixed value n. Specifically,

$$\begin{aligned} a_1 &= 1/10^0 = 1 \ , \\ a_2 &= 1/10 = 0.1 \ , \\ a_3 &= 1/10^2 = 1/100 = 0.01 \ , \\ a_4 &= 1/10^3 = 1/1000 = 0.001 \ , \end{aligned}$$

and so on to

$$a_n = 1/10^{n-1} = 0.000\ldots001 \ ,$$

where the final decimal digit 1 is preceded by $n-2$ zeros to the right of the decimal point. The sum of all terms of the sequence gives the series

$$S_n = a_1 + a_2 + \ldots + a_n \ .$$

If we set $n = 1, 2, 3$, etc., then

$$\begin{aligned} S_1 &= 1 \ , \\ S_2 &= 1 + 0.1 = 1.1 \ , \\ S_3 &= 1 + 0.1 + 0.01 = 1.11 \ , \\ S_4 &= 1 + 0.1 + 0.01 + 0.001 = 1.111 \ , \end{aligned}$$

and continuing this way

$$S_n = 1.11111\ldots1 \ ,$$

where the number of 1's following the decimal point is $n-1$.

Thus, if we allow n to increase indefinitely, the sum of all these terms is the decimal number 1 followed by an unending string of 1s after the decimal point. You probably know that we call this a *recurring decimal.* It is easy to verify by division that this is equal to the fraction (rational number) 10/9 or $1\frac{1}{9}$. If a series S_n takes some finite limiting value when $n \to \infty$, we say the series *converges* to that limit. In this example the limit is 10/9.

The sequence $1,\ 1/10,\ 1/10^2,\ \ldots$, has the property that each term is formed from the preceding one by multiplying by the same factor, 1/10 or 0.1. It is a special case of a more general sequence called a *geometric progression.* You may have met geometric progressions already. The series obtained by adding the terms in a geometric progression is a *geometric series.*

Generalization from a particular case to a wider context is an important way in which mathematics expands to cover many situations in one framework.

A geometric progression with n terms is a sequence with some given first term a. The successive terms are obtained by multiplying each preceding term by a common factor, r, giving the sequence

$$a,\ ar,\ ar^2,\ ar^3,\ \ldots\ ,\ ar^{n-1}.$$

It may help to be an artful dodger

There is an ingenious way to get a formula for the sum of these terms. Like many situations in mathematics the dodge seems obvious once we see it. In this and many other situations, spotting the trick *ab initio* is something of an art. In Sect. 4.5 we look at another way to verify the result we get here, using there an approach that is applicable to an even wider range of problems.

The school algebra trick for getting the sum S_n, where

$$S_n = a + ar + ar^2 + ar^3 + \ldots + ar^{n-1}, \tag{4.1}$$

is to multiply both sides of this expression by r giving

$$rS_n = ar + ar^2 + ar^3 + ar^4 + \ldots + ar^n. \tag{4.2}$$

Subtracting (4.2) from (4.1) we see at a glance that, on the right of the equality signs, all but the first term in (4.1) and the last term in (4.2) cancel out leaving

$$S_n - rS_n = a - ar^n\ ,$$

whence, by simple algebra

$$(1 - r)S_n = a(1 - r^n)\ ,$$

and

$$S_n = \frac{a}{1-r} - \frac{ar^n}{1-r}\ . \tag{4.3}$$

So far we have said nothing about a and r except implicitly that they represent numbers. In practice, these may be any real numbers except that r must not equal 1. The real numbers include all the rational numbers, and also numbers called *irrational numbers.* You may already be familiar with the latter. If not, don't worry, we meet them in the next chapter. If we let $r = 1$ the expression (4.3) for S_n would consist of two terms on the right each with zero in the denominator. The reason division by zero is not defined (i.e., it is not a permissible operation) in the real number system will become clearer in later chapters, but you can see the source of the difficulty here by looking at (4.1) and (4.2). If $r = 1$ then (4.1) and (4.2) become identical. Then if we subtract (4.2) from (4.1) we get only the trivial but true result $0 = 0$. When $r = 1$ we see from (4.1) that $S_n = na$.

Let 'thou shall not divide by 0' be your golden rule

In practice, we may also exclude the case $r = 0$, for then all terms after the first are zero, a situation of no practical interest.

What happens to S_n when $n \to \infty$? To simplify things for the moment assume a and r are both positive, because this is a practical case of immediate interest. In (4.3) the first term on the right does not depend upon n. When $r > 1$ the second term becomes larger and larger in magnitude, so that S_n will increase indefinitely as n increases. This behaviour is often described by saying the sum is *divergent* as $n \to \infty$.

If r lies between 0 and 1, but does not take either of these extreme values, often written as $0 < r < 1$, it is intuitively obvious that as n increases, r^n becomes smaller and smaller. But remember intuition is a good servant but a bad master. Here, however, intuition serves us well. We show formally in Section 4.5 that it is indeed true that r^n approaches zero when $n \to \infty$, if $0 < r < 1$. In fact, the difference between zero and the value of r^n can be made as small as we please (i.e., as close to zero as we please) by taking n sufficiently large. We express this as '$r^n \to 0$ as $n \to \infty$'. It now follows that as $n \to \infty$, (4.3) simplifies to

$$S = \frac{a}{1-r}\,. \tag{4.4}$$

We have dropped the subscript n from S_n to indicate that we are no longer considering a fixed n, but have allowed n to tend to infinity. S is sometimes called the *sum to infinity* of a geometric series. When r is positive, and less than 1, the sum is said to converge to $a/(1-r)$. Although we have not considered the situation where r is negative, it can be shown that (4.4) is still valid if $-1 < r < 0$. Trivially, it also holds when $r = 0$.

4.4 Resolving a Paradox

What, if anything, have geometric progressions to do with the 'walking to the shop' version of the Achilles and the tortoise paradox in Sect 4.1? My friend

Jock walks 5 km at 5 kph to reach the shop, while I, having a 1 km start, only have to walk 4 km at my speed of 4 kph.

As the problem is posed in Zeno's paradox, after Jock has walked 1 km he will be at my starting point. This takes him 12 minutes, or 1/5 of an hour, since he covers 5 km in one hour. In that time I have progressed 4/5th of a km, since, if I walk at 4 kph, it takes me 15 minutes to cover each 1 km. Therefore, in 12 minutes I cover $12/15 = 4/5$ km. To summarize, Jock in covering the first stage of the Zeno paradox, has covered a distance of 1 km in 12 minutes. In that time I have progressed to be 4/5 km ahead.

Since I am now 4/5 km ahead, in his next stage Jock must cover 4/5km. This takes him 4/5 of 12 minutes since he covers 1 km in the latter time, i.e., it takes him $(12 \times 4/5)$ minutes. While Jock is covering this second stage, I cover 4/5 of the distance he covered, i.e. a distance of $[(4/5)\times(4/5)]$ km. Jock's time to cover this new distance is, by arguments similar to those applied at the previous stage, $[12 \times (4/5) \times (4/5)]$ minutes.

Do you see a pattern emerging? You should verify that at the next stage of 'catching up' Jock has to walk $(4/5)^3$ km, and that this will take him $12 \times (4/5)^3$ minutes. Continuing this way, it is easy to see that the times in minutes taken by Jock to cover each stage form a geometric progression with terms

$$12,\ 12 \times (4/5),\ 12 \times (4/5)^2,\ 12 \times (4/5)^3,\ \ldots\ ,\ 12 \times (4/5)^m,\ \ldots$$

and that the distance in km he covers at each stage is a geometric progression with terms

$$1,\ (4/5),\ (4/5)^2,\ (4/5)^3,\ \ldots\ ,\ (4/5)^m, \ldots\ .$$

The corresponding infinite series sums are obtained respectively by substituting first $a = 12$ and $r = 4/5$, then $a = 1$ and $r = 4/5$, in (4.4), giving in the first case

$$S = 12/\left(1 - \frac{4}{5}\right) = 60\ ,$$

and in the second

$$S = 1/\left(1 - \frac{4}{5}\right) = 5\ .$$

The first of these results implies that in 60 minutes (1 hour) the distance between Jock and I will have shrunk to zero, (i.e., he will have caught up to me). The second implies that after he has covered 5 miles he will have caught up with me. These are the results we got in Sect. 4.1 using the simpler first reasoning, so there is no contradiction.

Perhaps you feel that the solution given here, when compared to the easier one given in Sect. 4.1, is using a sledgehammer to crack a nut. In this example introducing concepts of infinity and limits does create something like a sledgehammer. We shall see in later chapters that this sort of sledgehammer is useful for cracking objects larger than the nut we had here.

Geometric progressions help sort out the ambiguity about interest rates for the savings account mentioned in Sect. 1.1. This was about the advertisement that claimed an annual equivalent rate of 5 per cent corresponded to a gross rate of 4.89 per cent per annum.

The explanation hinges on the fact that interest is paid monthly, and the interest paid in any one month itself earns more interest in the following months. Monthly interest payments are made at a rate of 1/12 of the annual gross rate, i.e. at a rate of $4.89/12 = 0.4075$ per cent per month. This means that if we invest £100 (or dollars, euros or whatever is our local currency), in one month it will grow to £100.4075. This may be expressed as 100×1.004075. Next month we earn interest at the same rate on £100.4075. This increases our capital to $100.4075 \times 1.004075 = 100 \times 1.004075^2$. A pattern is emerging. At the end of the third month our original capital of £100 has increased to 100×1.004075^3. These terms form a geometric progression with $a = 100$, $r = 1.004075$. After 1 year the value of our accumulated capital is $100 \times 1.004075^{12} \approx 105.00$. This is indeed correct to 2 decimal places. Thus, our capital has increased from £100 to £105 over 12 months.

This is exactly what would happen if, instead of interest being added monthly at the rate given above, no interest was given until the end of the 12 month period, and it were then added at the *annual equivalent rate* of 5 per cent per annum.

Where algebra meets geometry

An attractive, and important, feature of modern mathematics is that by looking at things from different angles we often give them new interpretations. The introduction to geometric progressions was basically via arithmetic and algebra. One may also form a progression in a way that gives a geometric interpretation of convergence to a limit. Consider a portion of a straight line of unit length. Such a finite portion is often called a *segment*, and one of unit length a *unit segment*. We first split this unit segment into two equal parts. It is convenient to refer to these as the left and right segment as they are labelled in Fig. 4.1. The length of each will be 1/2.

We split the right segment in half once again. The left portion resulting from that split will have length $1/4 = (1/2)^2$. We repeat this process many more times, and at each step add the length of the left hand segment in the new split to those of all other segments to its left.

If we carry on indefinitely, there remains only a right hand segment that may be made as small as we please. If all is well, the sum of the lengths of the successive left hand segments should approach unity, the length of the original segment.

The sum is that of the geometric series

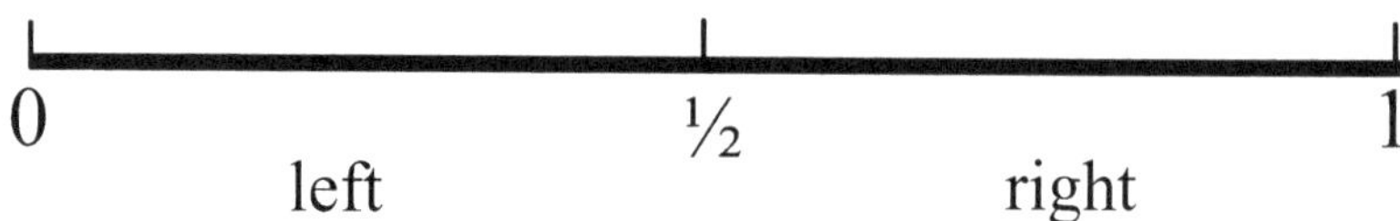

Fig. 4.1. Splitting a unit segment into half, forming left and right segments

$$S = \frac{1}{2} + \left(\frac{1}{2}\right)^2 + \left(\frac{1}{2}\right)^3 + \ldots + \left(\frac{1}{2}\right)^n + \ldots ,$$

which, by (4.4) has, as it should, the sum $S = (1/2)/(1 - 1/2) = 1$.

A necessary condition for convergence of an infinite series is that the terms of the sequence from which it is formed tend to zero, for otherwise the sum could not reach a limit. Intuition may suggest this is a sufficient condition. Here, however, intuition proves a bad master, as the following example shows.

Conditions may be necessary but not sufficient

The infinite series

$$1 + \frac{1}{2} + \frac{1}{3} + \frac{1}{4} + \ldots + \frac{1}{n} + \ldots$$

is the sum of the reciprocals of the natural numbers, where for any number r, the reciprocal is $1/r$. The series is called the *harmonic series.* As $n \to \infty$ then $1/n \to 0$, but the harmonic series does not converge to any finite limit. The sum increases indefinitely as n increases.

To show this we find a new series having all terms either the same or less than a corresponding term in the harmonic series. This new series turns out to be divergent. Since all terms in the harmonic series are positive, and the same or greater than those in the new series, it follows that the harmonic series must be divergent.

The first two terms in this new series are 1 and $1/2$, the same as the first two in the harmonic series. The next term is obtained by replacing $1/3$ by $1/4$ and adding it to the following term which is also $1/4$, giving $1/4 + 1/4 = 1/2$. We then take the next four terms ($1/5$ to $1/8$) and replace each by their smallest member $1/8$. Adding these four we again get $1/2$. The pattern is now becoming clear. We take the next 8 terms, and replace each by the smallest member of that group, which is $1/16$. The sum of these 8 replacements is again $1/2$. The process continues taking groups of the next 16, 32, 64, 128, and so on, terms to form a new infinite series

$$1 + \frac{1}{2} + \frac{1}{2} + \frac{1}{2} + \cdots .$$

This sum obviously becomes larger and larger as we add more terms, i.e., it diverges. Since this series diverges, so must any series whose terms are all either the same or greater. As aready pointed out, the harmonic series has that property.

In later chapters we meet other series of practical importance where we need to know whether they are convergent or divergent, or have some other limiting property.

4.5 Different Kinds of Proofs

We have used several unrelated dodges to establish results, or to confirm the truth of assertions. The indirect, or reductio ad absurdum, proof in Sect. 2.9 for there being an infinity of prime numbers was a different approach from that for obtaining the sum of a geometric series in Sect. 4.3. We used yet another dodge to show that the harmonic series is divergent.

If each fresh problem required a new 'dodge' for its solution mathematics would have made slow progress. Fortunately, there are many situations where the same approach may be used to establish a wide range of results. Indirect proof is one of these. Another is *mathematical induction.*

In science generally induction describes inferences based on observation or experiment. Observations over thousands of years give us confidence that the sun will rise each morning in the east, and set each evening in the west. This is not a proof that the sun will behave that way tomorrow. Indeed, I and many other people have on at least one occasion seen the sun rise in the West! Can you work out how this could come about. If you are stumped I'll let you into our secret at the end of this chapter.

In Mathematics empirical observations may suggest a formula or some other result, but this does not prove that it is always valid.

The analogue of an empirical argument in a mathematical context arises if we claim (without proof) that the sum of the first n integers $1, 2, \ldots, n$ is $n(n+1)/2$. This is true if we put $n = 1$, again if $n = 20$, again if $n = 100$, so we gain confidence in the formula. That does not prove it is universally true. It would be very tedious (for all practical purposes impossible without a powerful computer) to check by simple addition that the formula held for all integers n less than 1 000 000, say. Even if we did this, it would still leave a doubt, if only a small one, whether the formula were true if n exceeded one million.

If I claim that if we take any two positive integers a, b such that a is greater than b (written as $a > b$), then for any real r it follows that $ar > br$, is my assertion correct? If we assign any permissible (positive integer) values to a and b and take r to be any positive number (not necessarily an integer) I assure you the result will hold. Try it yourself for a number of cases.

Now set $a = 7$, $b = 5$ and $r = 0$. Both ar and br become zero, so one is no longer greater than the other. If we again take $a = 7$, $b = 5$ and set $r = -1$, we get $ar = -7$ and $br = -5$ and -7 is not greater than -5, indeed it is less. (If I am short of 7 cows and call this -7, I have fewer cows than if I had -5).

This counter example shows the original assertion is untrue. Had we specified that r must be a positive real number, the assertion would have been valid. Failure to realise simple truths about inequalities is the basis of many 'spoof' mathematical proofs. If you are not already familiar with the main rules for handling inequalities these are given in the Appendix, Sect. A.1.

Mathematical induction is, unlike its empirical counterpart, a rigorous logical method of proof. It works this way:

- *Step 1.* A proposition is asserted [e.g., the sum of the first n natural numbers is $n(n+1)/2$].
- *Step 2.* We assume the proposition is true for some fixed $n = N$. We check whether this implies it is also true if $n = N + 1$. If this is established we proceed to step 3. If it is not true one stops, because the proposition cannot hold.
- *Step 3.* We check if the proposition holds for $n = 1$. If it does not, the proof fails at this stage. If it holds the proof is complete, because step 2 implies it holds for $n = 2$, and by the inductive argument hence for $n = 3, 4, 5, \ldots$, and thus for all n.

We now use mathematical induction to prove that the sum of the first n natural numbers is $n(n+1)/2$. Having stated our proposition, we move to step 2 and assume the result holds for some $n = N$, i.e., that

$$1 + 2 + 3 + ... + N = \frac{1}{2}N(N+1) \, .$$

Adding $N + 1$ to each side gives

$$\begin{aligned} 1 + 2 + 3 + ... + N + (N+1) &= \frac{1}{2}N(N+1) + (N+1) = (N+1)(\frac{1}{2}N + 1) \\ &= \frac{1}{2}(N+1)(N+2) \, . \end{aligned}$$

This is exactly what we get by replacing N by $N + 1$ in the formula we are trying to establish. To complete the proof, we check whether the result holds for $n = 1$. It does, since $1 = 1 \times (1+1)/2$. From step 2 it follows that the result also holds for $n = 2, 3, 4, \ldots$. Thus it is universally true.

Unfortunately, mathematical induction does not tell us how we got the idea that the sum might be $n(n+1)/2$. That requires a different approach. The result is easily obtained directly in this example. The trick is to write down the terms in natural and reverse order, then to sum them in pairs, i.e.

$$\begin{aligned} S_n &= 1 + 2 + 3 + \ldots + (n-1) + n \\ S_n &= n + (n-1) + (n-2) + \ldots + 2 + 1 \end{aligned}$$

whence

$$2S_n = (n+1) + (n+1) + (n+1) + \ldots + (n+1) + (n+1) = n(n+1) .$$

The result follows on dividing both sides by 2.

A more important use of mathematical induction is to verify a result that might have been obtained by complicated manipulations, that may themselves be liable to error. The method is also useful if a result is suggested by intuition, perhaps as a potential generalization of some established result.

Here is an example where mathematical induction provides a useful tool to establish a result that empirical studies suggest is true, but where there is room for doubt that can only finally be removed by a rigorous proof. At first sight this result may seem little more than a curiosity, but one of its uses is to confirm an assumption we made in Sect. 4.3 when studying a geometric progression, i.e., that $r^n \to 0$ as $n \to \infty$, providing $0 < r < 1$.

We first prove by induction that if n is a positive integer, and x is any real number greater than -1, then

$$(1+x)^n \geq 1 + nx.$$

We assume the result is true for some $n = N$. We now multiply both sides of the inequality by $(1+x)$. This does not change the sign of the inequality since $(1+x)$ is positive providing $x > -1$. So

$$(1+x)^N (1+x) \geq (1+Nx)(1+x)$$

or equivalently

$$(1+x)^{N+1} \geq 1 + Nx + x + Nx^2.$$

The term Nx^2 is always non-negative, since N is positive and x^2 may only be zero or positive. Therefore, the inequality is made stronger by dropping this term. This implies

$$(1+x)^{N+1} \geq 1 + (N+1)x ,$$

which establishes that if the inequality holds for N, it also holds for $N+1$. Now, when $n = 1$, we have, almost trivially, $(1+x)^1 \geq 1 + x$, i.e., we have equality. Thus the assertion also holds for $n = 2, 3, 4, \ldots$.

In Sect. 4.3 we asserted that if r is positive and less than 1, then $r^n \to 0$ as $n \to \infty$. The proof requires some ingenious algebraic manipulation. If r is positive and fixed and less than 1, we may write $r = 1/(1+x)$, where x is now positive and fixed. Therefore, $1/r = 1 + x$ and so $(1/r)^n = (1+x)^n$. From the result we proved above that $(1+x)^n \geq 1 + nx$, since nx is positive when $x > 0$, it follows that $(1/r)^n > nx$. Using an easily established rule that inverting an inequality involving positive quantities changes the direction of the inequality, we get $r^n < (1/x).(1/n)$. Since x is positive and fixed, and $1/n$ becomes ever closer to zero as n increases, it follows that $r^n \to 0$ as $n \to \infty$.

It can be shown that the result also holds if r is negative and less than 1 in magnitude, i.e., if $-1 < r < 0$.

The formula for the sum of n terms in a geometric progression may also be verified by induction. We assume that providing $r \neq 1$ the sum of the first N terms is

$$S_N = \frac{a(1 - r^N)}{1 - r} \tag{4.5}$$

This is easily seen to be equivalent to (4.3). From (4.1) it is also clear that the $(N + 1)$th term is ar^N, so that if (4.5) holds, then

$$S_{N+1} = \frac{a(1 - r^N)}{1 - r} + ar^N = \frac{a(1 - r^N) + ar^N(1 - r)}{1 - r} == \frac{a(1 - r^{N+1})}{1 - r}.$$

This has the form of (4.5) with N replaced by $N + 1$. Thus, if the result is true for N it is true for $N + 1$. It is easily verified that (4.5) holds for $n = 1$, because then $S_1 = a$. So the result is universally true for positive integral n. While this confirms the result, as pointed out earlier, we need some other method to obtain (4.5) in the first place.

Mathematical induction is also useful if we want to check an assertion made by somebody else, when we ourselves are unable to obtain it directly. Maybe we don't know how it was obtained originally. There are also situations where we may have little more than a hunch that a result holds, perhaps because we feel it is an intuitively reasonable generalization of something known to be true. Here is an example. It only needs straightforward algebra to verify that

$$(a^2 + b^2)(c^2 + d^2) = (ac - bd)^2 + (ad + bc)^2. \tag{4.6}$$

Suppose now that a, b, c, d are all integers. Then (4.6) implies that the product of two integers, each of which is the sum of squares of two other integers, may itself be expressed as the sum of squares of two integers. For example, if $a = 3$, $b = 7$, $c = 4$ and $d = 9$, we see from (4.6) that

$$(3^2 + 7^2)(4^2 + 9^2) = (3 \times 4 - 7 \times 9)^2 + (3 \times 9 + 7 \times 4)^2 ,$$

or

$$5626 = (-51)^2 + (55)^2 .$$

It is easily verified that this is true.

An astute mathematician might conjecture that this result may have extensions. One possibility is that the product of any number, n, of integers, each of which is the sum of squares of two integers, might still be expressible as the sum of squares of two integers. Put algebraically, the assertion is that if we specify integers $r_i = (a_i^2 + b_i^2)$, $i = 1, 2, \ldots, n$, then the integer, q_n, where

$$q_n = r_1 r_2 \ldots r_n ,$$

is itself a sum of squares of two integers.

To establish this by induction, we first assume it is true for $n = N$. Then $q_{N+1} = r_1 r_2 \dots r_N r_{N+1} = q_N \times r_{N+1}$. This is of the form of the left hand side of (4.6), whence it follows that if the result is true for N it is true for $N + 1$. We know it is true for $n = 2$ [this is effectively (4.6)]. Thus it is also true for $n = 3, 4, 5, \dots$.

An interesting application of (4.6) is for the generation of triplets of integers satisfying the Pythagorean relationship $x^2 + y^2 = z^2$. If we choose number pairs such that the left-hand side is a perfect square, then using (4.6) we can express this as a sum of two squares of integers.

For example, setting $a = c = 7$ and $b = d = 5$ we find $ac - bd = 24$ and $ad + bc = 70$ while $(a^2 + b^2)(c^2 + d^2) = 74^2$. We can easily verify that

$$24^2 + 70^2 = 74^2.$$

4.6 Loose Ends

This chapter is something of a collection of loose ends that itself creates even more loose ends. I have touched upon ways to interpret infinity, and upon the notion of a limit. Both topics are met frequently in this book. In particular, aspects of infinity are further developed in Sect. 5.4, while Chap. 10 brings in other notions of infinity, and ties them to the wider concept of limits applied to sequences, series and algebraic functions. In Chap. 11 the fundamental role of limits in integral and differential calculus is explored. These ideas are expanded further in Chap. 12. As stated in Sect.4.2, sets are dealt with more fully in Chap. 16.

Piling them high

Something not so obvious. This loose end is a good example of the less obvious. Suppose we have a standard chess board with 64 squares arranged in an 8×8 array. On the first square we place a circular disk of thickness 1 mm. – roughly the thickness of a typical small-denomination coin. On the next square we place 2 such disks, on the next 4, on the next 8, on the next 16, doubling the number each time until we have piles on all 64 squares. Then, just for fun we collect the disks from each of the first 63 squares and place them on top of the pile on the 64th square. Remembering that each of the disks in 1 mm thick, approximately how high will the pile be? Which of the following do you think the best approximation:

- It will be almost as high as the Eiffel Tower

- It will be almost as high as Mount Everest.
- The top of the pile will be almost as far from the ground as the distance between the earth and the moon.
- The top of the pile will be further from the ground than the distance of the sun from the earth.

Think about this for a while before reading on.

Have you made any progress? The numbers on each square are

$$1,\ 2,\ 2^2,\ 2^3,\ \ldots\ ,\ 2^{62},\ 2^{63}.$$

Does something seem familiar? This is a geometric progression with $a = 1$, $r = 2$ and $n = 64$. The sum gives the total number of disks that we pile on top of each other on one square on the board. Using (4.3) it is easy to show that this sum is $2^{64} - 1$.

My pocket calculator told me that this number is approximately

$$1.8446744 \times 10^{19}.$$

If you have a calculator, or computer software that can handle very large numbers, you will find the exact value is

$$18\ 446\ 744\ 073\ 709\ 551\ 615\ ,$$

so the calculator approximation is correct to 8 significant figures.

What about the height of the pile of disks? Remember each disk is 1 mm thick, so the total pile is 18 446 744 073 709 551 615 mm high. Dividing by 1000 gives the height in metres. Dividing by 1000 again we get the height in kilometres, i.e., rounding to the nearest kilometre it is

$$18\ 446\ 744\ 073\ 710\ .$$

A slightly less accurate approximation rounding to 4 significant figures, which is sufficiently accurate for the purposes of answering the question we posed above, is 1.845×10^{13} km. The only other information we need, to see which was the correct option from the four given, is that the mean distance of the sun from the earth (it varies slightly depending on the time of year) is approximately 1.50×10^8 km.

Even if each disk only weighs 1 gm, don't try lifting the pile up single handed unless you feel confident about lifting weights of about 1.845×10^{13} tonnes.

Finally what about my assertion on p. 55 that some of us have seen the sun rise in the west? This can be experienced at certain times of the year if flying westward at supersonic speeds. When British Airways operated Concorde from London to New York they had an evening service that departed London after sunset at certain times of the year and arrived at New York when it was still daylight. In these circumstances crew and passengers saw the sun rise above the Western horizon during the journey.

5

More of the Numbers Game

5.1 More about Rational Numbers

In Sect. 2.7 we met rational numbers as a device for allocating equal shares when there is no integer solution. They also have a key role in increasing precision of measurement. As civilizations became more sophisticated, measurement acquired an importance akin to that of counting. Numbers were used to measure quantities like length, area, mass, temperature, or the passage of time.

Given a unit of measurement, it is easy to associate measures with counts of that unit. Many measurements cannot be made with sufficient accuracy by counting only complete units. One solution is to use smaller units. Another is to incorporate fractional parts of units into the count.

Taking a *foot* as the unit, the room in which I am writing this book is more than 10 ft, but less than 11 ft, wide. We may reduce this element of uncertainty by dividing the original unit into n equal parts. Conventionally, in the USA or the UK, if the foot is the original unit we set $n = 12$, and call the equal parts inches. Had I measured the room width in metres, I would conclude my room was more than 3 m, but less than 4 m, wide. For more refined measurements we may divide the metre into 100 equal parts, each called a *centimetre*, or into 1000 equal parts, each being a *millimetre*.

The size of each equal sub-unit formed by dividing the original unit into n equal parts is conveniently denoted by $1/n$. A measurement of r such sub-units is denoted by r/n in terms of the original unit. This is the familiar rational fraction. For example, my room is 10 ft 7 in wide. This may be expressed as $10\frac{7}{12}$ feet, this being a rational number expressing the width in feet. Alternatively, I could record the width as 127 in., or in metric units as 3.226 m or 3226 mm.

In Sect. 2.1 we mentioned that while the product of two rationals, a/b and c/d, is the comparatively simply $(a/b) \times (c/d) = ac/bd$, obtained by multiplying numerator and then denominator factors. For addition the rule is the more complicated

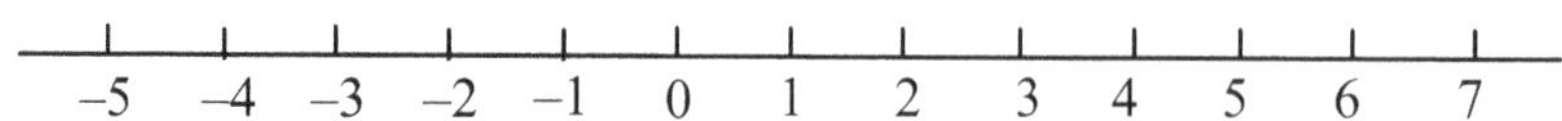

Fig. 5.1. Representation of integers by equally spaced points on a line

$$\frac{a}{b} + \frac{c}{d} = \frac{ad + bc}{bd}. \tag{5.1}$$

These and other rules such as $a/a = 1$, and that $a/b = c/d$ implies $ad = bc$, ensure a coherent system of measurements. For example, if a block of chocolate weighs $1/4$ kg and another block weighs $1/2$ kg using (5.1) we correctly conclude the total weight is $3/4$ kg.

Positive and negative rational numbers may be operated on using the rules (2.1)–(2.5), i.e., the commutative and associative laws of addition or multiplication, and the distributive law.

We can refine precision of measurement by introducing more rational numbers. We may do this by increasing n in our equal subdivisions of a unit into n segments each of length $1/n$.

Mathematicians often visualise numbers as points on a line. The positive and negative integers are equally spaced on that line in an order indicated in Fig. 5.1, which shows a finite segment of such a line.

Rational numbers of the form r/n, where $r < n$, may be represented in the segment from 0 to 1, usually called the interval (0, 1), on this line. The 'fraction' $1/2$ is represented by the midpoint of the interval (0, 1). The decimal fraction $0.666666666\ldots = 2/3$ corresponds to a point to the right of that corresponding to $1/2$, but closer to that point than to the point representing 1. For any chosen positive integer n, points may be inserted between 0 and 1 each corresponding to a rational number r/n, where r is an integer satisfying the condition $1 \leq r \leq n - 1$. We may do this for increasing n, eventually making n as large as we please, so there is no finite limit to the number of rationals between 0 and 1.

If $r > n$, then r/n can be expressed as an integer N, or as an integer N and a remainder expressible as a fraction s/n where $s < n$. It may also be expressed as a decimal consisting of an integer N, followed by a decimal fraction having the form $N.a_1a_2a_3\ldots$, where the a_r are each integers between 0 and 9. The decimal may terminate, or it may be a recurring decimal. We look at these forms in more detail later.

5.2 A Limitation of Rational Numbers

For practical measurement purposes the rational numbers appear to suffice, for they are densely placed on the line in Fig. 5.1. However, even the Greeks knew that there were other numbers. An example dating back to Euclid may be put in geometric terms. If we construct a square with each side of unit length, then the length of the diagonal cannot be expressed as a rational number. If x denotes the length of the diagonal, then by Pythagoras's theorem

$$x^2 = 1^2 + 1^2 = 2\ ,$$

and x is therefore given by the positive square root of 2 (written $\sqrt{2}$), i.e., $\sqrt{2}$ is the number which, when multiplied by itself, gives 2.

1.4142 is not the square root of 2

If $\sqrt{2}$ were a rational number, it must be expressible in the form $x = p/q$, where p and q are in their lowest terms, i.e., p and q have no common factor. This assumption leads to a contradiction. To show this, we note first that $p/q = \sqrt{2}$ implies that $p^2/q^2 = 2$, or $p^2 = 2q^2$. This in turn implies p^2 is even, and thus that p is also even, since the square of any odd number is necessarily odd. Pause a moment to be sure you see why. We may therefore find an integer r such that $p = 2r$. This implies that $p^2 = 2q^2$ is equivalent to $4r^2 = 2q^2$ or that $2r^2 = q^2$. This in turn implies that q^2, and hence q itself, is even. However, we already know that p must be even. Thus p and q are both even, and therefore divisible by 2. This means that the assumption that p and q have no common factor is untenable. Thus $\sqrt{2}$ cannot be expressed as a rational number.

The above argument is another example of an indirect proof like that used by Euclid to show that there are infinitely many primes (Sect. 2.9).

Despite our finding that $\sqrt{2}$ is not rational there is, nevertheless, a point on the straight line in Fig. 5.1 that corresponds to $\sqrt{2}$. This can be found by a standard Euclidean geometry construction using ruler and compasses only. It is easy to construct two lines of unit (or any other specified length) at right-angles to each other to form adjacent sides of a square. If you do not know how to do this, please take my word for it. We assume this has been done, and in Fig. 5.2 this construction is imposed with the initial side corresponding to the interval (0, 1) in Fig. 5.1. The line from 0 to the point A, where A is at unit distance above the point 1, is the diagonal of a unit square (i.e. a square with each side 1 unit). If we use compasses to form an arc with centre at 0 to pass through the point A (i.e., with radius equal to the diagonal length), where this arc intersects the real-number line is a point corresponding to $\sqrt{2}$.

Establishing that numbers like $\sqrt{2}$ cannot be expressed as rational fractions of the form p/q does not, in practice, cause major problems in measuring the length of the diagonal of a unit square, because we can get good rational approximations. This is easily illustrated by representing such approximations as decimal fractions.

Since $1^2 = 1$ it follows that 1^2 is less than $(\sqrt{2})^2 = 2$. Also since $2^2 = 4$, it follows that 2^2 is greater than $(\sqrt{2})^2$. Thus, 1 is an underestimate of $\sqrt{2}$, and

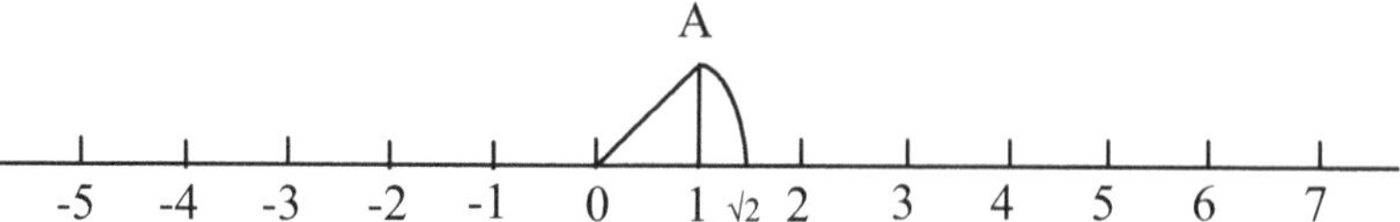

Fig. 5.2. Positioning an irrational number $\sqrt{2}$ on a line representing the real numbers

2 is an overestimate. Proceeding in this way we get a sequence of narrowing intervals that contain $\sqrt{2}$

$$\begin{aligned}
\mathbf{1}^2 &= 1 < 2 < \mathbf{2}^2 = 4 \\
\mathbf{1.4}^2 &= 1.96 < 2 < \mathbf{1.5}^2 = 2.25 \\
\mathbf{1.41}^2 &= 1.9881 < 2 < \mathbf{1.42}^2 = 2.0164 \\
\mathbf{1.414}^2 &= 1.999\,396 < 2 < \mathbf{1.415}^2 = 2.002\,225 \\
\mathbf{1.4142}^2 &= 1.999\,961\,64 < 2 < \mathbf{1.4143}^2 = 2.000\,244\,49\ ,
\end{aligned}$$

where successive interval end points are in **bold**.

It follows from the last line that $\sqrt{2}$ lies between 1.4142 and 1.4143. This is a tedious process, though it can be continued to give closer and closer approximations to $\sqrt{2}$. For many practical purposes in engineering and other technologies the approximation 1.4142, to four decimal places, is sufficient.

One reason approximating to more decimal places is tedious is because there is no rule for deciding what the next decimal digit might be. Starting with a 4-decimal place approximation all we know is that $\sqrt{2}$, expressed to five decimal places, must lie between 1.41420 and 1.41430. We may reasonably hazard a guess that it lies nearer to the former. This is because the square of the former given above, i.e., 1.99996164, is closer to 2 than is the square of the latter, i.e., 2.00024449. Indeed, we find

$$\mathbf{1.41421}^2 = 1.999\,989\,924 < 2 < \mathbf{1.41422}^2 = 2.000\,018\,208.$$

In the 21st century this monotonous search for a better approximation to an irrational number, or some alternative 'short cut' methods based on algebraic approximations, are somewhat academic. Most pocket calculators give at the press of a button, for any irrational square root, an approximation correct to some eight to ten decimal places. Mine gives to 9 decimal places, $\sqrt{2} = 1.414213562$.

The distinction between rational and irrational real numbers has implications for other aspects of modern mathematics. It is only in the last two centuries that many characteristics of the number system have been clarified by the development of modern number theory.

5.3 Number Theory

Like many facets of pure mathematics, a soundly based number theory requires the development of what to nonmathematicians may seem to be pin-pricking — some think plain boring — logical arguments. Nevertheless, it is worth looking at some aspects because, among other things, they throw more light on the nature of infinity. Many readers — particularly those who are not aspring to be pure mathematicians — may find the nitty gritty detail in the rest of this chapter hard to digest at first reading, so I suggest that if you find the detail hard to follow, perhaps even a little boring, you skim through it quickly and then move on to the next chapter.

Any rational number such as 47/211 or −323/101 or 4/5 or 20/5 can be expressed uniquely either as an integer or as an integer followed by a decimal fraction which may, or may not, terminate. The general form is

$$N.\, d_1\, d_2\, d_3\, \ldots\, d_r\, \ldots$$

where N is a positive or negative integer or zero, and d_r is a digit between 0 and 9 occupying the rth decimal place. Some examples are

$$\begin{aligned} 5\tfrac{1}{8} &= 5.125, \\ -7\tfrac{3}{7} &= -7.428571428571428571428571\ldots, \\ 22\tfrac{1}{11} &= 22.09090909\ldots, \\ \tfrac{1}{13} &= 0.076923076923076923\ldots, \\ 23\tfrac{23}{50} &= 23.46. \end{aligned}$$

All the above numbers either terminate after a finite number of decimal places, or else continue indefinitely with, after at most a finite number of decimal places, a repetitive pattern. In the second of the above examples this is a repeated '428571'. In the third it is a repeated '09', and in the fourth a repeated '076923').

It is not difficult to see that the decimal expansion of any rational number, expressed in the form r/s in its lowest terms (i.e. where r and s have no common factor), must either terminate, or have a recurring pattern. This is because the process for getting the decimal expression is essentially that of long division. This terminates if, at any stage, we get a remainder 0. If we never get a remainder 0, the decimal expression never terminates. In that case it must have a recurring pattern of length not greater than $s - 1$. This is because there are only $s - 1$ possible non-zero remainders at each decimal position. Thus, a remainder must be repeated either at one of these $s - 1$ positions, or at the sth decimal place. Thereafter the pattern repeats itself. We may regard an integer, or a terminating decimal, as a special kind of recurring decimal where we place zero in all decimal places after termination. The recurring pattern from then on is simply a string of zeros.

5.4 The Road to Infinity

Some infinities differ from others

In Sect. 4.2 we saw that the set of all even positive integers could be brought into one-to-one correspondence with the set of all positive integers. What is more surprising, is that the set of all rational numbers can be arranged to give a one-to-one correspondence with the set of all positive integers. This is described by saying the rational numbers are *countable* or *denumerable*. We outline an argument to establish this.

All positive rational numbers can be written as fractions of the form r/s where r, s are positive integers. It is not difficult to see that if we allow r, s to take all integer values less than or equal to n this gives rise to less than n^2 rational numbers. To show this, we first set $s = 1$ and allow r to take each value between 1 and n. This gives us, in effect, the integers 1 to n. We call this stage 1. Next, as stage 2, we set $s = 2$ and again allow r to take each value between 1 and n. This gives a sequence of n rational numbers of the form

$$1/2, 2/2, 3/2, \ldots, (n-1)/2, n/2.$$

However, if r is an even number, then the 'rational number' $r/2$ is an integer. Thus, it has already been recorded at stage 1. This means that at stage 2 we have only introduced $n/2$ additional rational numbers if n is even, and $(n+1)/2$ additional rational numbers if n is odd.

As an illustration, suppose we fix $n = 7$. Setting $s = 1$, and allowing r to take values between 1 and 7, gives the integers 1, 2, 3, 4, 5, 6, 7. Repeating the process with $s = 2$ gives the rationals 1/2, 2/2, 3/2, 4/2, 5/2, 6/2, 7/2. Since $2/2 = 1$, $4/2 = 2$ and $6/2 = 3$, the only new rationals at this stage are 1/2, 3/2, 5/2, 7/2.

We may continue to stage 3, setting $s = 3$ and again letting r take all integral values between 1 and n. This gives a sequence of n rationals of the form

$$1/3, 2/3, 3/3, \ldots, (n-1)/3, n/3.$$

Once again, some of these will be rationals already included as 'integers'. These correspond to any value of r that is exactly divisible by 3, i.e., for which $r \equiv 0 \pmod 3$.

At stage 4, we set $s = 4$ and form the sequence of n numbers

$$1/4, 2/4, 3/4, \ldots, (n-1)/4, n/4.$$

It is not difficult to see that any rational in this sequence having a common factor in numerator and denominator will be equal to a rational included at an earlier stage. For example, again taking $n = 7$, the sequence is 1/4, 2/4,

3/4, 4/4, 5/4, 6/4, 7/4. Here $2/4 = 1/2$, $4/4 = 1$, $6/4 = 3/2$, are already included at either stage 1 or stage 2.

We may continue in this way through stages $5, 6, 7, \ldots, n$. At each stage we form exactly n rationals. A little consideration makes it clear that if at any stage, q, there is a common factor between the numerator and q, then the corresponding rational number will already have been recorded in its lowest terms at an earlier stage. Mathematically, we are using the fact that any rational number r/s may be reduced to its lowest terms by cancelling out all common factors between r and s.

When we reach stage n we have recorded all possible rational numbers between 0 and n, where n may be made as large as we please. As we have seen many of these will be recorded more than once.

In generating each of the n stages, we have initially formed n sequences each of length n. The total number of rationals so generated is therefore $n \times n = n^2$. We have shown above some of these rationals are repeated at different stages. However, ignoring such repetitions for the moment, it is easy to show that for any n each of the n^2 rationals generated above can be matched with an ordered set of integers from 1 to n^2. One way to do this is to match those generated at stage 1 with the integers 1 to n, those generated at stage 2 with the integers $n+1$ to $2n$, those generated at stage 3 with the integers $2n+1$ to $3n$. Proceed in this way, we finally match those generated at stage n with the integers $(n-1)n$ to n^2. In the limit as n tends to infinity this matching means the set becomes a countable or denumerable set.

We still have a countable set if we remove all rationals that are repeated because r and s have a common factor. This is because each time we remove such an r/s, we simply need to reduce by 1 the integer corresponding to all successive entries in the enumeration.

Here is a simple illustration for $n = 4$. Then, $n^2 = 16$ giving for the 4-stage matching

1	2	3	4	5	6	7	8	9	10	11	12	13	14	15	16
↓	↓	↓	↓	↓	↓	↓	↓	↓	↓	↓	↓	↓	↓	↓	↓
1/1	2/1	3/1	4/1	1/2	2/2	3/2	4/2	1/3	2/3	3/3	4/3	1/4	2/4	3/4	4/4

Eliminating all elements except those where r/s have no common factor, we get a reduced matching of relevant rationals

1	2	3	4	5	6	7	8	9	10	11
↓	↓	↓	↓	↓	↓	↓	↓	↓	↓	↓
1	2	3	4	1/2	3/2	1/3	2/3	4/3	1/4	3/4

This concept generalizes and may be used to match the rationals with the positive integers however large n may be.

That is why, in the limit, as n tends to infinity, we say the rationals are *countably* or *denumerably* infinite. This type of infinity is like that of the positive integers, or that of all integers or that of all primes, so it may seem that there is only one kind of infinity.

We have considered only positive rationals in the above argument. We do not go into detail, but it is fairly clear that the argument can be extended to cover zero and the negative rational numbers.

What about irrationals? How do they fit into the scheme of numbers expressed. say, as decimal fractions? A feature of the decimal approximation to $\sqrt{2}$ is that there is no recurring pattern, This indeed is one way of distinguishing a rational from an irrational number. The successive approximations that we used in Sect. 5.2 to generate a non-terminating decimal that approaches the value $\sqrt{2}$ as we increase the number of decimal places indefinitely is only one way to generate an irrational number. Intuitively, we may feel that by slotting each such number between two rationals that all real numbers should still be denumerably infinite, but this is not so. This may be shown by an *indirect proof.* We find a contradiction if we assume the set of all real numbers, consisting of both rationals and irrationals, is denumerable.

If indeed all real numbers were denumerable, there must be some way to bring them into one-to-one correspondence with the integers. We may designate such a sequence this way:

$$\begin{array}{ll} \text{1st number} & N_1.d_{11}d_{12}d_{13}d_{14}\ldots \\ \text{2nd number} & N_2.d_{21}d_{22}d_{23}d_{24}\ldots \\ \text{3rd number} & N_3.d_{31}d_{32}d_{33}d_{34}\ldots \\ \ldots\ldots\ldots\ldots & \ldots\ldots\ldots\ldots \\ r\text{th number} & N_r.d_{r1}d_{r2}d_{r3}d_{r4}\ldots \\ \ldots\ldots\ldots\ldots & \ldots\ldots\ldots\ldots \end{array} \tag{5.2}$$

We have used what is called a *double suffix* notation for the decimal digits. This notation only became widely used during the 20th century. In the present context d_{ij} means the digit in the jth decimal place in the ith number. That is, the first suffix, i, refers to a digit in a specific number (the ith number), and the second suffix, j, to the position of the digit within that number. Thus d_{74} refers to the digit in the 4th decimal place in the 7th number. A complication arises if we want to refer specifically to, say, the 14th digit in the 17th number. The notation d_{1714} is then ambiguous, and indistinguishable from that for the 714th digit in the first number. To get round this difficulty either write $d_{17,14}$, using the comma to show the relevant separation, or write more fully 'd_{ij}, where $i = 17$ and $j = 14$', etc. Several examples in later chapters show the versatility of double suffixes.

We now show that however many numbers there are in the denumerable sequence (5.2), we can always find a new number that is not in it. This implies a contradiction, because we have assumed the set contains all real numbers. The contradiction implies the totality of real numbers is not denumerable.

To form a number not in the above sequence, we denote the digit in the ith decimal place in this new number by e_i. We choose for the first decimal place, a digit e_1 between 1 and 8 that differs from d_{11}. We then choose for the second decimal place a digit e_2 between 1 and 8 that differs from d_{22} (the second digit in the second number). We continue in like manner to select digits in the remaining places. This means that for all i, the digit e_i is different from d_{ii} (the ith digit in the ith number). Excluding 0 and 9 from the selection does not invalidate the argument, but it avoids potential minor technical difficulties.

The new number $N.e_1e_2e_3\ldots$ formed by this 'diagonal' selection process must differ from any number already in the ordering. It cannot equal the first, differing from it in the first decimal place, nor can it equal the second, differing from it in the second decimal place (as well as perhaps in the integral part N), and so on. Thus the denumerable table does not contain all real numbers. This contradiction means the real numbers are not denumerable. Thus, they represent a different type of infinity. This new type of infinity is described as an *uncountable* or a *nondenumerable* infinity.

To summarize: The rational numbers form a countable or denumerable infinite set. The complete set of real numbers, including the irrational numbers, form a noncountable or nondenumerable set.

These results are at the heart of number theory. They stem from work by the German mathematicians Georg Cantor (1845–1918) and Julius Wilhelm Richard Dedekind (1831–1916).

6

Complex numbers

6.1 The Next Step

We introduced the irrational number $\sqrt{2}$, via Pythagoras's theorem in the particular case of the length of the diagonal of a unit square. It is one solution to the equation $x^2 = 2$. That equation has two solutions written $x = +\sqrt{2}$ and $x = -\sqrt{2}$. Conventionally, if we write $\sqrt{2}$ without a sign, this is taken to be $+\sqrt{2}$. Irrational numbers make possible the solutions of many quadratic, and other, equations when these solutions are not expressible as rational numbers.

We face a new problem if we want to solve the equation $x^2 = -1$. There is no real number, rational or irrational, whose square is -1. More generally if a is a positive real number there is no real number solution to the equation $x^2 = -a$.

An extended number system that has many practical implications is possible if we introduce just one new number to represent the square root of -1. We use the symbol i for this, a symbol suggested by the name *imaginary*. That name is an unfortunate historical hangover, for i is no more imaginary than a rational or an irrational number! It has, however, a different purpose — one not directly related to counting or to physical measurements.

For a start it helps us solve quadratic equations. By itself, i solves the special equation $x^2 = -1$. By combining it with real numbers, to form what are called *complex* numbers (more unfortunate terminology), this enlarged system not only provides all possible solutions to any quadratic equation

$$ax^2 + bx + c = 0 \, ,$$

where a, b, c are real, but also to any equation of the form

$$a_n x^n + a_{n-1} x^{n-1} + \ldots + a_1 x + a_0 = 0 \, ,$$

where the coefficients a_r, $r = 0, 1, 2, \ldots, n$ themselves may be either real or complex numbers.

We have defined i as a number such that $i^2 = -1$. We call i the *imaginary unit*. It provides a useful extension of the number system not only for mathematicians. Electrical engineers and physicists would be virtually unable to work without complex numbers. Incidentally, engineers often write 'j' instead of i, because they use the symbol 'i' for electric current.

We define a complex number as one of the form $a + b\mathrm{i}$ where a and b are real numbers. If $b = 0$ the number is equivalent to the real number a. If $a = 0$ the number $b\mathrm{i}$ is referred to as an *imaginary number*. For the general complex number $a + b\mathrm{i}$, a is called the *real part* and b the *imaginary part*.

We add and multiply these numbers by the same rules as we use for real numbers, with the proviso that we replace i^2 by -1 whenever it occurs. Thus, for addition

$$(a + b\mathrm{i}) + (c + d\mathrm{i}) = (a + c) + (b + d)\mathrm{i} \ .$$

For multiplication

$$(a + b\mathrm{i}) \times (c + d\mathrm{i}) = ac + bc\mathrm{i} + ad\mathrm{i} + bd\mathrm{i}^2 \ ,$$

which, on setting $\mathrm{i}^2 = -1$, becomes

$$(a + b\mathrm{i}) \times (c + d\mathrm{i}) = (ac - bd) + (ad + bc)\mathrm{i}.$$

The rule for addition is simple. The real part of a sum is the sum of the real parts, and the imaginary part is the sum of the imaginary parts. For multiplication the rule is more complicated. The real part of a product is the product of the real parts minus the product of the imaginary parts. The imaginary part of the product is the sum of the products of the real part of each number with the imaginary part of the other.

In elementary algebra we learn that $(a^2 - b^2)$ factorises to $(a + b)(a - b)$ but that $(a^2 + b^2)$ cannot be factorized in real numbers. Complex numbers removes this restriction, since by the multiplication rule,

$$(a + b\mathrm{i})(a - b\mathrm{i}) = (a^2 - ab\mathrm{i} + ab\mathrm{i} - b^2\mathrm{i}^2) = a^2 + b^2.$$

The rule for subtraction of complex numbers is analogous to that for addition, i.e.,

$$(a + b\mathrm{i}) - (c + d\mathrm{i}) = (a - c) + (b - d)\mathrm{i} \ .$$

The rule for division may be deduced using the rules for multiplication in association with one of two tricks that are often valuable in algebraic operations. These 'tricks' are to remember that

- adding and subtracting the same number leaves an expression unaltered;
- multiplying and dividing by the same non-zero number leaves an expression unaltered.

Using the latter we get the following rule for division:

$$\frac{a+bi}{c+di} = \frac{(a+bi)(c-di)}{(c+di)(c-di)} = \frac{ac+bd}{c^2+d^2} + \frac{bc-ad}{c^2+d^2}\mathrm{i}\,.$$

Here we have multiplied and divided by the same number $(c - di)$.

A numerical example is

$$\frac{7+3\,\mathrm{i}}{2-\mathrm{i}} = \frac{(7+3\,\mathrm{i})(2+\mathrm{i})}{(2-\mathrm{i})(2+\mathrm{i})} = \frac{14+6\,\mathrm{i}+7\,\mathrm{i}-3}{4+1} = \frac{11+13\,\mathrm{i}}{5}\,.$$

You are probably familiar with a quadratic equation in x in the form

$$ax^2 + bx + c = 0.$$

where $a \neq 0$, and know that its solution is

$$x = \frac{-b \pm \sqrt{b^2 - 4ac}}{2a}. \tag{6.1}$$

If $b^2 - 4ac > 0$ there are two real roots. If $b^2 - 4ac = 0$ there is a unique real root, while if $b^2 - 4ac < 0$ there are no real roots. In this last case there are two complex roots.

We defined i to be the square root of -1 to help solve the equation $x^2 = -1$. This is a special case of the general quadratic with $a = 1$ and $b = 0$ and $c = 1$, so (6.1) gives the two roots $x = 0 \pm \sqrt{(-1)} = \pm\mathrm{i}$, indicating that both i and $-\mathrm{i}$ are square roots of -1. The latter is easily verified by multiplication since $(-\mathrm{i}) \times (-\mathrm{i}) = \mathrm{i}^2 = -1$.

Complex numbers satisfy the associative and commutative laws of addition and multiplication, also the distributive law. The results of addition, multiplication, subtraction or division of complex numbers all lead to complex numbers. Sometimes these may have either the real, or imaginary part, zero, e.g., $(a+b\mathrm{i})+(a-b\mathrm{i}) = 2a$. It can also be shown that processes such as taking square roots and cube roots give results that are also complex numbers.

Mathematical word theft

Given an algebraic entity such as a complex number (remember real and imaginary numbers are only special cases of such numbers), together with rules of operation that always lead to further entities of the same kind (here complex numbers) constitutes what mathematicians call a *field*. Here an everyday word is given a technical meaning, something that is common in mathematics. Other everyday words with special meanings in mathematics include *set*, *group*, *mapping*, *domain* and *manifold*. For *field*, the connotation is that of a field as an enclosure.

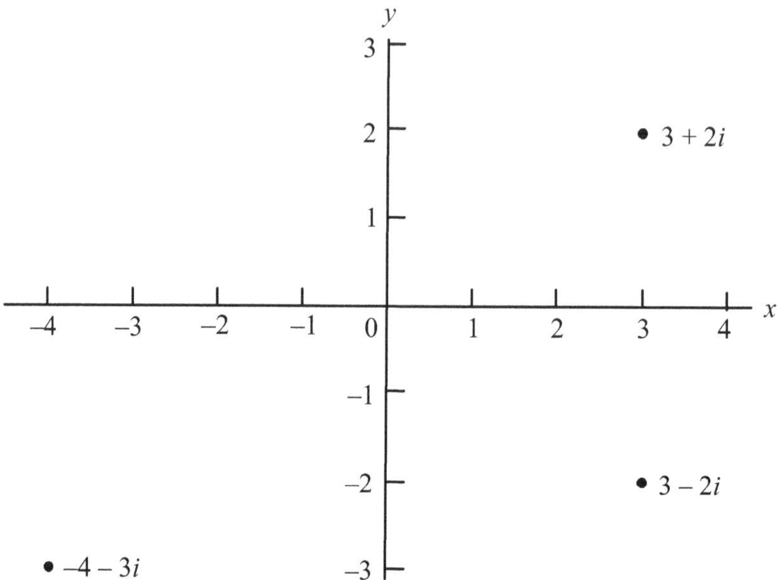

Fig. 6.1. Representation of complex numbers on an Argand diagram

We simplify multiplication and division by writing complex numbers another way. The key for doing this lies in extending the geometric representation for real numbers as points on a line as we did in Fig. 5.1.

About 200 years ago several eminent mathematicians including Carl Friedrich Gauss (1777–1855), Caspar Wessel (1745–1818) and Jean Robert Argand (1768–1822) all proposed similar extensions for a geometric representation of complex numbers.

Argand's work seems to have attracted most attention, for the representation is now usually called an *Argand diagram.* On it, all complex numbers are represented by points in a plane. The procedure is closely related to a Cartesian co-ordinate system. Readers not familiar with this notion may like to study Chap. 7 before continuing with this one.

To tie the concept of Argand diagrams more closely to Cartesian coordinates, we now replace a, b in $a + b\mathrm{i}$ by x, y and write $x + y\mathrm{i}$ for a general complex number, and refer to it as z, i.e., $z = x + y\mathrm{i}$. In an Argand diagram this complex number is represented by a point in a plane with Cartesian coordinates (x, y). Thus, any pure real number x lies on the x-axis, and a pure complex number $y\mathrm{i}$ lies on the y-axis. Fig. 6.1 shows the position of several complex numbers on an Argand diagram.

The points representing the numbers $z_1 = 3 + 2\mathrm{i}$ and $z_2 = 3 - 2\mathrm{i}$ are reflections of each other in the x-axis. They are a special case of pairs $z = x + y\mathrm{i}$, $\bar{z} = x - y\mathrm{i}$ with this property. We call $\bar{z}$ the *conjugate* of z and $z\bar{z} = (x + y\mathrm{i})(x - y\mathrm{i}) = x^2 + y^2 = r^2$, say, and r^2 is real and nonnegative.

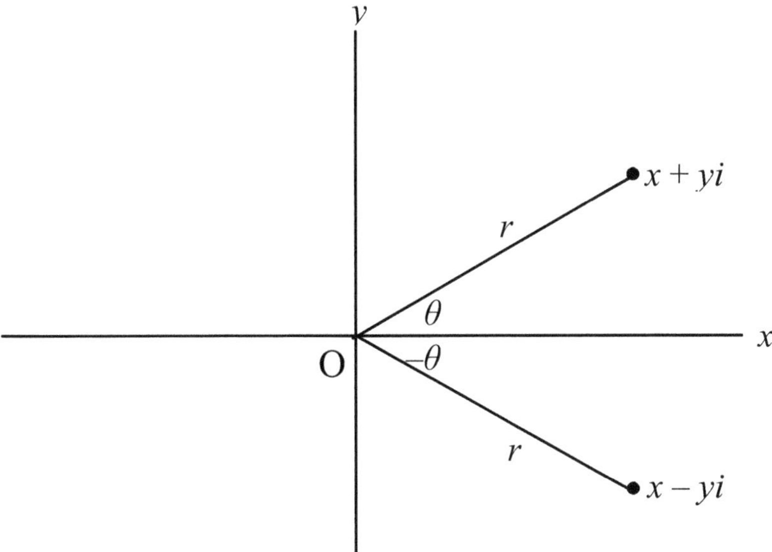

Fig. 6.2. Relation between modulus and amplitude for a conjugate pair of complex numbers

From the Argand diagram, Fig. 6.2, we see by Pythagoras's theorem, that r^2 is the square of the distance of the points $x \pm yi$ from the origin. The positive square root, r, is called the *modulus* of z.

The angle θ between the positive direction of the x-axis and the line from the origin to z, is called the *amplitude* of z. The modulus of $\bar{z}$ is the same as that of z, but the amplitude of $\bar{z}$ is $-\theta$. This accords with a convention that angles measured anticlockwise from the positive direction of the x-axis are positive, and those measured clockwise are negative.

You may know the definitions of the sine and cosine of any angle θ (these are conventionally written as $\sin\theta$ and $\cos\theta$) are $\cos\theta = x/r$ and $\sin\theta = y/r$, where r is always positive. This implies $x = r\cos\theta$ and $y = r\sin\theta$, so that

$$z = x + y\,\mathrm{i} = r(\cos\theta + \mathrm{i}\sin\theta).$$

Some common properties of sines and cosines relevant to what follows are described briefly in the Appendix, Sect. A.3.

If $z = \mathrm{i}$, this is represented by a point with coordinates (0, 1), and this implies $r = 1$ and $\theta = 90°$, whence $\mathrm{i} = 1(\cos 90° + \mathrm{i}\sin 90°)$. Also if $z = 1 + \mathrm{i}$ we have $r = \sqrt{2}$ and $\theta = 45°$, whence

$$z = \sqrt{2}(\cos 45° + \mathrm{i}\sin 45°)\ .$$

If $z = -\sqrt{3} + \mathrm{i}$ then $r = 2$ and $\theta = 150°$, so

$$z = 2(\cos 150^\circ + \mathrm{i} \sin 150^\circ) .$$

The above results for $z = \mathrm{i}$, $1+\mathrm{i}$, $-\sqrt{3}+\mathrm{i}$, are easy to verify if one is familiar with the values of the trigonometric ratios for angles of 0, 30, 45, 90 degrees.

On an Argand diagram an angle is not unique. The same angle is obtained by adding or subtracting multiples of 360°. It is convenient to introduce uniqueness by specifying that we always consider only values of θ lying between -180° and $+180^\circ$, or alternatively, between 0° and 360°.

6.2 De Moivre's Theorem

We now derive a result that makes multiplication (and division) with complex numbers easy. To obtain it we need two standard trigonometric results (see Sect. A2 of the appendix). These express the sines and cosines of sums of angles in terms of sums and differences of products of these trigonometric ratios for the individual angles. That sounds a mouthful, but the expressions are:

$$\cos(\theta + \phi) = \cos\theta \cos\phi - \sin\theta \sin\phi ,$$
$$\sin(\theta + \phi) = \sin\theta \cos\phi + \cos\theta \sin\phi .$$

If

$$z_1 = r_1(\cos\theta + \mathrm{i}\sin\theta) \text{ and } z_2 = r_2(\cos\phi + \mathrm{i}\sin\phi) ,$$

then multiplying and using the above results gives

$$\begin{aligned} z_1 z_2 &= r_1 r_2(\cos\theta\cos\phi - \sin\theta\sin\phi) + \mathrm{i}(\sin\theta\cos\phi + \cos\theta\sin\phi) \\ &= r_1 r_2[\cos(\theta + \phi) + \mathrm{i}\sin(\theta + \phi)] . \end{aligned}$$

Thus, to multiply two complex numbers, we *multiply* the moduli and *add* the amplitudes. The notation is sometimes further simplified by writing

$$\cos\theta + \mathrm{i}\sin\theta = \text{ cis } \theta ,$$

whence the multiplication rule becomes

$$z_1 z_2 = r_1 \text{ cis } \theta \times r_2 \text{ cis } \phi = r_1 r_2 \text{ cis } (\theta + \phi).$$

In particular, if $z = r \text{ cis } \theta$, then $z^2 = r^2 \text{ cis } 2\theta$ and $z^3 = r^3 \text{ cis } 3\theta$, and continuing in this way for any integer n

$$z^n = r^n \text{ cis } n\theta .$$

If $r = 1$ it is easy to see that the point z lies on the unit circle on an Argand diagram, and we have $(\text{cis } \theta)^n = \text{ cis } n\theta$, or more fully

$$(\cos\theta + \mathrm{i}\sin\theta)^n = \cos n\theta + \mathrm{i}\sin n\theta.$$

This result, due to Abraham de Moivre (1667–1754), is called *de Moivre's theorem.*

Some examples show its usefulness. If you are familiar with the binomial theorem, you will know that for any x, y

$$(x+y)^3 = x^3 + 3x^2y + 3xy^2 + y^3 .$$

If you do not know the binomial theorem, the result is easily obtained by first squaring $(x+y)$ and then multiplying the result by $(x+y)$.

Trigonometry without triangles

Using this binomial result, together with de Moivre's theorem, we get

$$\begin{aligned}(\cos\theta + \mathrm{i}\sin\theta)^3 &= \cos^3\theta - 3\cos\theta\sin^2\theta + \mathrm{i}(3\cos^2\theta\sin\theta - \sin^3\theta)\\ &= \cos 3\theta + \mathrm{i}\sin 3\theta .\end{aligned}$$

From the Argand diagram it is easily seen that two complex numbers are identical only if the real and imaginary parts are both equal, whence

$$\begin{aligned}\cos 3\theta &= \cos^3\theta - 3\cos\theta\sin^2\theta\\ \sin 3\theta &= 3\cos^2\theta\sin\theta - \sin^3\theta .\end{aligned}$$

Also, there is a well-known result in elementary trigonometry that for all θ

$$\cos^2\theta + \sin^2\theta = 1 .$$

Using that relationship the expressions for $\cos 3\theta$ and $\sin 3\theta$ reduce to

$$\begin{aligned}\cos 3\theta &= 4\cos^3\theta - 3\cos\theta ,\\ \sin 3\theta &= 3\sin\theta - 4\sin^3\theta .\end{aligned}$$

Most mathematicians, unless they are working on specialized problems involving trigonometry, may seldom use these by-no-means-obvious formulae for the sines and cosines of a multiple, 3θ, of θ in terms of the corresponding ratios for θ.

However, the above derivation is a good example of the power achieved by associating two or more techniques that are often looked at separately in elementary mathematics. Here we have established a trigonometric result, using a theorem derived from a geometric interpretation of a complex number as in an Argand diagram, and linking this to an algebraic result in the form of the binomial theorem. One may extend these ideas to express sines and cosines of, say, 4θ in terms of those for θ. Try this for yourself.

Our second example of an application of de Moivre's theorem is more important, and is relevant to solving algebraic equations. For the equation $x^n = a$, where n is a positive integer, any solution is described as an nth root of a. Thus, $x^2 = 1$ has two solutions, $x = 1$ and $x = -1$, among the real numbers. The equation $x^3 = 1$ has a real solution 1, and no other real solution. The equation $x^4 = 1$ has real solutions 1 and -1. It also has two imaginary solutions, namely i and $-$i, since both i^4 and $(-\mathrm{i})^4$ equal 1. Does the equation $x^3 = 1$ have other solutions among complex numbers?

Using de Moivre's theorem, it is not difficult to show that for any positive integer n, the equation $x^n = 1$ has exactly n distinct roots in the complex number system. Some, or all, of these may be either pure real or pure imaginary.

We illustrate this only when $n = 3$ and $n = 4$. If $n = 3$, since $x^3 = 1$ this implies $x^3 = \cos 0° + \mathrm{i} \sin 0°$. We pointed out above that if we add any multiple of 360° to the amplitude the value remains at 1. So we may replace $\cos 0° + \mathrm{i} \sin 0°$ by $\cos(360q°) + \mathrm{i} \sin(360q°)$ where q is an integer. Using de Moivre's theorem this gives

$$\{\cos[(360q°)/n] + \mathrm{i} \sin[(360q°)/n]\}^n = \cos(360q°) + \mathrm{i} \sin(360q°) = 1 + 0\mathrm{i}.$$

Putting $n = 3$ on the left-hand side, this implies the cube roots of unity are given by $\cos[(360q°)/3] + \mathrm{i} \sin[(360q°)/3] = \cos(120q°) + \mathrm{i} \sin(120q°)$. Setting $q = 0, 1, 2$ we get amplitudes $0°, 120°, 240°$. The last is equivalent to $-120°$. Substituting in the appropriate values of the sines and cosines, 3 cube roots of unity are

$$1, \ -\frac{1}{2} + \frac{\sqrt{3}}{2}\mathrm{i}, \ -\frac{1}{2} - \frac{\sqrt{3}}{2}\mathrm{i}.$$

If we substitute integral values of $q \geq 3$ we see that these values are repeated. Thus there are one real, and two complex, cube roots of unity. These are equally spaced on the circumference of the unit circle on the Argand diagram. If we set $n = 4$ and $q = 0$, 1, 2, 3, it is easily verified that the 4 fourth roots of unity are, as already indicated, 1, i, -1 and $-$i. Again, these are equally spaced on the circumference of the unit circle. We need not consider values of $q \geq 4$, for these roots are then repeated.

Where intuition works

You may feel that for any positive integer n there will be exactly n nth roots of unity, and that these will include the real value 1, and be equally spaced on the circumference of a unit circle on an Argand diagram. If so, you will be right. Intuition works here. I do not give a formal proof, though verification is straightforward.

6.3 The Fundamental Theorem of Algebra

Introducing complex numbers not only gives all solutions to equations of the form $ax^2 + bx + c = 0$ for all real a, b, c but, more importantly, leads to what is known as the *fundamental theorem of algebra*. This applies to *monic* polynomials, which are polynomials where the coefficient of the leading term is +1. The theorem asserts that

> Every monic polynomial of any degree n with real or complex coefficients, i.e., of the form
>
> $$f(x) = x^n + a_{n-1}x^{n-1} + a_{n-2}x^{n-2} + \ldots + a_2x^2 + a_1x + a_0$$
>
> has the property that $f(x)$ can be factorized into the product of exactly n factors
>
> $$f(x) = (x - \alpha_1)(x - \alpha_2)\ldots(x - \alpha_{n-1})(x - \alpha_n)$$
>
> where the α_i are complex numbers. Some may have complex part zero, and some may be equal.

We do not prove this.

Unfortunately, for those who have to solve equations of the form

$$x^n + a_{n-1}x^{n-1} + a_{n-2}x^{n-2} + \ldots + a_2x^2 + a_1x + a_0 = 0$$

the result is not particularly helpful. It does not tell us how to find the factors of $f(x)$, or the related solutions of the equation $f(x) = 0$. It only assures us the solutions do exist, and are given by the α_i.

When mathematicians are (almost) useless

Many fundamental results in mathematics are of a similar frustrating nature. This sometimes annoys workers in other disciplines who have specific problems to solve. Mathematicians are often able to tell them a problem has a solution, but are less good at telling them what that solution is. There is some truth in this. Finding a solution to a problem is often more demanding than simply establishing that one exists.

What is important about the fundamental theorem of algebra is that it tells us we may find all solutions to algebraic (i.e., 'polynomial') equations without having to invent or discover further kinds of numbers.

6.4 Loose Ends

Transcendental numbers. Having asserted that all algebraic (i.e., polynomial equations) have solutions in the field of complex numbers, you may feel we have all the numbers we need. But there are important numbers — many just real numbers — that are not solutions of algebraic equations. An example is π, familiar from schooldays as the ratio of the circumference to the diameter of any circle. We discuss the nature of π more fully in Chap. 12, where we meet another important irrational number that is also not a solution of any algebraic equation. Such numbers are called *transcendental numbers.*

Quaternions. The relation between real numbers and a straight line, between complex numbers and a plane, raises the question whether there might not be some new type of numbers associated with a three dimensional space. Or, if one wants to be more abstract, with any higher dimensional space. This possibility was explored by William Hamilton (1805–65). He found no useful generalization in three dimensions, but did find one in four dimensions which he called *quaternions.* These may be operated upon in the same way as real or complex numbers, except that multiplication is not commutative. There are close links between quaternions and what are known as *vector spaces*, a concept not covered in this book, but one that is important in many areas in physics, statistics and other fields of application.

Quaternions have a form which looks to be a generalization of a complex number. They may be written

$$a + b\mathrm{i} + c\mathrm{j} + d\mathrm{k}.$$

Here a, b, c, d are real numbers, and i, j, k are generalizations of the imaginary number, i, that satisfy the relationships

$$\mathrm{ij} = \mathrm{k},\ \mathrm{jk} = \mathrm{i},\ \mathrm{ki} = \mathrm{j},\ \mathrm{ji} = -\mathrm{k},\ \mathrm{kj} = -\mathrm{i},\ \mathrm{ik} = -\mathrm{j}. \tag{6.2}$$

The addition rule for quaternions is similar to that for complex numbers. The multiplication rule generalizes that for complex numbers. The complex number rule that i^2 is replaced by -1 gives way to the more complicated conditions (6.2). For example, the product ij is replaced by k, etc.

Diophantine equations. The fundamental theorem of algebra relates to a broad type of equation and establishes for these the existence of solutions in the complex number system. A different problem arises if we consider an equation in one or more variables that has integer coefficients and ask what, if any, integers satisfy the equation? Generally speaking, when we have one equation involving more than one unknown, there will be an infinity of possible solutions in the real number system. Among these, there may still be an infinity of solutions that are positive or negative integers, but this need not be the case. There may even be no such solutions. For example, for the equation

$$x^2 + y^2 = 5 ,$$

it is fairly easy to show that if p is any real number between -5 and $+\ 5$, and $q = \pm\sqrt{(5 - p^2)}$, then $x = p$ and $y = q$ is a solution. However, the only integer solutions are the eight solutions

(i) $x = 1, y = 2$, (ii) $x = 1, y = -2$, (iii) $x = -1, y = 2$, (iv) $x = -1, y = -2$, (v) $x = 2, y = 1$, (vi) $x = 2, y = -1$, (vii) $x = -2, y = 1$, (viii) $x = -2, y = -1$.

The equation $x^2 + y^2 = 3$ has an infinity of real solutions, but no integer solutions.

Equations with integer coefficients, where we consider only integer solutions, are called *Diophantine equations.* The term is a little misleading, as Diophantine really refers to this restricted type of solution rather than to the equation itself.

We do not prove it, but it can be established that equations of the form

$$ax + by = c$$

where a, b and c are integers, only have integer solutions if the highest common factor of a and b also divides c.

A good account of some of the properties of, and one might almost say some of the peculiar characteristics of, Diophantine equations is given in Devlin [14]. A slightly different approach is taken in the shorter treatment by Courant, Robbins and Stewart [13] (Chap. 1, Sect. 4). The term is sometimes extended to include solutions to equations that lie in more generally defined fields, using the term 'field' in the sense given on p. 73.

7

A Perfect Marriage

7.1 A Vital Link

In 1637 the French philosopher and mathematician René Descartes (1596–1650) linked geometry and algebra, introducing *Cartesian coordinates.* These let us move freely between, on the one hand lines, circles and other geometric figures, and on the other, algebraic equations. This is the basis of *analytic*, or *algebraic*, or *coordinate geometry.* All three names are used for what many regard as the perfect marriage of classical mathematics.

The germ of a Cartesian coordinate system lies in the association of real numbers with points on a straight line. We used that notion in Chap. 5. That simple idea predates the work of Descartes, and produces a one-dimensional Cartesian system, that in itself is only mildly interesting. If a and b are points on the line that represent two real numbers, about all we can say geometrically about them is that the difference $b - a$ is a measure of the distance between those points.

Things become more interesting with pairs of real numbers (x, y), called the coordinates, that represent points in a plane. Descartes represented these by drawing in a plane (e.g., on a sheet of paper) two special fixed straight lines perpendicular to each other. These are called the *axes.*

Each point on the first line, called the x-axis, represents a unique possible value of the real number x occurring in the pair (x, y). Each point on the second line at right angles to the first, and called the y-axis, represents some value of the real number y. Conventionally, we usually take the x-axis to lie horizontally across the page as in Fig. 7.1. We mark off on it points corresponding to the real numbers as we did in Fig. 5.1. Positive numbers lie to the right of zero, and negative numbers to the left.

The y-axis is a line obtained by rotating the x-axis anticlockwise through 90° about the point representing zero. That point is called the *origin.* On the y-axis positive values of y lie above, and negative values below, the origin. The origin, usually denoted by O, has co-ordinates $(0, 0)$.

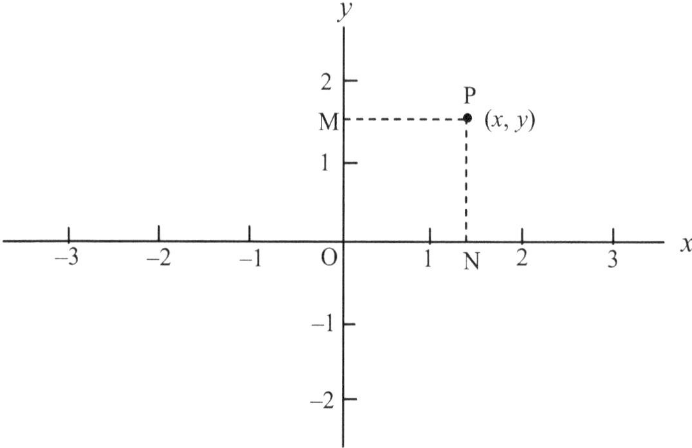

Fig. 7.1. A cartesian coordinate system

In the plane a point with coordinates (x, y) is a point P (Fig. 7.1) such that the length of the perpendicular MP from the y-axis to P is x and that of the perpendicular NP from the x-axis to P is y. In Fig. 7.1 it is easy to see that ON $=$ MP $= x$ and OM $=$ NP $= y$. Does this remind you of the Argand diagram in Sect. 6.1? In algebraic geometry, however, both x and y are real numbers in their own right, rather than real and imaginary parts of a complex number.

The axes divide the plane into four sectors called *quadrants*. The top right sector is the first quadrant. All points in it have both x and y positive. Moving anticlockwise to the top left, takes us to the second quadrant, where x is negative and y positive. A further movement anticlockwise takes us to the third quadrant, where both x and y are negative. In the fourth, and final, quadrant at the bottom right x is positive and y is negative. Fig. 7.2 indicates the position of several points and gives the numerical values of the coordinates of each.

7.2 Lines

Some simple algebraic equations are immediately interpretable in a Cartesian coordinate system. The equation $x = 0$ specifies all points on the y-axis; i.e., it is the equation of the y-axis. All points on it have coordinates of the form $(0, y)$. Similarly the equation $y = 0$ represents the x-axis. The equation $y = 1$ represents a line parallel to the x-axis that passes through the point (0, 1), since points with coordinates $(x, 1)$ lie on this line for all x.

The equation $x = y$ represents a line through the origin which bisects the angle between the positive directions of the x- and of the y-axes. Take a

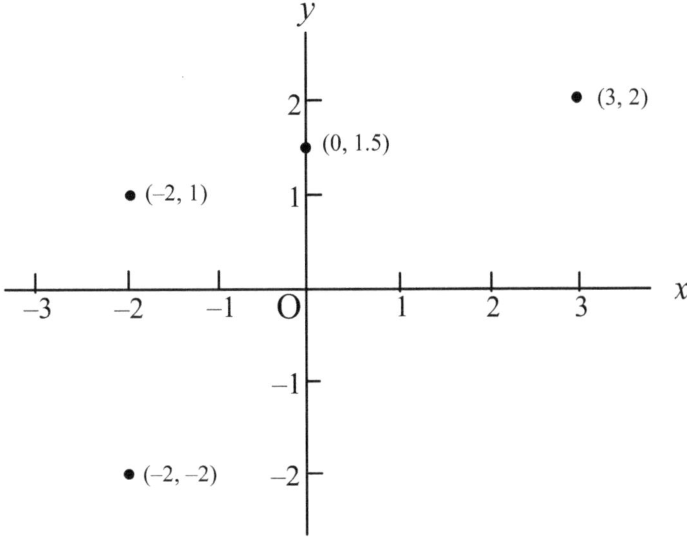

Fig. 7.2. Cartesian coordinate positions of four specific points

moment to convince yourself that the geometry of the coordinate system is such that all points with equal x and y coordinates lie on this line.

As you probably know already, we may associate any equation of the form

$$ax + by + c = 0 ,$$

where a, b, c are real numbers, with a straight line in the Cartesian coordinate plane. Conversely, any straight line in the plane has an equation which may be written in this form.

Fig.7.3 helps us see how to find the equation of a straight line that passes through any two points P, Q with known coordinates (x_1, y_1) and (x_2, y_2) respectively. To avoid some complications involving possible divisions by zero we exclude the special cases where either $x_1 = x_2$ or $y_1 = y_2$. Let R be any point on the line PQ having coordinates (x, y). QM and RN are parallel to the y-axis and with M, N lying on a line through P parallel to the x-axis. To obtain the equation to PQ we use Euclidean geometry. The triangles QMP, RNP are similar triangles. A well known property of such triangles is that RN/QM = NP/MP. As indicated on Fig. 7.3, QM $= y_2 - y_1$, RN $= y - y_1$, MP $= x_2 - x_1$ and NP $= x - x_1$, whence

$$\frac{y - y_1}{y_2 - y_1} = \frac{x - x_1}{x_2 - x_1} . \tag{7.1}$$

To find the line through the points $(2, 3)$ and $(-1, 4)$, we substitute these values for (x_1, y_1) and (x_2, y_2) in (7.1). Then the line through these points has the equation

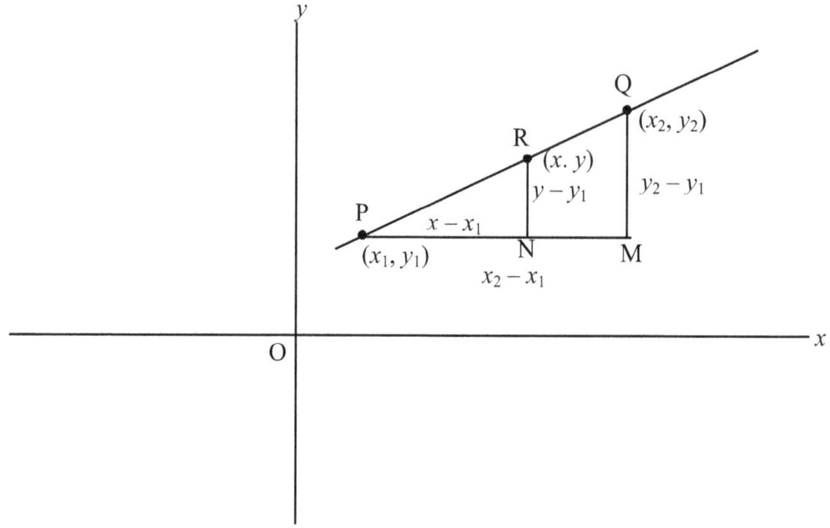

Fig. 7.3. Cartesian coordinate representation of a straight line through two given points P and Q

$$\frac{y-3}{4-3} = \frac{x-2}{(-1)-2} \; .$$

By simple algebra this reduces to

$$x + 3y - 11 = 0 \; . \tag{7.2}$$

This has the form $ax + by + c = 0$, where $a = 1, b = 3$ and $c = -11$. To check we substitute $x = 2$, $y = 3$ and then $x = -1$, $y = 4$ in (7.2). Both sets of values satisfy the equation, establishing that it passes through these points.

Uphill or downhill?

There are alternative, but equivalent, ways of writing equations like (7.1) and (7.2). We may recast (7.1) in the form

$$y = mx + c \; .$$

Doing this we find that

$$m = \frac{(y_2 - y_1)}{(x_2 - x_1)} \; .$$

By looking at Fig. 7.3 we see that m is the tangent of the angle QPM. This angle is equal to the angle between the positive direction of the x-axis and the line PQ. Thus m is also the tangent of that angle. It is called the *slope* (sometimes the *gradient*) of PQ. If the slope is negative, the line runs 'down hill' as we move from left to right across the graph.

The expression for c in terms of the coordinates of P and Q is the more complicated

$$c = \frac{y_1 x_2 - y_2 x_1}{x_2 - x_1} \; . \tag{7.3}$$

This has a simple geometric interpretation. If we substitute $x = 0$ in the equation $y = mx + c$ we have $y = c$. This implies that $(0, c)$ is the point where the line cuts the y-axis. It is called the y-intercept. Fig. 7.4 is a graph showing the line through the points $(2, 3)$, $(-1, 4)$ considered above. Starting from (7.2) this may be rewritten

$$y = -\frac{1}{3}x + 3\frac{2}{3} \; ,$$

from which we easily confirm that

$$m = \frac{(4-3)}{(-1-2)} = -\frac{1}{3} \; .$$

From Fig. 7.4 we see that the intercept on the y-axis is indeed at the point $(0, 3\frac{2}{3})$. This may be confirmed by substitution in (7.3).

7.3 Lines and Intersections

Two straight lines intersect at only one point unless they are parallel or coincide. In Euclidean geometry parallelism means the lines do not intersect, or as some people rather loosely say 'they intersect only at infinity'. If the lines coincide this implies they have the same equation. If two lines intersect at a unique point, and their equations are respectively

$$ax + by + c = 0$$

and

$$a'x + b'y + c' = 0 \; ,$$

their solution determines a point that satisfies both equations, i.e., the point of intersection of the lines represented by these equations. Remembering that we may write the two equations in an alternative form such as

$$y = mx + c \text{ and } y = m'x + c',$$

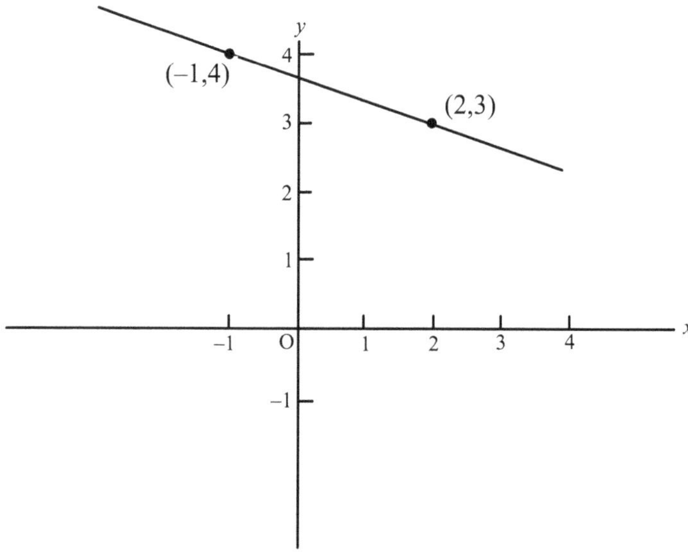

Fig. 7.4. Graph of the line through (2,3) and (−1,4)

it is easy to establish by the usual method for solving a pair of straight line equations that the solution is

$$x = \frac{c - c'}{m' - m}, \; y = \frac{cm' - mc'}{m' - m},$$

with the proviso that this solution is not defined if $m' - m = 0$, i.e., if $m' = m$. Remembering that m defines the slope, $m' = m$ implies both lines have the same slope. Therefore they are either parallel or identical. If, as well as $m' = m$, we have $c = c'$, the equations are identical, i.e., they represent the same line.

Possible situations when we have equations for three lines are easily visualized in the Cartesian coordinate system. They might intersect pairwise in three distinct points,these being the vertices of a triangle having sides formed by segments, or parts, of these lines. Another possibility is that two of the equations represent parallel lines, and that the remaining line intersects them. It is also possible that the third line might be parallel to the other two, or identical with one of them. The remaining possibility is that all three lines meet at a common point.

If two intersecting lines have equations

$$ax + by + c = 0 \text{ and } a'x + b'y + c' = 0\,,$$

then for all nonzero values of the real number k the equation

$$ax + by + c + k(a'x + b'y + c') = 0$$

represents another line that passes through the common intersection. This is easily verified by noting that any such equation is satisfied when both $ax + by + c = 0$ and $a'x + b'y + c' = 0$.

For example, it is easy to check that $x - 2y + 1 = 0$ and $4x - 3y - 1 = 0$ represent straight lines intersecting at the point $(1, 1)$. Thus, any equation of the form $x - 2y + 1 + k(4x - 3y - 1) = 0$ will pass through $(1, 1)$. For example, if $k = -5$ this equation is $-19x + 13y + 6 = 0$. By substitution, we see that this is satisfied when $x = y = 1$.

7.4 Other Ways to Find a Straight Line Equation

In Sect. 7.2 we found the equation to a straight line passing through two given points.

If we know the slope, m, of a straight line (i.e., the tangent of the angle between its direction and that of the x-axis), and also the co-ordinates of one point (x_1, y_1) that lies on the line, then the equation is

$$y - y_1 = m(x - x_1). \tag{7.4}$$

This line passes through the required point since the equation is satisfied if $x = x_1$ and $y = y_1$. Also the slope is m, since this equation is of the form $y = mx + c$ where $c = y_1 - mx_1$. The equation of the line of slope m that passes through the origin is $y = mx$. This follows immediately from (7.4) on setting $x_1 = y_1 = 0$; i.e., the fixed point determining the line is the origin.

Perpendicularity of lines is often of interest. If two directions differ by 90°, i.e., if $\phi = \theta + 90°$ then $\sin\phi = \cos\theta$ and $\cos\phi = -\sin\theta$ (see Appendix, Sect. A2). It follows that $\tan\phi = \sin\phi/\cos\phi = \cos\theta/(-\sin\theta) = -1/\tan\theta$. Thus, if the slope of a line is $m = \tan\theta$, then that of a line perpendicular to it is $m' = -1/m = \tan\phi$. To avoid possible difficulties with attempted division by zero we exclude the case where $\theta = 0°$ or $\theta = 90°$. These cases correspond to lines parallel to one or other of the axes, and are best dealt with separately.

There is a well-known ruler and compass construction in Euclidean geometry for drawing a line that passes through the midpoint of a segment PQ and is perpendicular to the line PQ . In the corresponding problem in analytic geometry we are given two points P, Q with coordinates $(x_1, y_1), (x_2, y_2)$. We seek a line that passes through the midpoint of that segment and is perpendicular to the line PQ. Referring to Fig. 7.3, and using the properties of similar triangles, it is easy to establish that the mid-point of the segment PQ has coordinates $x = (x_1 + x_2)/2$, $y = (y_1 + y_2)/2$. We established in Sect. 7.2 that PQ has slope $m = (y_2 - y_1)/(x_2 - x_1)$. Thus the slope of any line perpendicular to PQ is $m' = -(x_2 - x_1)/(y_2 - y_1)$.

Here is a numerical example. If P, Q have coordinates (1, 2) and (3, 6), the mid-point of the segment PQ has coordinates $x = (1 + 3)/2 = 2$ and $y = (2 + 6)/2 = 4$. The slope of PQ is $m = (6 - 2)/(3 - 1) = 2$. Thus any line

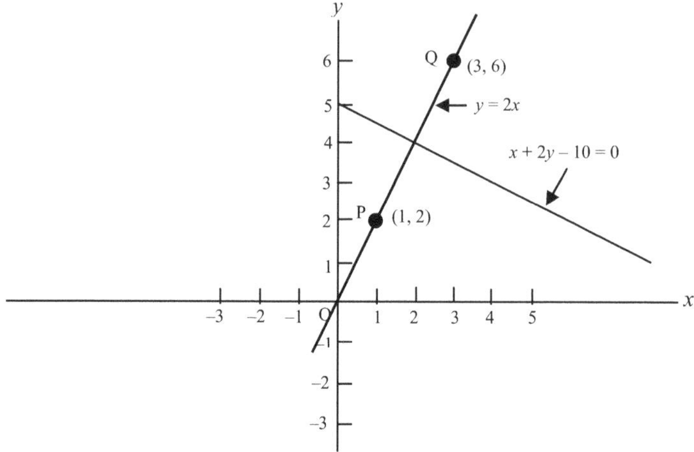

Fig. 7.5. Line through the mid-point of the join of two points P, Q that is perpendicular to that join

perpendicular to PQ has slope $m' = -1/2$. The line through the mid-point (2, 4) with this slope has, by (7.4), the equation

$$y - 4 = -\frac{1}{2}(x - 2)$$

which may be rewritten

$$x + 2y - 10 = 0 \, .$$

It is easy to verify by methods given earlier that the equation of the line joining P and Q is $y = 2x$, a line through the origin.

The line PQ, and the required perpendicular, are shown in Fig. 7.5.

Another property associated with a segment PQ, where the coordinates of P, Q are given, is its length, d, say. From Fig. 7.3 it is obvious if we apply Pythagoras's theorem to the triangle QMP that

$$d^2 = (x_2 - x_1)^2 + (y_2 - y_1)^2 \qquad (7.5)$$

7.5 Second Degree Equations

Going round in circles

Let P be a fixed point with coordinates (x_1, y_1). Though not essential to our argument, to emphasise this 'fixed' property we rename the coordinates

(a, b). Consider all points Q with coordinates (x, y) such that the length d of the segment PQ has constant value r. A little thought confirms that as the coordinates of Q change, the possible points Q all lie on the circumference of a circle with centre (a, b) and radius r satisfying the equation

$$(x - a)^2 + (y - b)^2 = r^2 . \tag{7.6}$$

This follows immediately from (7.5). Equation (7.6) involves terms of the second degree in both x and y. In the special case when the centre is at the origin, i.e., when $a = b = 0$, the equation simplifies to

$$x^2 + y^2 = r^2 .$$

Equation (7.6) may be expanded and the terms regrouped to give

$$x^2 + y^2 - 2ax - 2by + a^2 + b^2 - r^2 = 0 .$$

This is a particular case of the most general equation involving terms in x and y of the second degree. A widely used notation for the most general second degree equation is

$$ax^2 + 2hxy + by^2 + 2gx + 2fy + c = 0 . \tag{7.7}$$

In Sect.7.2 we saw that any first degree equation $ax + by + c = 0$ represents a straight line. Does (7.7) always represents a particular family of curves? We have shown that sometimes, at least, it represents a circle. At school most of us learn from another example that such equations do not always represent a circle. It is usually drilled into us that the familiar equation

$$y = ax^2 + bx + c ,$$

which can easily be re-expressed as a special case of (7.7), is the equation for a parabola.

Slicing through cones

We shall not consider in detail the analytic geometry of curves represented by (7.7). Subject to various conditions on the nature of the coefficients, (7.7) covers all possible curves known as *conic sections.* These are the possible curves formed if we take a cone (more specifically what is called a right circular cone) and study the curves formed by the intersection of the cone with a plane that slices through it. If the plane is at right-angles to the axis of the cone the intersection is a circle. Tilt the plane slightly, and the curve of intersection becomes an ellipse. The family of conic sections includes the following curves — circles, ellipses, parabolas, hyperbolas and as an extreme case, pairs of straight lines. It is easy to visualize these last arising if the plane passes through the vertex of the cone.

Which of these curves (7.7) represents is reflected in a simple function of the coefficients. If we write $D = ab - h^2$ then, if $D > 0$ the equation represents an ellipse. The circle is included as a special case of the ellipse. If $D < 0$ then (7.7) represents a hyperbola. If $D = 0$ (7.7) represents a parabola. We say more about functions of the type D, called *discriminants*, under the entry for that term in Sect.25.2.

Many fascinating properties of conic sections are only derived easily by using analytical geometry, since the curves take us well beyond the constructs of Euclidean geometry.

The parabola, represented by the equation

$$y = ax^2 + bx + c , \tag{7.8}$$

is often studied graphically in school algebra in connection with solution of the quadratic equation

$$ax^2 + bx + c = 0 , \tag{7.9}$$

and in a more general context of picturing the shape of parabolas.

To plot the graph of a straight line given an equation $ax + by + c = 0$, or in any other forms such as $y = mx + c$, we only need to determine two points on the line. To plot, even approximately, the curve corresponding to (7.8) for given values of a, b and c, we must work out a fairly extensive set of corresponding values of x and y representing points on the curve. Having marked these points on a sheet of graph paper, we trace out a smooth curve joining them.

There are a few general rules that help determine the shape of a parabola represented by (7.8). If a is positive the values of y become larger and larger when x takes larger and larger positive or negative values. We express this by saying $y \to \infty$ as $|x| \to \infty$. Here $|x|$, called the *modulus* of x, means the value of x ignoring sign, i.e., x and $-x$ have the same modulus, always taken to be non-negative. It can also be shown that if a is positive, then y has its minimum value when $x = -b/2a$. This minimum is $y = -(b^2 - 4ac)/4a$. This is most easily established using the differential calculus which we introduce in Chap. 11. A method known as *completing the square* is also widely used.

A useful trick of the trade

Completing the square uses the trick of adding and subtracting the same appropriate constant from an expression essentially equivalent to the right-hand side of (7.8). This constant is chosen so that all terms involving x can be written in a form $(x + k)^2$, where k is a constant.

If you have not met the method before, here is an outline of how it works. We first rewrite (7.8) in the form

$$y = a\left(x^2 + \frac{bx}{a} + \frac{c}{a}\right).$$

Adding and subtracting $(b/2a)^2$ inside the bracket, followed by a little algebra gives

$$\begin{aligned} y = a\left(x^2 + \frac{bx}{a} + \frac{c}{a}\right) &= a\left[x^2 + \frac{bx}{a} + \left(\frac{b}{2a}\right)^2 - \left(\frac{b}{2a}\right)^2 + \frac{c}{a}\right] \\ &= a\left[\left(x + \frac{b}{2a}\right)^2 - \frac{b^2 - 4ac}{4a^2}\right]. \end{aligned} \tag{7.10}$$

For real x, b and a we know that $(x + b/2a)^2$ is nonnegative. It takes its minimum value, zero, when x is chosen to make $x + b/2a = 0$, i.e., when $x = -b/2a$. Then $y = -(b^2 - 4ac)/4a$, and when a is positive this minimum value of y will be negative if $b^2 - 4ac > 0$.

If we now set $y = 0$, after a little algebraic manipulation we see from the last line in (7.10) that the roots of the equation $ax^2 + bx + c = 0$ are given by the well-known formula

$$x = \frac{-b \pm \sqrt{b^2 - 4ac}}{2a}.$$

Discriminants discriminate

If $b^2 - 4ac > 0$, it is not difficult to see that the roots are symmetrically placed about $x = -b/2a$, at which x-value the minimum occurs. If $b^2 - 4ac = 0$ there are two equal roots $x = -b/2a$, while if $b^2 - 4ac < 0$, as we saw in Sect. 6.1, there are two complex roots.

The quantity $b^2 - 4ac$ is another example of a discriminant. If it is positive (7.9) has two distinct roots, while if it is zero the two roots merge to a unique value. If it is negative there are no real roots.

If, in (7.8), a is negative then the parabola is tipped upside down, and the point corresponding to $x = -b/2a$ is a maximum.

Fig. 7.6(**a**) shows the shape of a graph of a parabola (7.8) when a is positive and $b^2 - 4ac > 0$. Fig. 7.6(**b**) shows a parabola with a negative and $b^2 - 4ac > 0$. Fig.7.6(**c**) shows a parabola with a positive and $b^2 - 4ac < 0$. In this last case the fact that the curve does not cut the x-axis shows that there are no real roots, but again there are two complex roots.

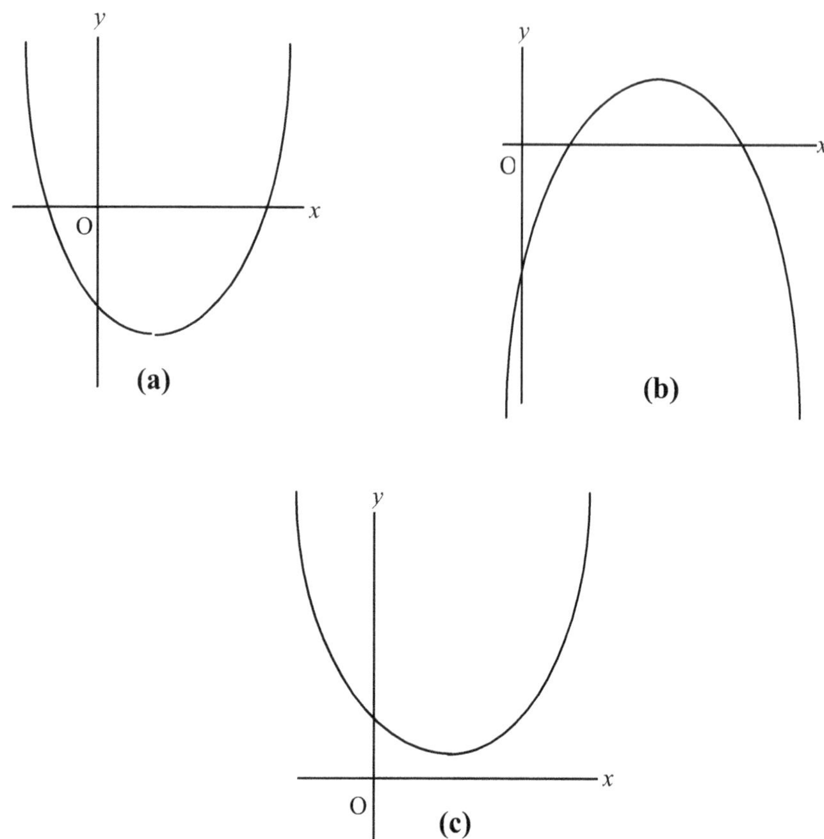

Fig. 7.6. Parabolas $y = ax^2 + bx + c$. In (**a**), $a > 0$ and $b^2 - 4ac > 0$. In (**b**) $a < 0$ and $b^2 - 4ac > 0$. In (**c**) $a > 0$ and $b^2 - 4ac < 0$

7.6 Cartesian Coordinates in Three Dimensions

An obvious extension to the two dimensional Cartesian coordinate system is to three dimensions. It helps if we think of a rectangular Cartesian system with the picture of a typical room in mind. The origin is at one corner at floor level, and we visualize the x-axis running in one direction along the junction of the floor and a wall. The y-axis runs from the same initial corner along the floor-wall junction at right-angles to the first. The z-axis rises vertically at the same corner. Points in the three dimensional space are represented by three coordinates reflecting lengths measured along these axial directions. They are written (x, y, z).

The three dimensional generalization of the equation $ax + by + c = 0$ is

$$ax + by + cz + d = 0 .$$

Again beware of intuition

A naïve intuitive thought is that this equation may represent a straight line in space. Looking at one or two special cases shows this is not so. What about the equation $z = 0$? This is satisfied by points with coordinates $(x, y, 0)$, where x and y may be any points in the plane containing the x- and y-axes. In our 'room' example this is the plane of the floor. Thus, the equation $z = 0$ is the equation to this plane. Similarly, the equation $x - y = 0$ is satisfied by all points with equal x and y coordinates, and any z coordinate. In our 'room' example this specifies a plane at right-angles to the plane of the floor that includes the line bisecting the angle between the x- and the y-axes. As an exercise, try to picture the plane represented by the equation $x+y+z=1$. You may find it helpful to consider the coordinates of the points at which this plane intersects each of the axes.

I do not prove it, but in three dimensions any linear equation of the form $ax + by + cz + d = 0$ represents a plane. To determine the equation to a plane we need only know the coordinates of any three points in that plane that do not all lie on one straight line. Alternatively, it suffices to know the direction of one line in the plane, and one point in the plane that is not on that line. Another possibility is to know one point in the plane and the 'direction' of a line at right angles to the plane. Such a line is called a *normal* to the plane. A little thought only is needed to see that all lines at right angles to a given plane have the same direction. One way of specifying that direction is by some function of the angles between that direction and each of the axial directions. We do not pursue here this important practical, but slightly more sophisticated, approach to determining the equation to a plane.

In Sect. 7.3 we saw that unless two straight lines are parallel, or identical, they met in a point. Similarly, unless two planes are parallel or identical, it is not hard to see that they intersect in a straight line.

Algebraically, if we have two non-parallel planes with equations

$$ax + by + cz + d = 0 \tag{7.11}$$

and

$$a'x + b'y + c'z + d' = 0\ , \tag{7.12}$$

these equations together determine the line of intersection.

Thus, a line in three dimensional space is specified by two equations to planes which must be satisfied simultaneously. If we fix the value of one of the coordinates on this line, setting, for instance, $z = 1$, we may determine a unique point on the line by substituting $z = 1$ in each of (7.11) and (7.12), and solving the resulting equations for x and y.

If the equations to three planes have a unique solution, that solution determines a common meeting point. There will be no unique solution in cases where the planes meet pairwise in a set of three parallel lines, or where two or all three planes are parallel.

Another practical concept that generalizes easily to three dimensions is that of *Euclidean distance*, or more precisely the square of that distance. This follows from a straightforward extension of Pythagoras's theorem. If the two points have coordinates (x_1, y_1, z_1) and (x_2, y_2, z_2), then the Euclidean distance, d, between them is given by

$$d^2 = (x_2 - x_1)^2 + (y_2 - y_1)^2 + (z_2 - z_1)^2 ,$$

where conventionally the positive square root is taken as a measure of distance.

By analogy with the derivation of the equation for a circle in two dimensions, a moment's reflection shows that if we replace (x_1, y_1, z_1) by a fixed point (a, b, c), and consider all points (x, y, z) that are at a fixed distance, r, from (a, b, c), then these points will lie on the surface of a sphere with equation

$$(x - a)^2 + (y - b)^2 + (z - c)^2 = r^2 .$$

7.7 Other Coordinate Systems

Analytical, or algebraic, geometry was made possible by Descartes' proposed coordinate system, but there are useful modifications of, or alternatives to, his system.

One modification that is not widely used in practice is to relax the restriction that the axes be at right-angles to each other. Such axes are referred to as *oblique axes*. The angle xOy between them is denoted by ω and the x, y coordinates are measured parallel to the axes. Fig. 7.7 shows the positioning of a point with co-ordinates $(-2, 3)$ in an oblique system.

Many of the properties associated with a rectangular system carry over to an oblique coordinate system with only minor modifications. For instance, the squared distance between two points (x_1, y_1), (x_2, y_2) becomes

$$d^2 = (x_2 - x_1)^2 + (y_2 - y_1)^2 + 2(x_2 - x_1)(y_2 - y_1)\cos\omega .$$

This reduces to (7.5) when ω is a right-angle since $\cos 90° = 0$.

A different coordinate system is one where the coordinates (x, y) are replaced by coordinates (r, θ), where these are derived from Cartesian coordinates using the relationship

$$r = \sqrt{x^2 + y^2} .$$

We determine θ knowing that $\cos\theta = x/r$ and $\sin\theta = y/r$. A little thought should convince you that in this system the equation

$$r = a ,$$

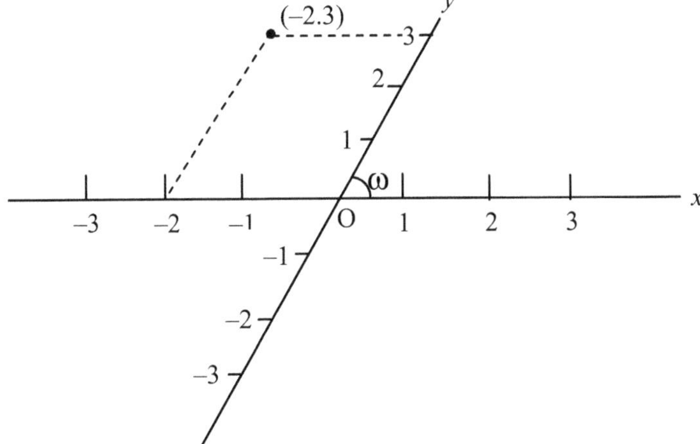

Fig. 7.7. The point with coordinates $(-2, 3)$ in an oblique coordinate system

when a is constant, represents a circle with centre at the origin and radius a.

Does the concept of these coordinates — called *polar coordinates* — remind you of something we have seen before? What about the Argand diagram we used to derive de Moivre's theorem?

7.8 Loose Ends

The many results that we have quoted without proof in this chapter themselves leave some loose ends. More important perhaps, we have only considered the graphical representation of a few simple functions, mainly straight lines, planes, circles and parabolas.

Trigonometric functions. The graphs of the trigonometric functions are interesting. If you are not familiar with the graph of functions like, say,

$$y = \sin(x + 45°) ,$$

where x is measured in degrees, you will find these are given in most books on elementary trigonometry. Indeed, this graph is simply a displaced version of that of the graph of $y = \sin x°$ which is shown in Fig. A.5 in the appendix.

Scaling. I have implicitly assumed so far in this chapter that the same 'scale' is used on both x- and y-axes. This is not essential, but it means care must be taken when drawing a graph to indicate the scales that are used. Generally speaking, scales should be chosen to make as much use of the available space as possible to show interesting features of whatever function or functions are being shown on the graph Again, Fig. A.5 provides a good example. There

y-values of interest are confined to the interval $[-1, 1]$, while x-values slightly exceed the range from 0 to 360. Valuable detail would be lost if the same scale were used for both axes. Using different scales on each axis may distort certain features. For example, if the graph of $y = x$ is plotted using different scales on the x- and y-axes, the drawn line will not make an angle of 45° with the x-axis, but the slope of the line, as defined by the coefficient of x, is still unity.

Dimensions. In the Cartesian system we think of a straight line as having one dimension, and of a plane as having two dimensions. A rectangular solid (e.g., a cube) has three dimensions.

Dimensions – some new notions

More generally, a curve is usually regarded as one dimensional no matter whether it lies on a flat surface such as a plane or on some curved surface such as that of a sphere. I am speaking loosely here for later — specifically in Chap. 22 — we show this is not always true! The reason a curve is regarded as one dimensional is associated with the concept of motion. A particle confined to a curve has only one possible direction of motion (along the curve). A real-world example is that of a train that moves only along a single track.

On the other hand motion on a surface, which may often take the form of an area enclosed by a curve, is two dimensional, since two components of direction are needed to specify possible motion. Typically, for a ship at sea we specify its position by latitude and longitude, and its direction of motion by changes in latitude and longitude. If motion away from a surface is allowed this is three dimensional. An aeroplane operates in a three dimensional space. We need three components to specify its position or motion, e.g., latitude, longitude and altitude.

Higher dimensions. Just as we extend an equation (i) $ax + by + c = 0$ to (ii) $ax + by + cz + d = 0$, by including an additional variable z, we may add yet another variable, say, t, to give an equation of the form

$$ax + by + cz + dt + e = 0 \,. \tag{7.13}$$

In Cartesian coordinate systems we represent (i) as a line in two dimensions and (ii) as a plane in three dimensions. For (i) we can draw simple diagrams representing the corresponding geometric entities (straight lines) on a plane such as a sheet of paper, or a flat-screen computer monitor.

When we move to three dimensional systems it is still relatively easy to depict the plane represented by an equation of the form (ii) by a model in three dimensional space. We did this when we took the origin at floor level in the corner of a room.

What about (7.13)? Our geometric intuition is limited to three dimensional space. Yet geometric concepts can be extended by introducing the notion of four and higher dimensions. In that situation (7.13) represents what is called a *hyperplane*. We look at some of the algebraic implications in Chap. 13.

Coordinate geometry gives us the power to define algebraically concepts of a geometrical nature in more than three dimensions, even though we cannot visualise these directly in the three dimensional space in which we live.

Without the mathematical tool of dimensions greater than three Albert Einstein would not have been able to formulate the theory of relativity.

8

Shape and size

8.1 Geometry in its Own Right

In Chapters 1 to 6 we looked at the development from integers as counting tools to real and complex numbers. We saw how major developments were triggered either by practical needs or by man's curiosity. In Chap. 7 we explained how René Descartes, who lived some 1900 years after Euclid, gave us a link between geometry and algebra.

We have touched briefly on some fundamentals of algebra — the use of symbols to specify any number, and said a little about algebraic equations, because these motivated work that led to the development of complex numbers. The approach so far has been largely 'geometry free' — just brief mentions of Pythagoras's theorem, of the association between the ordered real numbers and the points on a line, and between complex numbers and the points in a plane. We have seen how geometric concepts such as lines, circles and planes can be viewed algebraically by the use of algebraic or analytic geometry.

Now we look at geometry in its own right, because the concepts of form, shape and size are as basic to mathematics as are the natural numbers.

8.2 Axioms

A lot of mathematics is based on an *axiomatic* approach. Until recently many school pupils had their first encounter with an axiomatic system when studying geometry, more fully designated *Euclidean geometry.*

Teaching of elementary geometry has changed greatly over the last half-century, but the axiom-based system introduced by Euclid dominated for more than 2000 years.

In modern mathematics — ranging from different algebras, and several geometries (referred to collectively as non-Euclidean geometries), to probability theory and other topics — the foundation stones for many branches are specific, but different, axioms.

Axioms are sometimes regarded erroneously as obvious truths. Many axioms seem intuitively reasonable, but mathematicians have developed systems based on axioms that may appear to laymen to be somewhat esoteric, if not useless. Any set of axioms must be *self-consistent*, i.e., one axiom in a system must not imply that some other axiom may not, or cannot, hold. All mathematical properties of a system must be deducible from the axioms for that system.

By modern standards, the axioms laid down by Euclid for his geometry were somewhat shaky foundations for a large edifice. Useful as that edifice is, Euclidean geometry has limitations in the real world. Many of Euclid's results that we take as more or less obvious — for example that the sum of the three angles of a triangle is always 180° — are not universal truths. That they sufficed, and appeared to be universally true to Euclid and his followers, may not be entirely divorced from thinking in terms of a flat earth. Some astronomers in Euclid's day had collected evidence that the earth was approximately a sphere, but many people still believed in, or at any rate thought in, flat earth terms.

Much Euclidean geometry works completely in a plane, and also quite well locally, on the surface of a large sphere. That the latter is so, is only because small even expanses like the surface of a lake or a level playing field appear to be effectively flat. When we consider the surface of a whole sphere, or a large part thereof, we must part company from many Euclidean concepts.

On a plane or flat surface we readily accept that a straight line segment joining two points A, B defines a unique shortest distance between those points. We also have no qualms about asserting that, if we walk from A to B, and continue along the straight line beyond B, we will get ever further from A. To get back to A we must retrace our steps.

What about real world travel on the surface of a sphere, or something like a sphere? I say 'something like' because our Earth is only approximately spherical. It is somewhat flattened at the poles. We ignore that detail here and regard the Earth as a sphere. In the situations we are about to look at, we also ignore surface irregularities such as hills and mountains.

'Go west young man' may not be the best advice

Look at a map or a globe and you will see that the latitude 39°55′ N passes both through Philadelphia, USA, and Beijing, China. Most of us intuitively argue that if we travel east from Philadelphia we are travelling in a straight line and will eventually pass through Beijing. That we will do the latter is true, but if we want to get from Philadelphia to Beijing by the shortest route this is not the way to go. We do better travelling due west. Even so, there is a still shorter route.

We can demonstrate this with a globe and a piece of string. We place one end of the string on Philadelphia, then stretch it taut to pass through Beijing. The string then traces the shortest route between the two points if we are confined to the Earth's surface. Remember, we are ignoring irregularities such as mountains or valleys that might lie on the path. The route takes us more or less north over Greenland, across the Arctic circle and quite close to the North Pole. This is the so-called *great circle* route between these two cities.

What is a great circle? A great circle passing through any two given points on the surface of a sphere is a circle on that surface that passes through those two points, and has its centre at the centre of the sphere. There is only one great circle through two given points, unless these points are the opposite ends of a diameter of such a circle. In that case there are an unlimited number of great circles through the two points. A particular case of such diametrically opposite points are the north and south poles.

If two points, A and B, are not at the extremities of a diameter of a great circle, then if we keep to the surface of the sphere, the shorter arc of the great circle determined by those points is a unique shortest distance between them. This means that when we move from a flat surface (plane) to a spherical one, the straight line property of shortest distance applies only to an arc of a great circle.

It is easy to see that all points having the same longitude lie on a great circle determined by that longitude. Conversely, a fixed longitude determines a particular great circle, which passes through the north and south poles.

What about latitudes? Latitude 0°, the equatorial latitude, is the only latitude that corresponds to a great circle. Think carefully to be sure you understand why.

Some straight line properties in planes transfer to great circles on a sphere. There are also some differences between great circles and plane surface lines. On a straight line in a plane if one walks from A to B the shortest way back to A is to retrace one's step. On a great circle the shortest way back may be to continue in the same direction. This will be so if one has travelled more than half way round the relevant great circle. This is an obvious consequence of great circles having a finite length — the circumference — as distinct from the infinite extent of a planar straight line.

What about the rest of Euclidean geometry when we move from plane to sphere? What happens, for example, to triangles?

8.3 A Non-Euclidean Triangle

We know from experience that Euclidean geometry works quite well locally on the surface of a sphere, e.g., on the Earth over small distances, or over areas of a few square miles or kilometres, where the curvature of the surface is for most practical purposes insignificant.

What about large distances and areas? You will find it easy to follow this illustration using a piece of string on a globe. If you have no globe take an orange, a grapefruit, a soccer or a tennis ball, and imagine it is a globe.

Suppose the point A is at latitude 0° (i.e., on the equator) and longitude 0° (i.e., on the Greenwich meridian). We set off from A and travel east along the equator to a point B that is one quarter of the way round the world. Our latitude is still 0°, but our longitude is 90° east. Being on the equator we have traversed one quarter of a great circle. At B we turn left through a right angle, and proceed north for the same distance (a quarter way round the Earth) to get to C. This is the north pole. We have proceeded north along longitude 90° east. This again is an arc of a great circle. What is the shortest surface route back to A? You should soon be able to convince yourself that we have to travel due south along the Greenwich longitude of 0°. We do this if we turn left through 90° at the north pole.

Up the pole – where there is only one way to go

There is something special about the north pole. When we leave it every possible direction we may take is due south. That is why we specify that we cover the final leg from C back to A along longitude 0°, the Greenwich meridian. When we arrive back at A, we approach that point in a direction at right angles to that from which we left it. This 'triangle' ABC on the surface of the globe has all its angles 90°, so the sum of the 3 angles is 270°. The directions we have travelled are 'straight lines' in the sense that they have each been either on the equator, or along a specified longitude. Each sector is therefore the shorter arc of a great circle. Thus, each is the shortest distance on the surface of the sphere between the relevant points.

Further, ignoring the slight flattening of the Earth at the poles, we have travelled the same distance east, then north, then south. So this triangle on the earth's surface, called a *spherical triangle*, has all sides the same length, and is thus equilateral.

We have seen that in moving from a plane to the surface of a sphere, some, but not all, properties of a straight line convert to those we associate with great circles. It is this partial transfer that helps explain why the triangle we considered in our journey round quarter segments of the Earth generated a spherical triangle with angles summing to 270°.

You should spend a few minutes to see if you can construct, or at least visualize, some spherical triangles where all three sides are arcs on great circles, but where the sum of the angles is other than 270°.

Just as a triangle in the plane distorts when we move to the surface of a sphere, distortions occur to other plane figures when we try to place them on the surface of a sphere. If you have a globe handy, cut out a large square sheet of paper with side length roughly equal to the radius of the globe and try to place it to the surface of the globe so that it touches closely at all points. You will have a problem if you are not allowed to stretch, compress or tear the paper.

This difficulty is familiar to people who draw maps of the world for atlases. Most school atlases contain maps of the whole earth on a flat page that distort the spherical features in one or more ways. They may maintain the relative shapes of land masses, but distort their areas, making Greenland look larger than Australia. They may preserve relative size, but distort the shape. The difficulty arises from the impossibility of placing a flat surface onto a rigid sphere to touch everywhere. Conversely, we cannot open out the surface of a sphere to fit everywhere onto a plane without tearing or distorting it in a way that changes shapes or sizes. We look more closely at such matters in a wider context in Chap. 9.

8.4 Geometry and Art

Concepts like congruence (equality of size and shape of objects) and parallelism (straight lines are parallel if they do not meet no matter how far they are produced in either direction) present problems when we try to make a two dimensional picture of a three dimensional space.

Artists do this when they draw a 'true' or 'in perspective' representation of a landscape. If two objects are the same size in a three dimensional landscape, then in the two dimensional picture the nearer one occupies a larger area.

Fig. 8.1 is a simple drawing representing a straight railway track crossing a level plain. Looking at it as a 'picture', we would accept that the two rails forming the track are a constant distance apart, and that the sleepers are all the same size. Yet the rails meet at the top of the picture, and from bottom to top of the picture the sleepers look to decrease in size.

If we travel along the real line we would not fear that the engine would derail because the rails are getting closer. Nor would we expect the more distant sleepers to be getting closer to each other and smaller.

The notion of perspective has foundations in a different geometry to that of Euclid. It is called *projective geometry.*

Mathematical interest centres on which properties of points, lines and angles are preserved when we move from Euclidean to projective geometry. To answer a few of these questions we need to explain what is meant by projections and projective geometry. It is clear from Fig. 8.1 that parallel lines, such as real railway tracks, which never meet in the Euclidean sense, no longer retain that property when projected onto a flat picture. The rails do retain the property that they are straight lines. Similarly, the top surfaces of real sleepers are all equal rectangles, but they become trapeziums of different sizes in the pictorial representation.

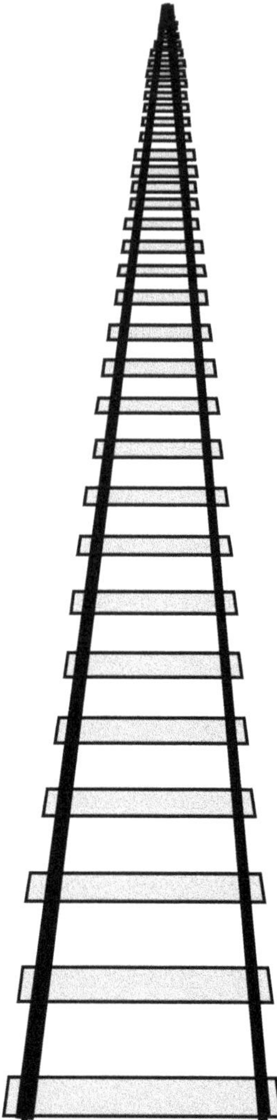

Fig. 8.1. A representation of a straight railway track on level ground

'A point has position but no magnitude' ***Euclid***

–

But not every point is a vanishing point

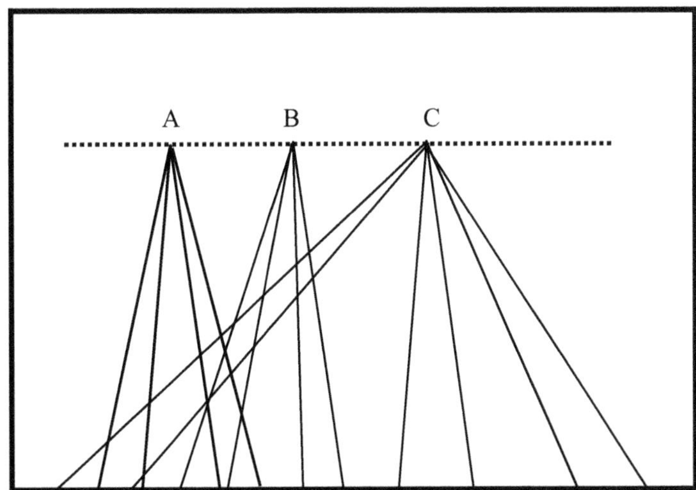

Fig. 8.2. A picture of sets of parallel lines in a plane, each set being in different directions, so the lines converge to different points on the horizon denoted by A, B, C on the dotted line representing the horizon

In Fig. 8.1 the point at the top of the picture where the 'parallel' lines appear to meet is the most distant point we can see. Artists call it a *vanishing point.* If the railway ran across a completely flat plain, this most distant point would, in artistic terms, be on the horizon.

Figure 8.1 represents a projection of rails and sleepers in a real world plane onto the draftsman's paper. The projection is a recording on a flat surface of the impression in an artist's eye. In a two dimensional perspective drawing or photograph the horizon is usually represented by a straight line running horizontally across the picture. Anything below that line represents the surface of the plane. Everything above is the sky. The two rails of the single track railway, appear to meet at a point on the horizon. This is true no matter what the gauge of the track (the distance between the rails) may be. In the picture any number of lines that are parallel to one another, irrespective of distance between them, will meet in one point on the horizon.

What happens if we take a second set of parallel lines in a different direction to those in the first set? In the picture those in the second set will meet at a point on the horizon, but at a different point from that where the first set meet.

Fig. 8.2 is a simple two dimensional drawing of three different sets of parallel lines. The outer rectangle is just a picture frame. The dotted line joining the points A, B, C where the sets of parallel lines meet represents the horizon. The four lines meeting at A provide the artistic representation of a set of four lines parallel to one another. The four lines meeting at B represent another parallel set having a different direction to the first set. A third set of six parallel lines in a different direction from those of the first two meet in C. The different directions are highlighted by some of the members of this third set crossing one or more members of each earlier set.

Geometrically, the line that is the horizon is composed of the set of 'points at infinity' (in a Euclidean sense), at which parallel lines in the real-world plane meet in the picture. There is an infinity of such points making up the last line. Each point is associated with a different direction for sets of parallel lines that 'meet' at that point. I am being vague about what infinity means, but think of it here as an unlimited number.

Perspective fascinated Leonardo da Vinci and other artists of his era. They realised that an accurate two dimensional drawing of a three dimensional landscape required a 'transfer' of what was perceived by an artist's eye to a sheet of paper. We get an idea of what this means if we look out of a window, regarding the window as the paper on which the picture we see is etched. If two objects outside are of the same size in the real three dimensional world, that which is closer occupies a greater window space. If we stand 1 m (one metre) behind a window 2 m wide and 1.2 m high, we will see only part of the wall of a shed that is 5 m long and 3 m high, if it is only 1 m in front of the window. If it were 200 m away, we would see the whole shed wall, and it would occupy a relatively small amount of window space — even less if it were 500 m distant. The picture will also change in perspective if we move away from, or closer to, the window, or sit rather than stand. This is because our eyes are in different positions in each case.

In projective geometry an artist's eye corresponds to a point known as the *centre of projection*. Basically, projective geometry is concerned with the projection of points in one or more planes onto another plane, using a centre of projection that lies outside those planes.

Fig.8.3(**a**) gives an illustration. The planes $\mathbf{P}$ and $\mathbf{P}'$ are any two planes in three-dimensional space. In general they are not parallel to each other. A point O outside these planes is the centre of projection. The central projection from O of any point M in the plane $\mathbf{P}$ onto the plane $\mathbf{P}'$ is the point M′ where the line OM intersects the plane $\mathbf{P}'$.

If A, B, C are any points on a straight line in $\mathbf{P}$ that project onto points A′, B′, C′ in $\mathbf{P}'$ it is not hard to see that these lie on a straight line in $\mathbf{P}'$. It is easily seen that any straight line in $\mathbf{P}$ projects onto a straight line in $\mathbf{P}'$. The length of straight line segments is not preserved under projection. For example, $\mathrm{AC} \neq \mathrm{A'C'}$.

Triangles project onto triangles, but properties such as equality of angles, similarity and congruence may not be preserved under projection. Projections of parallel lines are in general no longer parallel. This is consistent with our comments on perspective.

Key features of projections are that a point projects onto a point, and a straight line onto a straight line. From Fig. 8.3(**a**) we see that the line through A′C′ is the line of intersection of the plane $\mathbf{P}'$ and the plane containing the triangle OAC defined by the line AC and the centre of projection O.

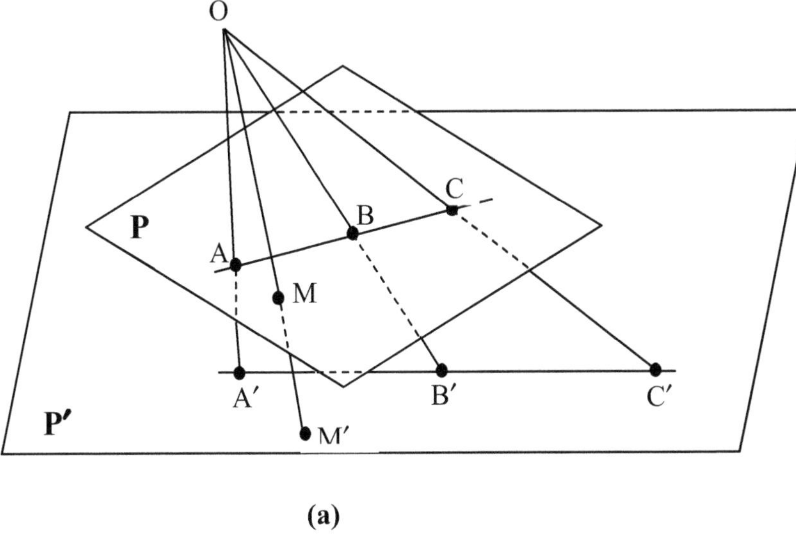

(a)

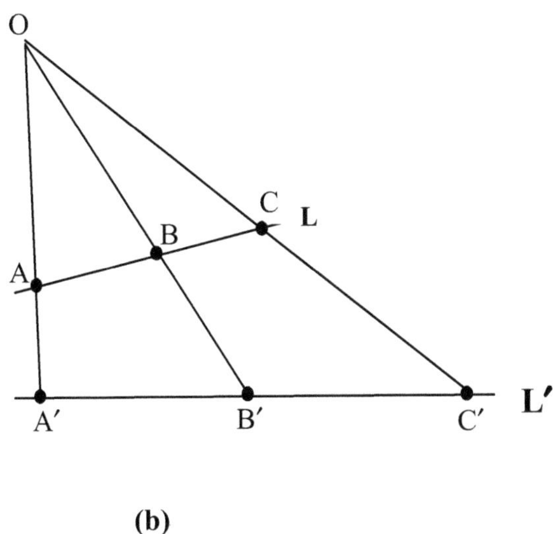

(b)

Fig. 8.3. (**a**) Projections from a centre of projection O of points and lines in a plane onto another plane and (**b**) a simpler projection in two dimensions

A point and a line are said to be *incident* if the line goes through the point, or if the point is on the line. If a point B and a straight line AC are incident, then after projection the corresponding point B′ and the projected line A′C′ will also be incident. There are a few 'special' cases that appear to be exceptions. For example, when the point M is such that the line OM is parallel to the plane $\mathbf{P}'$, or the plane OAC is parallel to the plane $\mathbf{P}'$. These difficulties can be removed by defining special concepts to cover them.

We have considered projections from one plane to another in three dimensional space where the centre of projection lies outside either plane. A simplified concept of projection is one in a two dimensional space. The point of projection lies within a two dimensional plane and the projection is that of points on one line **L** to another line $\mathbf{L}'$, both lines lying within the plane.

This is illustrated in Fig. 8.3(**b**). It is a special case, for we see on closer inspection that this two dimensional plane is in essence a representation, or copy, of the plane through O, A, C in Fig. 8.3(**a**).

8.5 Desargues' Theorem

A famous theorem in projective geometry was first stated by the French mathematician Girard Desargues (1591–1661). In a diagram like that in Fig. 8.4(**a**) through any point O, we draw three lines OA, OB, OC where A, B, C are any points on those lines. On each of these lines we mark further points A′, B′, C′ (which need not all be on the same sides of O as A, B, C). We temporarily exclude the case where any or all of the lines A′B′, B′C′, C′A′ are respectively parallel to AB, BC, CA. We extend as necessary, as shown in Figure 8.4(**b**), AB and A′B′ to meet at F, BC and B′C′ to meet at D and CA and C′A′ to meet at E. Desargues' theorem states that D, E and F lie on a straight line.

Unlike most theorems in Euclidean geometry, Desargues' theorem is not about equalities of lengths or angles, features not preserved in projective geometry. It is a truly 'projective' theorem, where the centre of projection is the point O. It uses only notions about incidences of straight lines and points.

The theorem has a remarkable symmetry. We may relabel the figure by first choosing as a new centre of projection, O, any of the remaining nine labelled points in Fig. 8.4(**b**). Then we can, by suitable relabelling of the rest of the points, find a new arrangement for which the theorem still holds with the same wording as we used above.

One possibility is given in Fig. 8.5 where O is now the point formerly labelled A′. You should follow through the statement given above to see that the theorem now holds with these new labels. As a further exercise you should verify that we might, with the same choice of O, interchange ABC and A′B′C′. We may also 'relabel' the triangles by interchanging A, B, C in any of six possible ways, and making corresponding interchanges in A′, B′, C′. There are in all 120 different ways of labelling the points in diagrams like Figs. 8.4(**b**) or 8.5 that all give rise to the same statement of Desargues' theorem. This follows because there are 10 choices for the position of O. Each of these gives rise to two possible triangles with which we may associate the undashed and dashed symbols A, B, C. Within these two choices the symbols A, B, C can be arranged in 6 orderings, giving in all $10 \times 2 \times 6 = 120$ different labellings.

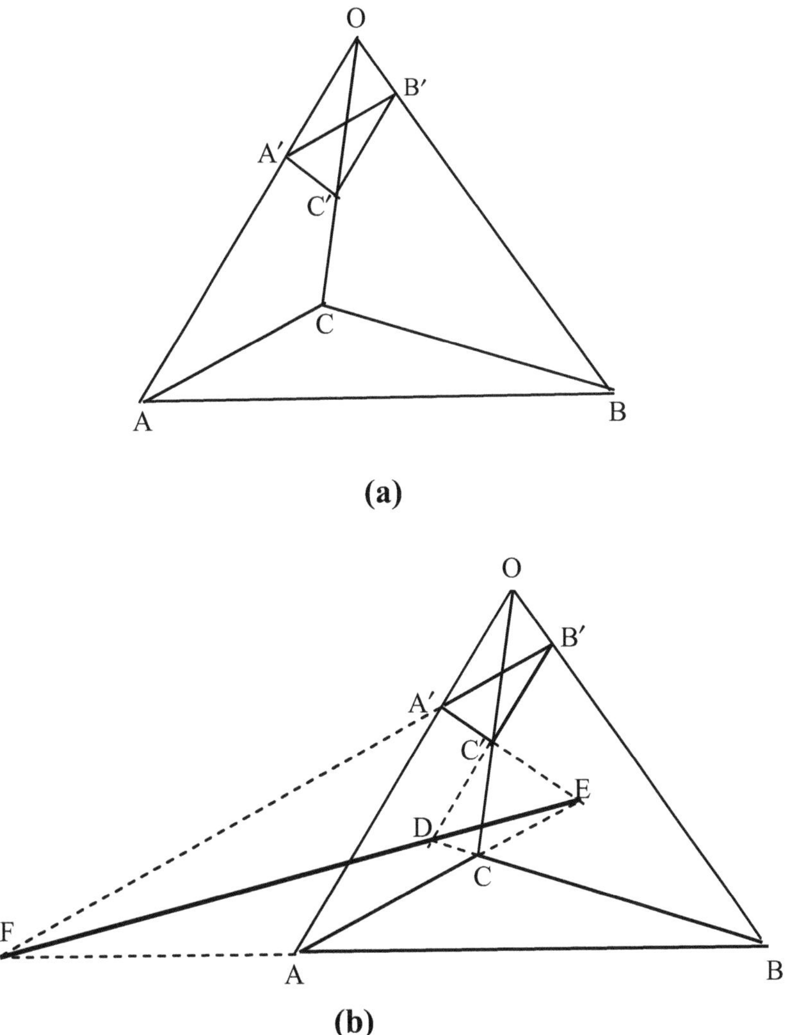

Fig. 8.4. Diagrams relevant to Desargues' theorem

The theorem can be proved in this two dimensional case by Euclidean methods, but that proof is complicated. What is surprising is that there is a three dimensional analogue that is easy to prove by projective geometry. Only a little further work is needed to extend this proof to the two dimensional case.

It is sometimes easier to think in three dimensions than in two

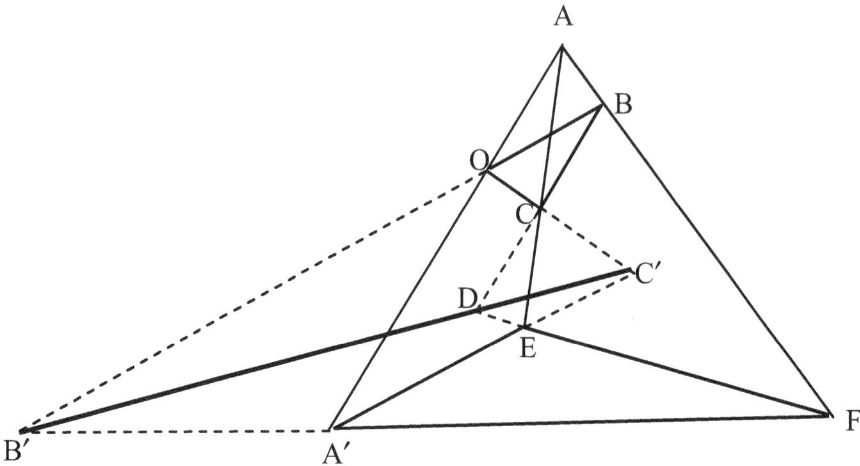

Fig. 8.5. Symmetry property of Desargues' theorem illustrated by appropriate relabelling of ten points

The three dimensional version of Desargues' theorem is described using Fig. 8.6. In Fig. 8.6(**a**) O,A,B,C are the vertices of a tetrahedron. This can be any tetrahedron, but it helps to visualize it as one with the triangular 'base' ABC lying in the horizontal plane (i.e., the plane of the page), with O a point above that plane. As in the two dimensional case, there is no restriction on the lengths OA, OB, OC. We may think of the edges OA, OB, OC, AB, BC, CA as being six thin rigid rods. We now take 3 more rods to form the triangle A′B′C′ where A′, B′, C′ lie at arbitrarily chosen positions on OA, OB, OC respectively. We temporarily apply the restriction that the points are so chosen that the triangle A′B′C′ does not lie in a plane parallel to that of the triangle ABC. In Fig. 8.6(**b**), the sides AB and A′B′ are extended to meet in F. Since the planes that include the triangles ABC and A′B′C′ are not parallel they intersect one another in a straight line. Since AB lies in one of those planes and A′B′ in the other, those two lines must meet at some point on that common line of intersection, i.e., F lies on the common intersection. Similarly, the point D where the lines BC and B′C′ intersect, and the point E where CA and C′A′ intersect, both lie on the common intersection line. Thus the three points F, D, E are collinear, i.e. they lie on the same straight line.

Look now at Fig. 8.6(**b**) and Fig. 8.4(**b**). These are identical apart from the fact that some of the lines differ in thickness. This was simply an artefact to make the descriptions a little clearer. Effectively, Fig. 8.4(**b**) is a two dimensional photograph of the real three dimensional system. Thus Desargues' theorem holds for photographs or drawings using a centre of perspective O. With a little *i*-dotting and *t*-crossing this is the basis for a proof of Desargues' theorem in two dimensions.

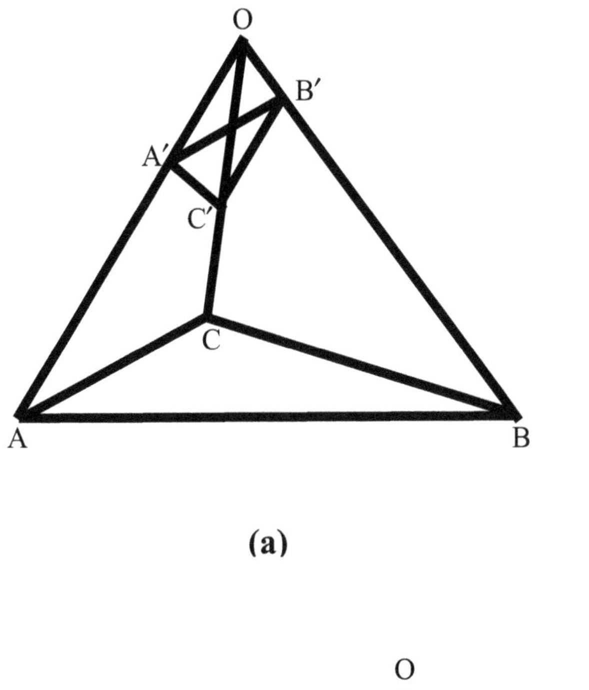

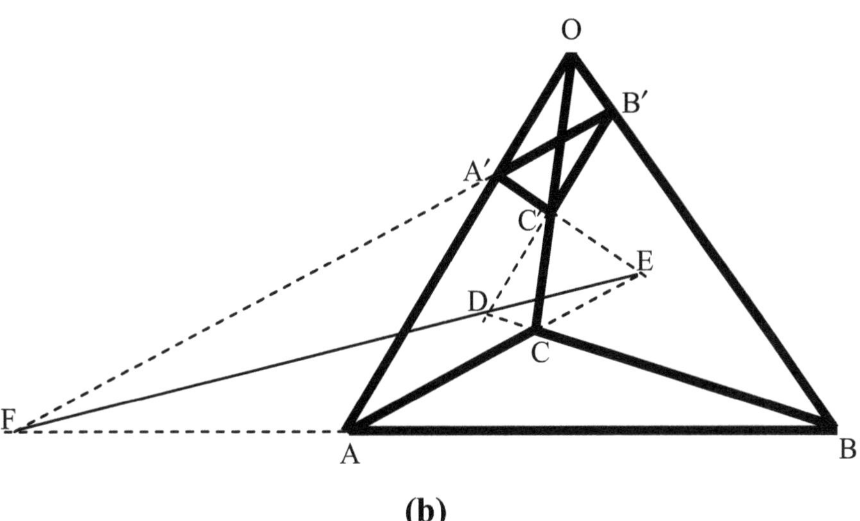

Fig. 8.6. Desargues' theorem applied to tetrahedrons

In some special cases in projective geometry there is an apparent breakdown. In particular, this may happen where parallelism means that some lines do not meet, or that lines do not intersect some plane. Projective geometry would be very complicated if these situations had to be treated separately.

Fortunately, they may be brought into one general framework by introducing the additional concepts of *ideal points* and *ideal lines*.

We look at what is involved only in general terms. Consider a situation like that in Fig. 8.3(**a**). We take two lines in the plane P that meet at some point B, say. Then if B′ is the projection of that point on the plane P′, the projections of those two lines onto P′ will also intersect at B′. Lines that are not parallel intersect by definition. Consider two lines that are 'nearly' parallel. We fix one and rotate the other about some fixed point in a way that makes them even closer to being parallel. The point of intersection will become more distant, eventually receding to what is popularly referred to as 'infinity'. That is why even in Euclidean geometry parallel line are sometimes described as meeting only at infinity.

In projective geometry this idea is usually encapsulated in the concept of an ideal point. This is an abstract notion like infinity itself. All lines that are mutually parallel are regarded as meeting at a common ideal point. We met this idea in the discussion of parallel lines in a drawing or photograph. There, we gave a physical representation of such an ideal point as a point on the horizon. Similarly, sets of mutually parallel lines in a different direction to the first set meet at some other ideal point as in Fig. 8.2. These ideal points in turn are said to lie on an ideal line. We represent this in a drawing or picture by the straight line 'horizon'. The concept of ideal lines and ideal points simplifies some proofs of theorems in projective geometry by avoiding the need to consider a number of special cases.

Another type of projection of interest to artists is what are called parallel projections. These cover, e.g., an artist's concept of parallel light beams from the sun. We do not pursue this further here.

8.6 Loose Ends

Map making. In Sect. 8.3 I alluded to the difficulties of map making when we attempt to translate, or map, points on the surface of a sphere to a plane. Some distortion is inevitable, either in shape or in representing distances or areas.

The distortion is negligible when we map only small areas, for then the curvature involved is minimal. If we want to project the whole spherical surface of the globe onto a plane surface, such as a sheet of paper, distortion is inevitable. The basic procedure for doing this is projection. Points on the surface of the sphere are projected onto another surface which can be opened out to give a plane surface.

The simplest such surface is a cylinder with diameter equal to that of the globe made by rolling up a rectangular sheet of paper. The map is obtained by unrolling the cylinder after recording the projected image. Different choices of a centre of projection, or method of projection, as well as different choices of the height of the cylinder and its positioning relative to the almost spherical globe representing the Earth, result in a range of potential planar maps. Appropriate choices will depend on what properties one wants to retain with little or no distortion.

For whole-world or large area maps one of the most commonly used and best known projections is Mercator's projection. It was proposed by the Flemish cartographer Gerardus Mercator (1512–1594).

Mercator wrapped a long paper cylinder round the globe to touch it at all points on the equator. His centre of projection was the centre of the globe. The projected lines of latitude and longitude form a rectangular grid. Lines corresponding to each degree of longitude difference are parallel vertical lines all at equal distances apart. Lines corresponding to each degree of latitude difference are parallel horizontal lines, but the distance between each increases as the latitude increases. There is little distortion of distance near the equator but major distortions as the parallel lines representing each additional degree of latitude move further apart. The poles are not represented on the map since the parallel lines representing each specific longitude then meet only at infinity. The north pole is effectively moving off to a line at infinity, or the top edge of a sheet of paper of infinite height, but finite width. An appealing feature of maps based on Mercator's projection is that bearings read from them are true bearings. This makes it a useful navigational tool.

Another cylindrical projection is one where projection is onto a cylinder touching the globe at the equator and the cylinder is of height equal to the diameter of the globe. A different type of projection to any described so far is used. If we consider the planes corresponding to each latitude, these form a series of parallel planes, and if these are extended (projected) to meet the cylinder they do so in parallel lines. These lines become closer for each degree of difference in latitude as we approach the poles. If we now consider the planes corresponding to each different longitude, and project these to meet the cylinder, they will do so in a series of parallel lines (equidistant for any given difference in longitude). This grossly distorts the shape of regions far removed from the equator, but relative areas are preserved by the mapping.

Conical projections are used to minimize distortions where it is important to do so at specific latitudes, especially if these are far removed from the equator. The simplest conical projection is one where the surface of a cone is tangential to the globe (i.e., touches the globe) at the latitude of interest, and the vertex of the cone lies on an extension of the polar diameter of the globe. With the centre of the globe as a centre of projection, there is little distortion in either distance or area near the tangential latitude.

Modifications of the conical projection include cones that make radial contact at two different latitudes, say 60° N and 70° N, effectively passing inside the globe between these two latitudes. Providing the chosen latitudes are not too different, there is little distortion in distance or area between these latitudes with any reasonable projection method. This is because the surfaces of globe and cone are reasonably close to one another in the projection region.

Fig. 8.7. A taut string with ends pinned at F_1, F_2 for mechanical construction of an ellipse

Many other projections are used, often for special purposes. These include, for example, zenithal projections in which some point on the surface is selected as a central point and great circles radiating from that point are represented by straight lines.

Cross ratios. An important concept in projective geometry omitted earlier because it would take too long to discuss in detail is that of the *cross ratio.*

If A, B, C, D are four points in that order on a straight line, L, then the cross ratio, c, is defined as

$$c = \frac{\mathrm{CA}}{\mathrm{CB}} \Big/ \frac{\mathrm{DA}}{\mathrm{DB}} .$$

If $\mathrm{A}', \mathrm{B}', \mathrm{C}', \mathrm{D}'$ are the projections of A, B, C, D on another line L′, then

$$c = \frac{\mathrm{CA}}{\mathrm{CB}} \Big/ \frac{\mathrm{DA}}{\mathrm{DB}} = \frac{\mathrm{C'A'}}{\mathrm{C'B'}} \Big/ \frac{\mathrm{D'A'}}{\mathrm{D'B'}} .$$

Thus, the cross-ratio is preserved under projection. This result holds for both central projection about a point O, and for parallel projection. For central projection it is not hard to prove using elementary trigonometry. On the basis of the definition alone, this may look a somewhat esoteric property, but the invariance of the cross ratio under projection is important. For an introductory account of the main properties of the cross ratio, and some of its uses, a good reference is [13] (Chap. IV).

Ellipses. The only non-Euclidean geometry we have discussed in even moderate detail so far has been projective geometry. Another limitation in Euclidean geometry arises when one wants to create curves that cannot be drawn with the restriction to ruler and compasses for construction. In broad terms, this restricts us to drawing straight lines and circular arcs. We cannot in general draw ellipses, hyperbolas. parabolas, spirals or most other curves with just a ruler and compasses.

A simple mechanical device for drawing ellipses consists of two drawing pins and a piece of string. The drawing pins are inserted in a plane surface such as a sheet of paper at some chosen distance, d, apart, where d is less than the length, l, of the string. The ends of the string are fixed, or held down by

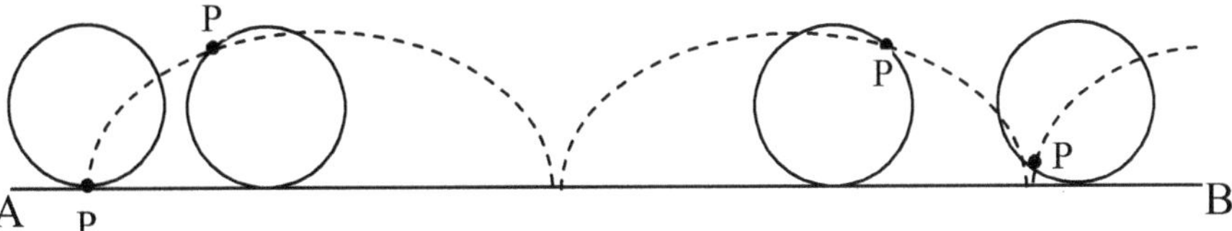

Fig. 8.8. A simple cycloid formed by a fixed point P on the circumference of a circle that rolls along a line AB

the drawing pins, at their points of insertion. A curve is traced out by keeping the string taut as we move a pencil. The shape we get is like that in Fig. 8.7. Different ellipses are obtained by altering the distance, d, between the fixed end points while keeping l fixed, or by altering l while keeping d fixed.

The characteristic property that generates the curve is that if the drawing pins are at points F_1 and F_2, and P is any point on the ellipse, then $F_1P + F_2P = l$. Also, if A, B are the points where the line F_1F_2 cuts the ellipse, then AB has length l. The 'diameter' AB is called the *major axis* of the ellipse and the points F_1, F_2 are called the *foci*. We do not pursue here the many interesting properties of this curve. It is easily seen to collapse to a circle of radius $l/2$ with centre F if F_1 and F_2 coincide at F.

Cycloids. Another interesting curve is that traced out by a fixed point on the circumference of a circle when the circle rolls along a straight line without slipping. We might picture it as the path of a piece of chewing-gum fixed to a bicycle tyre as one pedals along a straight level road. Fig. 8.8 shows four positions of such a fixed point, P, as it rolls from left to right along a line AB. We see that P traces out a series of arches (the dashed curve). These are easily seen to be of width equal to the circumference of the circle, and height equal to the diameter of the circle. The curve is called a *cycloid.* If the point is not on the circumference of the circle, but lies either inside or outside the circle, more general curves of the 'cycloid' family are formed.

Rolling around in circles

Related curves are obtained when smaller circles rotate inside (or outside) larger circles. It is not too difficult to see that if we trace the locus of a fixed point P on the circumference of a circle of radius r that rolls internally round a fixed circle of radius $R = 3r$ that we get the dotted curve illustrated in Fig. 8.9. This is an example of a *hypocycloid.* You might like to work out for yourself what form the hypocycloid would take if the radius of the larger circle is twice that of the smaller circle, i.e. $R = 2r$. See if you can work it out before reading on.

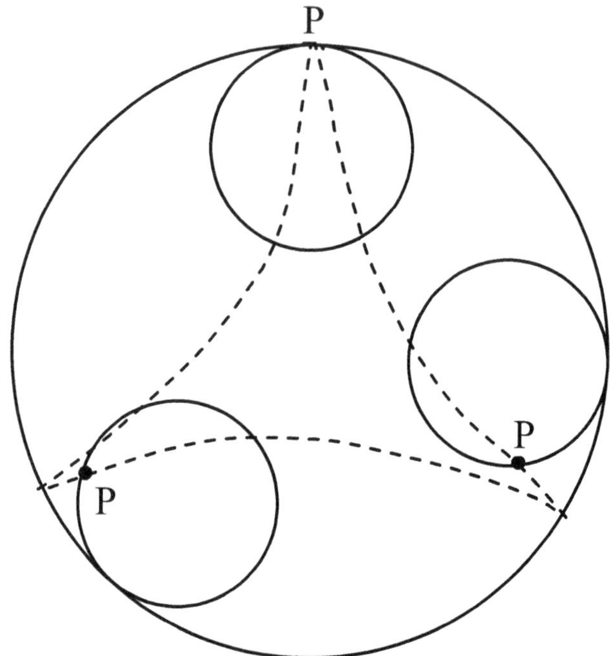

Fig. 8.9. Hypocycloid traced by a fixed point P on the circumference of a circle of radius r which rolls inside a fixed circle with radius $R = 3r$

The fixed point always lies on a diameter of the larger circle. It takes just a bit of careful thought to see why.

9

Stretch and Squeeze Geometry

9.1 Stretching and Squeezing

Basic Euclidean concepts — parallelism, similarity, congruence and certain other properties relating to lines, triangles, rectangles, polygons, polyhedra and circles — apply to rigid bodies. Only limited transformations are allowed. Items may be rotated, or moved to different positions, providing this does not change their shape.

In Chap. 8 we saw that projective geometry allowed some distortions, as defined by the projection. These included certain shape changes, providing points transformed to points, and straight lines remained straight lines. Directions of lines and lengths of segments could change, and parallel lines could be redirected to meet at points that did not comply with Euclidean notions of infinity. Triangles that were similar before projection need not be similar afterwards, and the area of a triangle could, and usually would, change after projection. Projective geometry is relevant to the artistic concept of perspective.

In Sect. 8.3 we pointed out that one cannot paste a plane sheet of paper on the surface of a sphere so as to make contact everywhere unless we can stretch and squeeze, or perhaps even tear or cut, the paper. What happens if we allow such activities?

Fig. 9.1(**a**) and Fig. 9.1(**b**) are closed figures in a plane that look to have little in common. That is true if we think only in terms of Euclidean geometry. They are not congruent or similar. While (**a**) can be described as a polygon split into smaller polygons, including rectangles and triangles, (**b**) is devoid of the straight lines, triangles, quadrilaterals, polygons or circles that are the life-blood of Euclidean geometry.

One property that Fig. 9.1(**a**) and (**b**) share is that each is divided into the same number of parts. Looking a little closer we see a further correspondence between the parts. Those labelled with each letter in the two diagrams either have, or have not, got a length of common boundary between them. For example, in each figure the region A has common boundaries with both B and

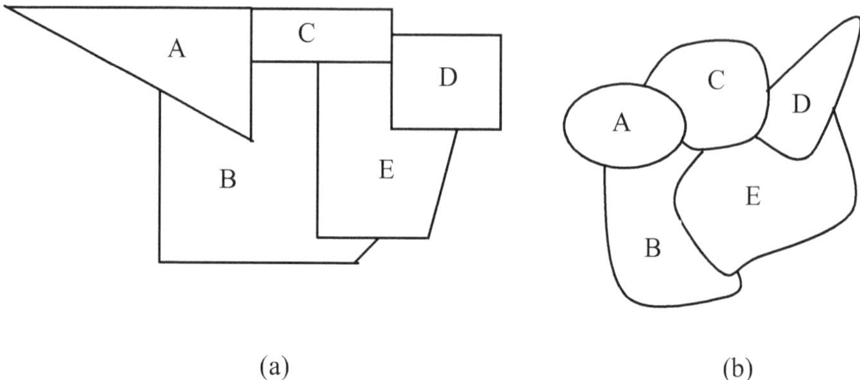

Fig. 9.1. From the Euclidean geometry viewpoint figures (**a**) and (**b**) have little in common but they are topologically equivalent figures

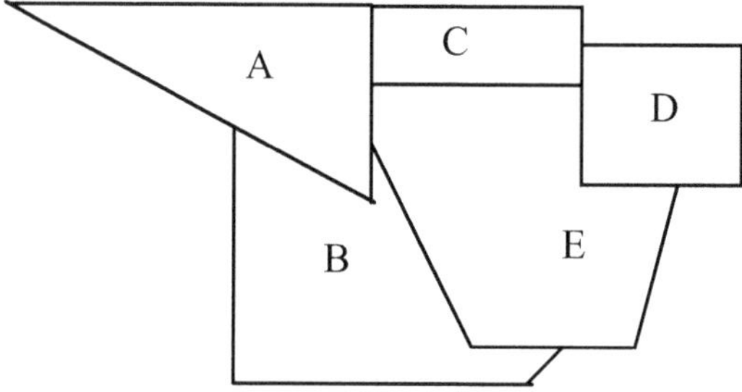

Fig. 9.2. A figure not topologically equivalent to those in Figures 9.1 (**a**) or (**b**)

C, but not with D or E. The two diagrams are said to be *topologically equivalent*, a term we explain more formally later. Broadly speaking it means that if we drew diagram (**a**) on a strong (in the sense that it would not easily tear) but pliable sheet of rubber we could, by stretching and squeezing it carefully and patiently, distort it to the shape in (**b**) without tearing the rubber. On the other hand, we could not so distort either Fig. 9.1(**a**) or (**b**) to Fig. 9.2 without some tearing or cutting and pasting. This would be needed to create a common boundary between A and E that does not exist in Fig. 9.1.

Geometry on a rubber sheet

A physical explanation of the impossibility of transforming either Figs. 9.1(**a**) or (**b**) to Fig. 9.2 by the rubber stretch analogy, is that the common boundary between C and B in Figs. 9.1(**a**) and (**b**) implies the existence of a large set of points in either region that are close together. These are neighbouring points on either side of the boundary. It is impossible, without tearing the rubber sheet, to eliminate the 'closeness' of such points, as has happened in Fig. 9.2, where there are no points in C and B that are close together. The diagram in Fig. 9.2 is described as *not* topologically equivalent to those in Fig. 9.1.

Topology is the study of properties of figures that may be transformed from one to another by transformations which, on the rubber sheet analogy, may be described as *without cutting or tearing.* For a complete understanding we need some provisos that we introduce later.

The subject is now a major branch of mathematics, although early interest in it grew from some tantalizing problems that in themselves were not of immediate practical importance. One of these is associated with maps and is known as the *four colour problem.*

9.2 The Four Colour Problem

In the 1850s a young graduate from the University of London, Francis Guthrie, started thinking about the minimum number of colours needed to produce a map showing all the English counties. The only constraint was that the same colour must not be used for any two counties that had a length of common boundary.

As a mathematician often does, Guthrie soon turned his thoughts to a more general question. What is the minimum number of colours required to cover any conceivable map on a plane or spherical surface if the same colour must not he used for two regions having a common boundary? The boundaries might be between administrative regions such as counties, or between areas of distinct geological structure, or just arbitrary geometric closed shapes.

A simple counter example like that in Fig. 9.3 shows that three will not suffice. In that diagram each region A, B, C, D needs a different colour.

Saving the printer's ink

It was conjectured, probably first by the English mathematician Augustus de Morgan (1806–1871), that four colours might suffice. It was not until 1890 that it was established that five would certainly be enough. In 1976 two American mathematicians, Kenneth Appel and Wolfgang Haken, proved that four colours would always suffice. Their proof broke new ground mathematically. It would not have been possible without a powerful computer and ingenious programming. There is no pencil and paper version of their proof. Indeed, there never can be one based on the method they devised. A pencil and paper proof may be possible, but it would need an approach logically different from that used by Appel and Haken. Those authors' views on their proof are given in [3].

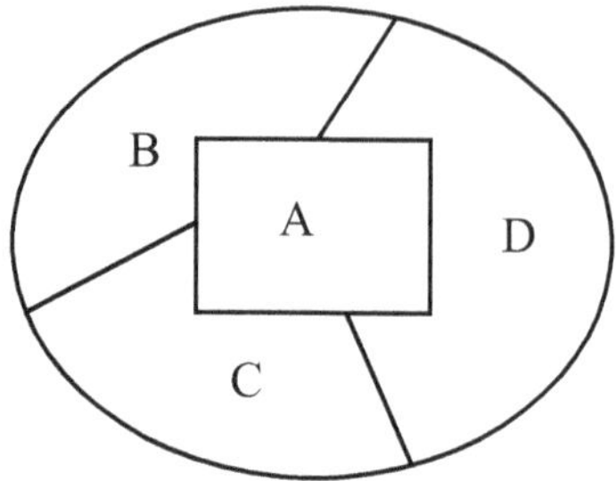

Fig. 9.3. A map of only four regions that requires four colours

The question central to the four colour problem is not whether all real maps in a plane or on the surface of a sphere can be so coloured. It is one of showing that it is impossible to produce a map that needs more than 4 colours. Because there are so many different possible ways of colouring a map, it may require ingenuity to actually produce a scheme that will do the job for a particular map. Again, that is not the question at issue.

A straightforward exposition of the historical steps, and the nature of the hurdles that had to be cleared in proving the four colour problem, is given by Devlin [14] (Chap. 7). A fuller account relating the topic to other work is that by Wilson [46]. We deal here only with some facets that link the problem to other aspects of topology.

Cutting back the deadwood

All topologically equivalent maps may be reduced to simple formats that retain only the relevant topological features. Clearing away the clutter to reduce a problem to its bare bones is often a vital preliminary to making progress in mathematics.

To start, we give the problem a more 'mathematical' flavour by replacing each colour by a number. The problem then reduces to whether we can allocate just four numbers 1, 2, 3, 4 in such a way that no two regions that have a common boundary are given the same number. Fig. 9.4 shows a diagram (map) with 6 regions and a possible allocation of four numbers that meets this requirement. Regions that meet at only one point are declared to have no common boundary. That this restriction is needed is obvious if we consider a map in which all regions are sectors of a circle and therefore share a common 'one point' boundary at the centre of that circle.

Another simplification we can make in representing topologically equivalent maps (i.e., those with the same disposition of common boundaries) is to select in each region a designated point. This point might be the position of the capital city, or the highest mountain peak. In more abstract mathematical

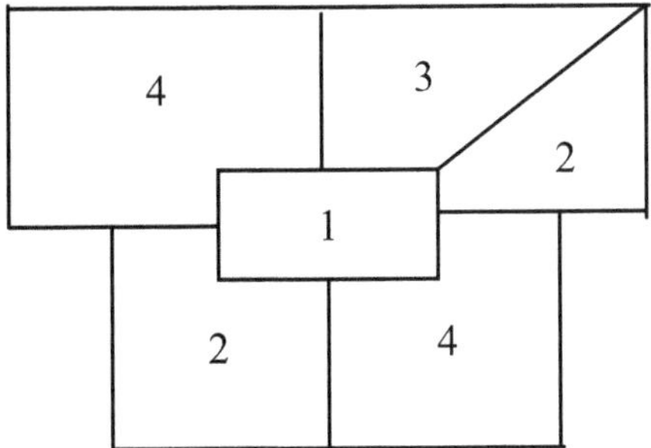

Fig. 9.4. A map where 6 regions are coloured by 4 colours; each number represents a different colour

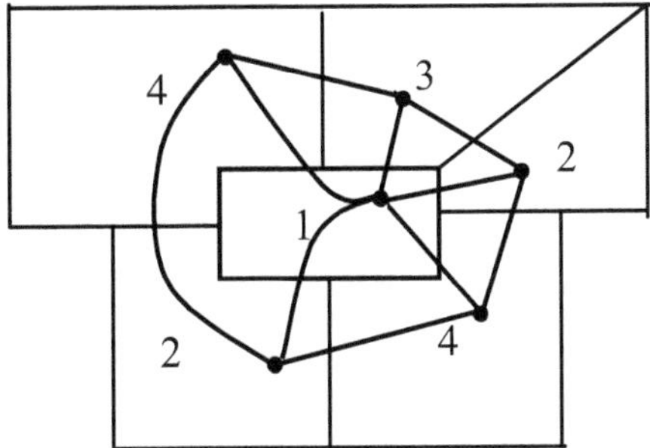

Fig. 9.5. Curves joining selected points in regions in Fig. 9.4 that have a common boundary and that do so without passing through any other region

terms it may be any point chosen at random, or at our discretion, within that region. If we now join *only* those chosen points that are in regions having a common boundary that we may cross without going through any other region we get a situation like that in Fig. 9.5. Regions that can be so joined are called *neighbouring regions.*

In Fig. 9.5, to comply with the condition that we only cross one boundary with any join, we use curved rather than straight lines for some joins. Careful reflection shows that in any map none of these joining lines need cross each other.

9.3 Graph Theory

We may assign each region number in the four colour problem to the point we have selected within that region. For reasons that will soon become apparent, these points are usually called *vertices*. We call a line joining two vertices an *edge*. In the four colour problem, if two vertices are joined by an edge, this implies that the regions corresponding to those vertices have a common boundary of nonzero length. Remember, we exclude the case where there is only one boundary point in common. There is no edge joining vertices in regions that have no common boundary.

The numbers associated with the vertices, and the information conveyed by the edges, tell us everything relevant to the colouring of topologically equivalent maps. One implication is that we cannot give the same numbers to vertices that are joined by an edge. We can, however, give the same number to vertices not joined by an edge. As already explained, some of the edges are curved in Fig. 9.5 to avoid passing through more than one boundary. In Fig. 9.6(**a**) we show the vertices and edges ignoring the rest of the original map information. We may then 'straighten out' the curved edges. This is done in Fig. 9.6(**b**), which is topologically equivalent to Fig. 9.6(**a**), because the same pairs of vertices are joined by nonintersecting edges.

The diagrams in Fig. 9.6 are a special case of a kind of graph first studied in a different context by Leonhard Euler (1707–1783). Diagrams like these are basic to the subject of *graph theory.* The graphs here are a different concept to those in Chap. 7 associated with coordinate geometry.

The graphs in this chapter are often referred to as *networks*, or more specifically as neighbouring networks, because the edges connect vertices in neighbouring regions. This approach is of interest in the four colour problem because it forms the basis for the eventual computer-generated proof of that conjecture.

Graph theory provides valuable tools for solution of many topological problems, and we look at some examples. These are indicative only, and are far from exhaustive.

A key formula particularly relevant to networks stems from Euler's work. He was not aware of it, but he was taking the first steps in developing topology. Euler obtained his original formula in a study of polyhedrons — three dimensional figures having vertices joined by straight lines called edges. These edges define plane polygons known as *faces.* The simplest polyhedron is a tetrahedron. It has four faces. Each is a triangle, a triangle here being classed as a three sided polygon.

A formula that holds for any polyhedron

Fig. 9.7 is a two dimensional picture of a tetrahedron. For visual purposes think of the triangle ABC as lying in the plane of the page, and the vertex D as a point above the page. The tetrahedron has 4 faces (the triangles ABC, ABD, BCD and CAD), 4 vertices (A, B, C, D), and 6 edges (AB, BC, CA, DA, DB, DC). Denoting the number of edges by E, the number of faces by F and the number of vertices by V, we have $F = 4$, $V = 4$ and $E = 6$. Therefore $V - E + F = 2$. This is a topological property of any tetrahedron. It does not

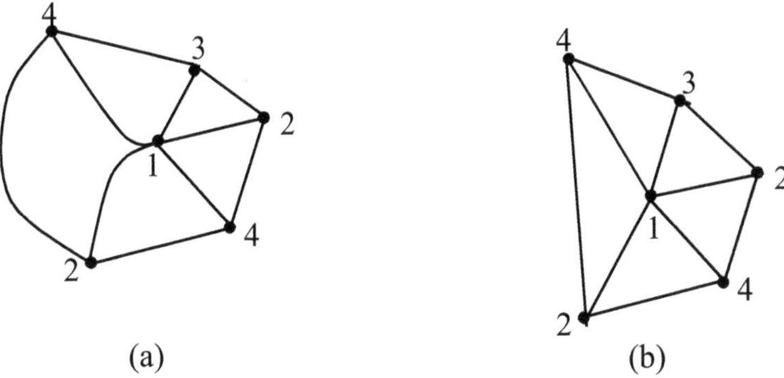

Fig. 9.6. Figure (**b**) is a neighbouring network topologically equivalent to (**a**)

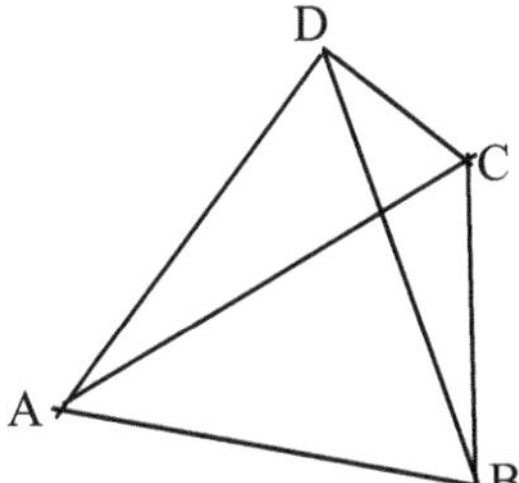

Fig. 9.7. A two dimensional picture of a tetrahedron

depend on the lengths of the edges, or on the shape or area of the triangular faces.

The important result Euler obtained was that not only for tetrahedrons, but for all polyhedrons,

$$V - E + F = 2.$$

This is usually referred to as the *Euler formula.*

There is one restriction. The polyhedron must be simple. By this we mean there are no holes in it. An example of a polyhedron with a hole is a small section of brickwork, like a wall, with one interior brick taken out. A more sophisticated example, which we examine at the end of this section, is illustrated in Fig. 9.12.

For any simple polyhedron Euler's formula is easily verified with the help of diagrams like Fig. 9.8 where there are 12 vertices (labelled A to L) and 26 edges (AB, BC, CD, DE, EF, FG, GH, HA, FD, FC, GC, HB, IA, IB, IC, ID, IJ, JK, JL, KD, KE, KL, LF, LG, LH, LA) and 16 faces (LHA, LAIJ, LJK, LKEF, LFG, LGH, IAB. IBC, ICD, IDKJ, KDE, DEF, FDC, FCG, GHBC, HAB). Thus $V - E + F = 12 - 26 + 16 = 2$.

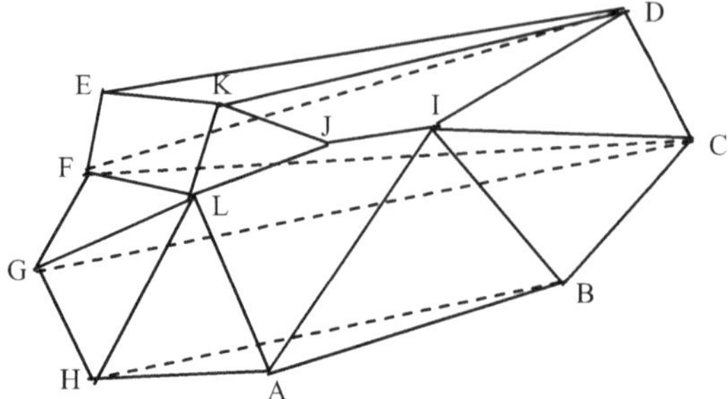

Fig. 9.8. A simple polyhedron useful for verifying Euler's formula.

It is a useful exercise to draw other examples of polyhedrons with differing numbers of faces, edges and vertices, and for each to verify the above equality. If we found just one case when the equality did not hold this would mean that it was not universally true. Failure to find one such case among a large number of polyhedrons still does not prove the result holds universally.

However, Euler's formula is easily shown to hold for all simple polyhedrons. We show first that a related formula holds for neighbouring networks. This provides a link to establishing the formula for simple polyhedrons.

Consider the network in Fig. 9.6(**b**). There are 6 vertices (V), 10 edges (E) and 5 faces (F), so $V - E + F = 6 - 10 + 5 = 1$. This equality holds for any neighbouring network.

Here is an outline of the proof. Fig. 9.9 is a modified version of Fig. 9.6(b) that is topologically equivalent to the latter. Each vertex label has been replaced by a letter. We remove one outer edge of the polygon ABCDE, say, AB. Inspecting Fig. 9.9 shows that this decreases E by 1 and also decreases F by 1. The number of vertices is unaltered. Thus, there is no change in the value of $V - E + F$. This remains true if we continue to remove each outside edge. After all outside edges are removed Fig. 9.9 reduces to the network in Fig. 9.10.

In Fig. 9.10 there are no longer any faces, i.e., $F = 0$. Each of the edges AF, BF, CF, DF and EF are referred to as 'dangling' edges. Removing a dangling edge and the associated dangling vertex reduces both the number of vertices and the number of edges by 1. So again $V - E + F$ is unaltered. If we remove all 5 edges in turn the same is true. We are left with one vertex only, so that $V - E + F = 1$. [Caution: do not confuse the vertex F (roman) with the number of faces F (italic)].

We can rebuild the original figure by reversing these steps without altering this value of $V - E + F$, thus confirming that $V - E + F = 1$ for the original network.

A moment's consideration will show that any neighbouring network may be reduced to a single vertex by successive removal of dangling or outside edges. This establishes that $V - E + F = 1$ holds for all such networks.

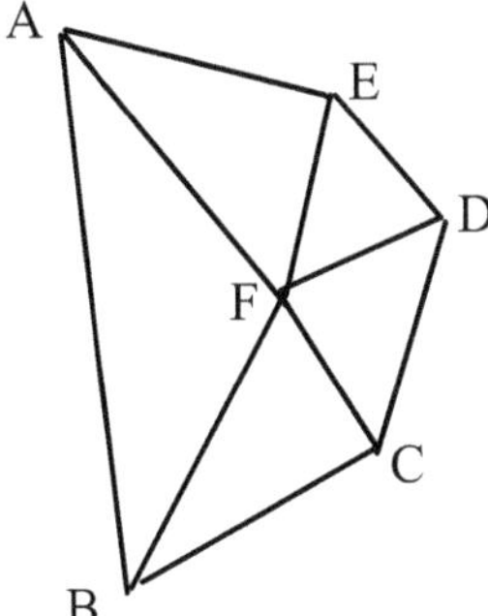

Fig. 9.9. The network used in the text to outline the derivation of Euler's formula

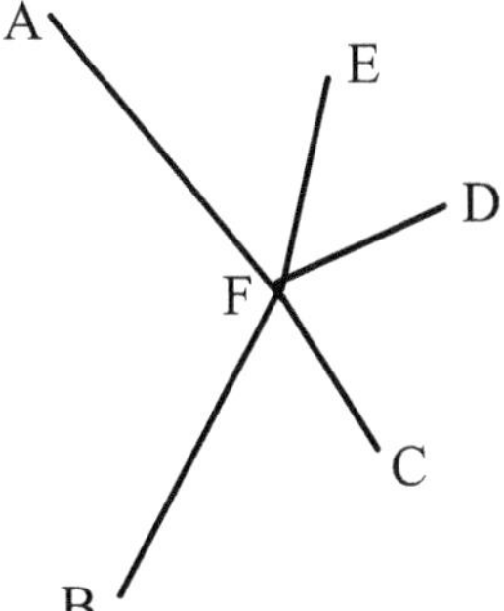

Fig. 9.10. The network in Fig. 9.9 after removal of outside edges

You should follow through an appropriate reduction for the network in Fig. 9.11. An obvious starting point is to remove the dangling edge AB.

One case we exclude from the above argument is that of any isolated vertex with no path (edge) connecting it to any other vertex. Such a vertex is not part of a neighbouring network.

I have asserted that for any simple polyhedron $V - E + F = 2$. To prove this we use a topological property. This is that any neighbouring network without dangling edges that is drawn on a perfectly elastic rubber sheet can be transformed into a polyhedron with one missing face. The converse is also true.

To see that this is so universally requires both a good three dimensional imagination and an appreciation of the 'elastic' properties inherent in a more formal definition of topological equivalence.

It is easy to see that such a transformation is possible in a simple case like that in Fig. 9.9. Imagine that figure drawn on a rubber sheet lying on a table. If we peg down all the vertices except F, we can make the required transformation by lifting the vertex F above the table surface by stretching.

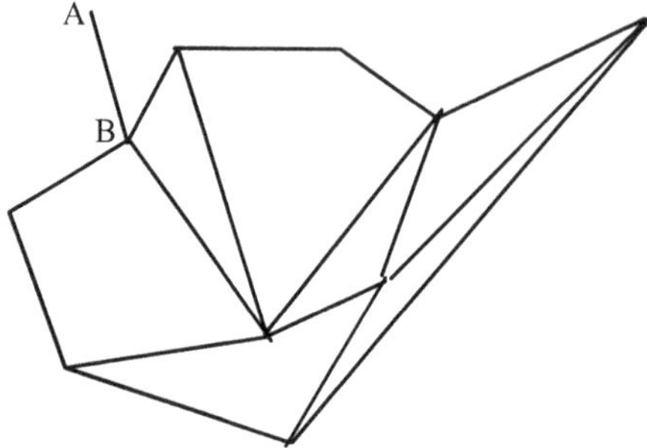

Fig. 9.11. A further network that may be used to demonstrate the validity of Euler's formula

We may visualize Fig. 9.9 as a two dimensional view (e.g., a photograph) of a polyhedron with the vertex F situated above the plane of the page, and the remaining vertices all lying in the page. If we 'freeze' the rubber sheet in the stretched condition and pick it up, we have a hollow polyhedron with the polygon face ABCDE 'missing'. Adding this face increases the number of faces by one, but leaves the number of edges and vertices unchanged. So for the polyhedron $V - E + F = 2$.

This argument extends to any simple polyhedron where, after removal of one face, it is topologically equivalent to some neighbouring network, for which we know $V - E + F = 1$.

I mentioned earlier that polyhedrons may have one or more holes. Fig. 9.12 gives an example. The polyhedron specified by the vertices ABCDEFGH is drilled out (i.e., removed) from a simple polyhedron to create a 'hole'. It is easy to check that the resulting polyhedron with that hole has 16 vertices. These are shown by the dots at the end of each edge in Fig. 9.12. There are 4 faces at the front, 4 at the back, also 4 outer faces, and a further 4 surrounding the hole at its sides, giving a total of 16 faces. Counting all the lines in the figure, each of which represents an edge, we find there are 32 edges. Thus, for this polyhedron with one hole $V - E + F = 16 - 32 + 16 = 0$.

I do not prove it, but this is a characteristic of any 'one hole' polyhedron. The value, n, of $V - E + F = n$ is called the *Euler number*. For a simple polyhedron the Euler number is 2. For a polyhedron with one hole it is zero. For a polyhedron with two holes it may be shown that $V - E + F = -2$, so the Euler number is -2.

9.4 Topological Equivalence

To see why the Euler number is important, we look more closely at what is meant by topological equivalence. The rubber sheet analogy is useful, but we need a clearer idea of what we have loosely describe as achievable 'without

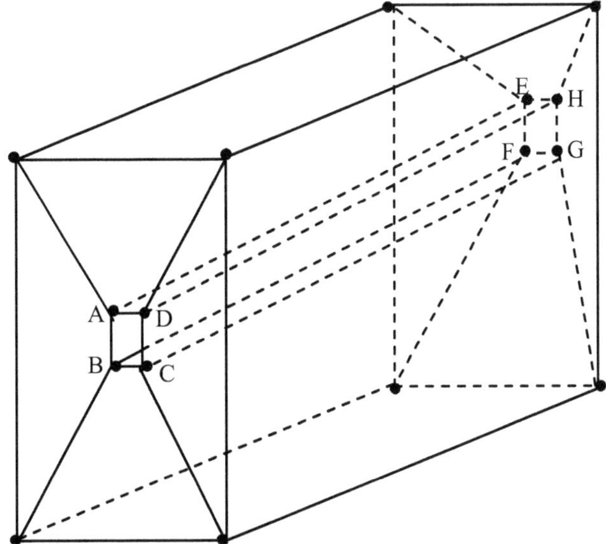

Fig. 9.12. A polyhedron with one hole obtained by removing the polyhedron with vertices A, B, C, D, E, F, G, H

Fig. 9.13. Some simple topologically equivalent curves

cutting and pasting'. Indeed, these latter operations may, in a restricted sense, sometimes be allowed.

Simple examples involving lines and curves illustrate key ideas. Fig. 9.13 shows some 'one dimensional' (i.e., line) figures that are topologically equivalent. It is easy to see that if we visualise a line as a thin broken rubber band, or as a piece of string, we can by stretching, bending or compressing, transform any one of these curves to one of the others.

We could bend a broken rubber band into the form of a circle, providing we hold or glue the broken ends together. In the light of our, so far incomplete, account of topological equivalence, this makes it tempting to think that the circle in Fig. 9.14 is topologically equivalent to the lines in Fig. 9.13.

Conversely, it might be argued that by stretching a circle at opposite ends of a diameter we can collapse it first to an ellipse, and eventually to a straight line, as in Fig. 9.15. If we allowed this, again we have the intuitively unsatisfying inference that a circle and a straight line appear to be topologically equivalent.

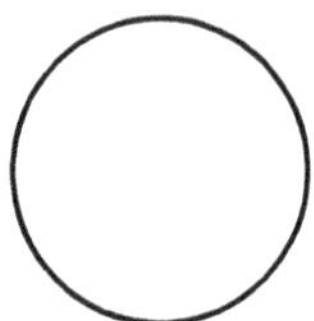

Fig. 9.14. A curve that is not topologically equivalent to those in Fig. 9.13

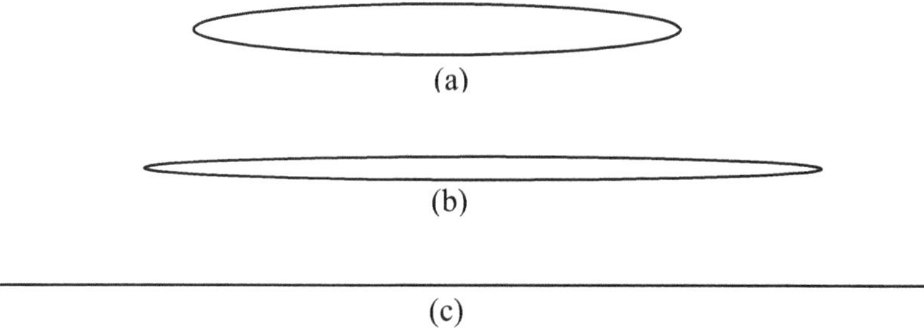

Fig. 9.15. Collapsing a circle (**a**) to an ellipse (**b**) and eventually to a straight line (**c**) by stretching

If a circle were indeed topologically equivalent to a straight line the concept would seem unreasonably wide. Our unease would arise largely because a circle is a closed curve with both an inside and an outside, whereas a line, or curve, in Fig. 9.13 is not closed.

A more formal definition of topological equivalence gets rid of this anomaly. We impose a further condition hinted at on p. 121. The condition is that any two points which are arbitrarily near to each other must remain close after the transformation.

This means that a circle cannot be topologically equivalent to a straight line, because, if we cut a circular rubber band at any position, neighbouring points on each side of the cut are now liable to wide separation after bending or stretching. Similarly, if we reverse the process illustrated in Fig. 9.15 and try to get from the line (**c**) to the circle (**a**) by prising apart the two strands of rubber in the line (**c**), points that were close together will now be far apart if they are on different strands as we move from (**c**) to (**b**) to (**a**).

Cutting, pasting and forceful prising apart are forbidden. There is an exception. We may cut and manipulate a surface providing we restore the cut in exactly the same configuration after the manipulation.

All this can be put more formally in mathematical terms, but the terminology required is not covered here.

Bearing in mind what is allowed, we can explore more fully the idea of topological equivalence and give a clearer indication of what properties may properly be called *topological*.

We look again at the Euler formula. We derived it for simple polyhedrons assuming straight edges and plane faces. It has nothing to do with the length of the edges, or the size and shape of the faces.

Indeed, we may go further, and make deformations to a rubber polyhedron to convert it to a sphere. The edges become curves on the surface of the sphere, and the faces are parts of the spherical surface bounded by those edges. The Euler formula $V-E+F=2$ still holds for this 'figure' on the spherical surface. We can reverse the process, converting any such figure on the spherical surface back to a polyhedron for which the Euler formula still holds.

Thus the formula may be associated with a sphere — specifically with the surface of a sphere. In effect, it tells us that if we divide the surface of a sphere in any way by noncrossing lines joining a set of vertices, then in every case $V-E+F=2$.

Mathematical doughnuts

We saw for a particular polyhedron with one hole that $V-E+F=0$. I asked you to accept that the result held for any one-hole polyhedron. By similar reasoning to that for transforming a polyhedron to a sphere, the equality still holds if straight edges and plane surfaces are allowed to become curved and stretched. By such transformations a one-hole polyhedron can always be changed to a map on a whole ring resembling a single-hole doughnut. Mathematicians call this shape a *torus*.

The four-colour theorem no longer holds on the surface of a torus. It takes more than four colours to depict quite simple maps drawn on a figure like a doughnut. It is worth spending a little while to devise an example of a map on a torus that requires at least 5 colours. You should not find this too difficult. In fact, the minimum number of colours for all possible maps is 7 for a torus. This is not easy to establish.

Polyhedrons with two, three or more holes are all feasible configurations. It is hardly surprising that the value of $V-E+F$ remains constant for all polyhedrons with 2, 3, 4 or more holes, However, that constant is different for each specified number of holes. By arguments similar to those used for topological equivalence of a simple polyhedron to a sphere, and of a one hole polyhedron to a torus, it is easily deduced that all two hole polyhedrons are topologically equivalent both to a two-hole doughnut like that in Fig. 9.16, or to a sphere with 2 handles as in Fig. 9.17, so far as relevant surface properties are concerned. Remember that the Euler numbers associated with $V-E+F$ are concerned with surface properties.

Closely related to the concept of the Euler numbers that take different values for topologically distinct surfaces is that of the *genus* of a surface. The genus is determined by the effect of cutting each type of surface in a particular manner. A circle drawn in a plane divides the plane into two parts, that which is inside, and that which is outside, the circle. If we draw a circle on a sheet of

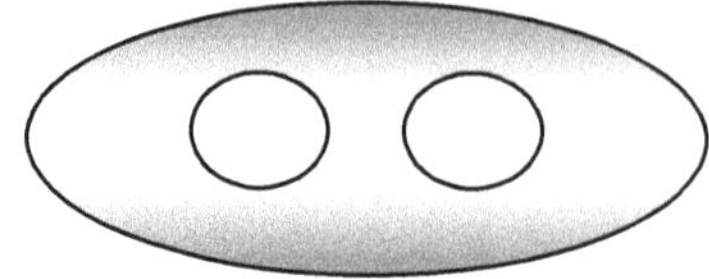

Fig. 9.16. A 'two-hole' doughnut is topologically equivalent to a 'two-hole' polyhedron

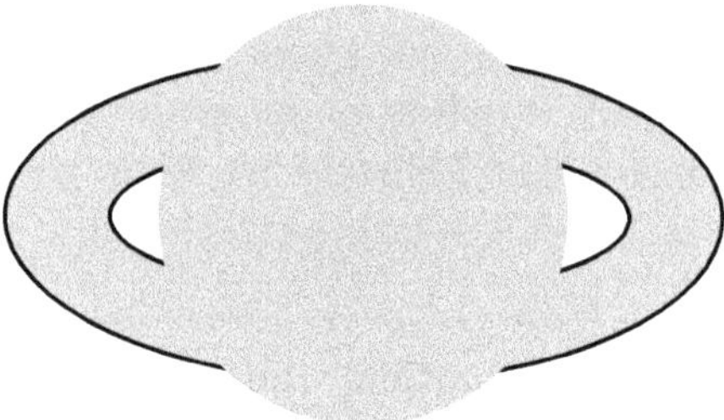

Fig. 9.17. A sphere with two handles is topologically equivalent to a two-hole doughnut

paper and cut it out, we have two distinct entities. One consists of all points inside, the other of all points outside, the circle.

If, as in Fig. 9.18, we inscribe a circle on the surface of a sphere (such as a tennis ball), and cut it out, something similar happens. The spherical surface separates into two unconnected parts.

However, if, on the surface of a torus we mark a circle in the way depicted in Fig. 9.19 and cut round the circle, the torus does not divide into two separate parts. If we draw a second such circle at some other part of the torus and make a further cut on this circle, the torus will now be in two parts. The figure resulting from making the cut is, as a result of that operation, no longer topologically equivalent to the torus. Think carefully about what property for topological equivalence breaks down here.

If we cut a two-hole surface such as that in Figs. 9.16 or 9.17 at two nonintersecting circles, such as those in Fig. 9.20, it is evident that these circles will not separate the surface into two disconnected parts. However, if we draw a third non-intersecting circle of the same type anywhere else it will.

The maximum number of cuts by circles, or by figures topologically equivalent to circles, that are possible without causing separation of a given topological figure is called the *genus*. The genus of a sphere is zero, since one circular cut will always do the job. The genus of a torus is 1, since two cuts may be needed to make a separation. That of a two-hole surface is 2, because we may need three such cuts to get separate surfaces.

We do not prove it, but an important result in topology is that for a surface of genus g, the following relationship holds:

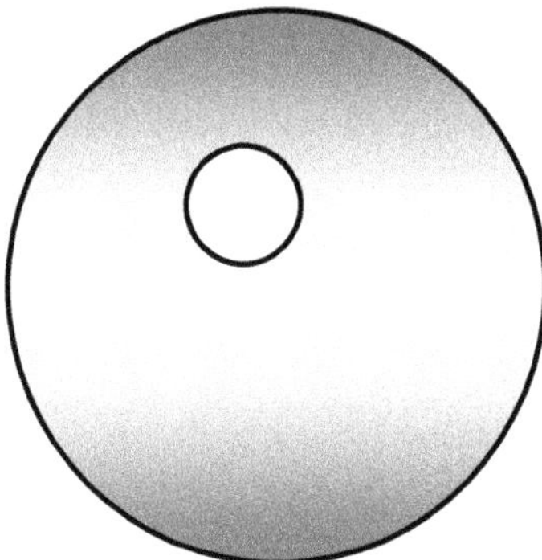

Fig. 9.18. A circle on the surface of a sphere divides the surface into two distinct parts

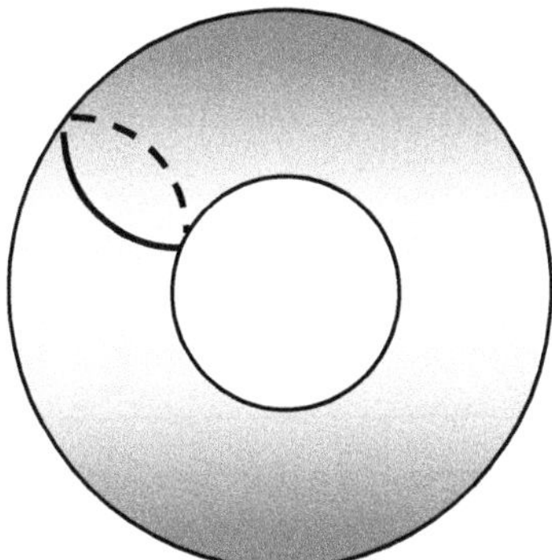

Fig. 9.19. Cutting a torus at the dotted circle does not produce two distinct parts of the surface

Fig. 9.20. Cutting a two-hole doughnut on the indicated circles will not result in two disconnected parts

$$V - E + F = 2 - 2g ,$$

where now we may generalize slightly the definitions of V, E, F given in the Euler formula, and apply it to a complete map of the surface where V is the number of vertices, E is the number of arcs forming the regional boundaries, and F is the number of regions. Thus, the basic Euler formula for the sphere is $V - E + F = 2$, as we have already proved. For the torus we have $V - E + F = 0$, as we have already indicated without a formal proof. This is consistent with our finding that $g = 1$ for the torus. For the two hole surface, where $g = 2$, we get $V - E - F = -2$. Thus, the value of $V - E + F$ provides a useful tool for recognising topologically equivalent groupings of surfaces.

9.5 One-sided Surfaces

The topological properties discussed so far have applied to what might be described as typical, or familiar, surfaces. A characteristic of these is that they are two-sided. We may distinguish between the two sides of a sheet of paper by painting each a different colour. We may discriminate between the outer and inner surfaces of a sphere or a torus by painting each surface a different colour.

On a surface that is not closed, like a sheet of paper, the colours meet only at the edges. If the surface is closed, like that of a sphere or a torus, the two colours never meet.

When both sides are the same side

The German mathematician, August Ferdinand Möbius (1790–1868), is best known for his demonstration of a simple surface that has only one side. It is universally known as a *Möbius strip* or *band.*

Before doing what Möbius did, we consider an even simpler band formed by pasting (or stapling) together the two ends of a strip of paper as illustrated in Fig. 9.21, where there are no 'twists' in the band so formed. This band, which we might think of as part of the curved surface of a cylinder, has an inside and an outside surface that we could paint in different colours. We might try doing various things with this cylinder-like strip. We could draw a pencil line on one face or the other parallel to, and equidistant from, the two edges. If we cut along that line the band will separate into two parts. None of this is surprising. Commonsense and experience tells us the strip should behave this way when we perform these operations.

What Möbius did was to make what may seem a trivial difference to form his band. He gave the strip a half twist before joining the ends. We can make ourselves a Möbius strip using paper and paste pot or stapler. After doing this, starting at any point, try painting one side in some colour. Intuitively the strip seems to have two sides, but what happens with our painting? We don't reach an edge at which we must stop before we get back to the starting point. The complete strip is by then painted. Similarly, if we draw a line down the centre equidistant from the edges, we get back to our starting point

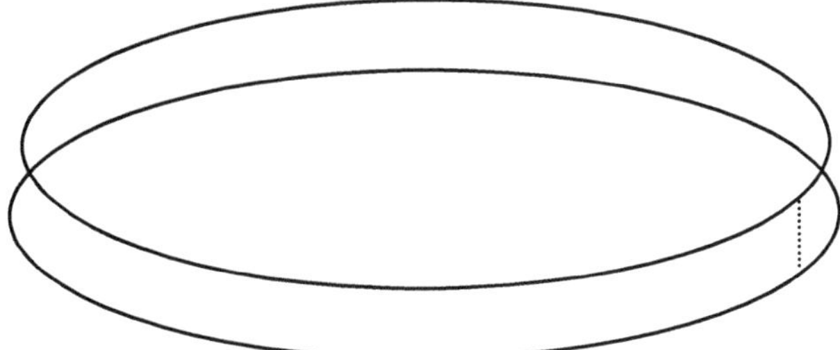

Fig. 9.21. A band formed by joining the ends of a strip of paper without any twists

without meeting an edge. The line appears to cover what were originally the two sides of the strip we started with.

The Möbius strip has only one surface. There is not an inside and an outside, or upper surface and lower surface, as there was on the strip of paper before twisting and joining. Cut along the centre pencil line and see what happens to the Möbius strip. Is it what you expected?

It is amusing — perhaps even educational — to see what happens with simple variants of what we have just done with a Möbius strip. For example, what happens if instead of cutting the strip along a line in the middle of the strip, we start cutting on a line one-third of the way in from an edge, and continue cutting parallel to the edge? Try again, starting one quarter of the way in from an edge.

We formed a Möbius strip by giving a strip of paper a single half-twist before joining the ends. What happens if we give the paper a full twist? Try it and see.

9.6 Loose Ends

The Klein Bottle. The concept of one sided surfaces is not confined to twisted strips of paper. A bottle that has neither an inside nor an outside is harder to visualize, but the Klein bottle has this property. It has no edge and only one side, the surface in effect being allowed to pass through itself. It is not easy to visualize this surface intuitively, and two dimensional drawings of it are not always convincing. The diagram in [37] (p. 250) is more convincing than many.

Getting knotted

Knot theory. Knot theory — a topic in its own right — relies heavily on topological concepts. Boy scouts are expected to tie reef knots rather than granny knots, but most scouts would be hard pressed to describe the difference in mathematical terms. Most people think of tying knots in a piece of string. If the ends are then joined the result is what mathematicians call a knot diagram. These form what are sometimes called entwined circles. A useful introduction to knot theory, and the relevant topological aspects, is given by Devlin [14], (Chap. 10).

Neat packaging. The concept of folding and packing is closely allied to stretch and squeeze geometry.

The Japanese art of Origami, the shaping of paper into forms and structures such as dragons or butterflies, was long regarded as an artistic, rather than a mathematical, skill.

It is only recently that computer power has made possible a study of the geometry of complex folding problems. These studies are relevant to many packing problems, where equipment must be folded to fit into a confined space in such a way that unpacking or unfolding is safe and quick. A typical example is the folding of parachutes so that they are virtually certain to open without entanglement at the tug of a cord. Yet they must be fitted into containers so compact that they do not endanger the parachutist when jumping from the aircraft. Instruction in folding and packing parachutes is a vital part of the training of potential jumpers and sky divers. The strict controls on how the packing is done have evolved as a matter of experience, rather than by studying the mathematics of folding.

That folding is an important part of any packing problem is familiar to all of us when packing our holiday luggage. One can cram more into a suitcase if clothes are folded neatly rather than being simply bundled in. More seasoned travellers usually learn from experience that ‘folded neatly’ is a vague concept. Some methods of folding a shirt or blouse are more useful than others to maximise the potential carrying power of a piece of luggage.

A practical problem that stimulated research into the geometry of folding was the development of safety airbags for cars. These must pack neatly into a small space on the dashboard, steering column, or door of a vehicle. They must inflate in about one second after crash impact to protect driver or passenger before their head makes contact with steering wheel or windscreen. Engineers have applied computer studies of the geometry of folding to this problem.

A more sophisticated application of the geometry of folding concerns the design of large lenses consisting of nearly flat sheets, where ridges and valleys in the sheets change the direction of the incident light, and focus it in a similar way to conventional lenses. Such lenses were first developed in 1822 by the French physicist Augustin Fresnel (1788–1827) and bear his name. This approach is being further developed by replacing the ridge and valley feature by fan-like structures, which it is hoped will be useful for orbital telescopes. Solar panels used as power sources on satellites have also been made more efficient by sophisticated studies of folding patterns.

Devising suitable folding mechanisms and patterns is important to enhance road safety for cars, by ensuring that they will crumple on collision in a way less likely to injure the occupants. Similar considerations apply to designing buildings in regions subject to earthquakes. The aim is to design these so that if they collapse they do so in a way least likely to injure occupants or passers-by.

Folding problems first interested the Chinese some 2000 years ago, even before their study was made an art by the Japanese. Modern computers have made possible a mathematical study of all the diverse aspects of folding and packing, with potentially important terrestrial and extra-terrestrial applications.

These problems provide good examples of application of a combination of art and geometric skills to physical and social problems. Many examples are given by Lang [32].

10

Limits

10.1 Some Simple Limits

In Chapter 4 we looked at a modern version of the paradox of the tortoise and the hare. We showed how applying the notion of a *limit* to a geometric series resolved Zeno's paradox.

Limits are important whether one is considering a sequence, a series or some function of one or more variables. Early studies of limits were largely curiosity driven. Later formal studies gave us one of the most powerful mathematical tools — the *calculus* — which we look at in the next chapter.

In a sequence we are interested in what happens to the nth term when n becomes large. For a series we often want to study the behaviour of the sum, S_n, of n terms of a sequence as n becomes large. The S_n themselves, of course, form a sequence $S_1, S_2, S_3, \ldots, S_r, \ldots$.

We saw in Chapter 4 that the terms of a sequence may either tend to a finite limit or diverge when n becomes large. They may also oscillate. For example, the sequence

$$\frac{1}{2}, -\frac{2}{3}, \frac{3}{4}, -\frac{4}{5}, \ldots, (-)^{r+1}\frac{r}{r+1}, \ldots$$

oscillates between values that approach $+1$ from below and -1 from above as r becomes large.

For a function of a variable x we may want to know how it behaves when x becomes large. We are often even more interested in what happens when x approaches some finite value a at which the function is not defined because it takes the form $c/0$, where c is some constant, including the case where $c = 0$, when the undefined form is $0/0$. Limits of this sort are at the heart of *differential calculus.*

We saw in Sect. 4.3 that the sum of the terms in a geometric progression with first term a and common ratio r tended to a limit $a/(1-r)$ for positive values of r less than 1. We did not prove it there, but it also tends to this limit for negative r providing $|r| < 1$.

We also saw in Sect. 4.4 that if

$$S_n = 1 + \frac{1}{2} + \frac{1}{3} + \ldots + \frac{1}{n} ,$$

and we allow n to increase indefinitely this sum becomes increasingly large, even though $1/n$ becomes ever closer to zero. This is formally expressed by saying that the series S_n is *divergent*. This means that S_n may be made as large as we please by making n sufficiently large.

In practice, a sequence or a series which is divergent, or which oscillates, is usually of less interest mathematically than one that tends to a finite limit when n becomes large. This does not mean that divergent or oscillatory behaviour may not be important in real-world problems. To the engineer it may mean that a particular design for a bridge or a building is unstable, and that the structure is almost certain to collapse!

Here is an example of a series where the limiting behaviour is not immediately obvious. Consider first the sequence whose rth term is $a_r = 1/(r-1)!$ where $r!$, called factorial r, is, for any integer r, the product of all the integers between 1 and r, i.e.,

$$r! = 1 \times 2 \times 3 \times \ldots \times (r-3)(r-2)(r-1)r .$$

We also define $0! = 1$.

If we consider the sum of the first $n+1$ terms of the sequence $\{a_r\}$, then

$$S_{n+1} = 1 + \frac{1}{1!} + \frac{1}{2!} + \frac{1}{3!} + \ldots + \frac{1}{(n-1)!} + \frac{1}{n!} . \tag{10.1}$$

It is easy to see that the following inequality holds for $n > 2$:

$$\frac{1}{n!} = \frac{1}{2} \times \frac{1}{3} \times \frac{1}{4} \times \ldots \times \frac{1}{n} < \frac{1}{2} \times \frac{1}{2} \times \frac{1}{2} \times \ldots \times \frac{1}{2} = \frac{1}{2^{n-1}} .$$

Applying this repeatedly, in (10.1) we get:

$$S_{n+1} < 1 + 1 + \frac{1}{2} + \frac{1}{2^2} + \ldots + \frac{1}{2^{n-1}} = 1 + \frac{1 - \left(\frac{1}{2}\right)^n}{1 - \frac{1}{2}} .$$

This is because the terms after the first in the series $1+1+1/2+(1/2)^2+\ldots$ form a geometric series with $a = 1$ and $r = 1/2$ (see Sect. 4.3). Since the fraction $(1/2)^n$ tends to zero as $n \to \infty$, it follows that $S_n < 3$, no matter how large n is. Also, $S_n > 2$ because $S_2 = 2$ and for $n > 2$ all remaining terms are positive. This implies that S_n must tend to some limit between 2 and 3 when n becomes large. Mathematicians denote the value of this limit by e, so we may write

$$\mathrm{e} = 1 + \frac{1}{1!} + \frac{1}{2!} + \frac{1}{3!} + \ldots + \frac{1}{r!} + \ldots , \tag{10.2}$$

where the second set of dots ... means that the sequence goes on 'without end'. A less verbose way of putting this is to say that $S_n \to \mathrm{e}$ as $n \to \infty$.

We may use (10.2) to compute an estimate of e to any required degree of accuracy. For example, if we include only the first 7 terms, up to and including $1/(6!)$, and denote this sum by s, then

$$\begin{aligned}\mathrm{e} - s &= \frac{1}{7!} + \frac{1}{8!} + \frac{1}{9!} + \ldots = \frac{1}{7!}\left(1 + \frac{1}{8} + \frac{1}{8 \times 9} + \ldots\right) \\ &< \frac{1}{7!}\left(1 + \frac{1}{8} + \frac{1}{8^2} + \ldots\right) = \frac{1}{7!}\left(\frac{1}{1 - 1/8}\right) = \frac{8}{7! \times 7} \approx 0.000227\end{aligned}$$

This means that the 'error' in approximating to e by summing only the first seven terms in (10.2) cannot exceed 0.000227. Thus, the approximation is correct to at least 3 decimal places, i.e., to 3 decimal places $\mathrm{e} = 2.718$ since, as we can easily check, the sum $s = 2.71805555\ldots$ where the decimal is now recurring.

What kind of number is e?

The number e is called the *exponential constant.* We explain its importance in mathematics in later chapters. Like π, e is both irrational and transcendental. Some readers may know that it is the 'base' for the so-called *natural logarithms* which we discuss in Sect. 12.5.

A limit that is far from obvious

In Sect. 12.3 we shall look at what happens to the sequence $(1 + 1/n)^n$ when $n \to \infty$. You probably know, and it is easy to show, that if we raise a fixed number greater than 1 to a high power, n, the resulting number increases as n increases. We also know that the number 1 raised to any positive integral power n, however large n may be, remains 1. In the sequence $(1 + 1/n)^n$ we have a situation where, as n becomes larger and larger, the number we raise to the nth power does not stay fixed, but decreases, and comes closer and closer to 1. Finally, it becomes virtually indistinguishable from 1. Finding the limiting value of $(1+1/n)^n$ is a matter of weighing up competing influences. Is the important thing that the number we are raising to the power n is always (even if only just) greater than 1, or is it more important that it becomes effectively indistinguishable from 1 when n is large? In the former case we might expect $(1+1/n)^n$ to tend to infinity, in the latter case we might expect it to tend to 1. We find in Sect. 12.3 that it does neither. Perhaps more surprisingly, the limit is the exponential constant e. At this stage just take this as a warning that it is often not obvious what a limit will be.

10.2 Limits of Functions

We now turn to functions of any real variable x. A notation commonly used, and one you may already have met, is to denote any function, f, of x by $f(x)$. An example is

$$f(x) = x^3 + 3x^2 - 7x + 3.$$

The notation $f(x)$ for a specified function, f, of x is useful when we may be interested in the value of that function for different values of x. For example, when $x = 0$ the function above takes the value 3, and when $x = 10$ it takes the value 1233 (shown by substituting $x = 10$). We summarize these results by writing $f(0) = 3$ and $f(10) = 1233$. The notation $f(x)$ is also useful for describing particular classes of functions, using phrases such as 'Suppose $f(x)$ is a polynomial of degree n ...'

The function above, $x^3 + 3x^2 - 7x + 3$, is defined for all real values of x. If we wish we might also define it for complex values, but in this chapter, indeed in most of this book, we consider only real values. Other examples of functions of x are

$$f(x) = \sin x, \quad f(x) = \frac{x^2 - a^2}{x - a}, \quad f(x) = \frac{1}{x}, \quad f(x) = \frac{\sin x}{x}.$$

I use these later to clarify several points about limits.

If we want to speak about, or manipulate, several different functions of x we distinguish between them by using letters additional to f. For example, if we are interested in both $\sin x$ and $1 - x^2$ we might write $f(x) = \sin x$ and $g(x) = 1 - x^2$. If we want to work with a function $\sin(1 - x^2)$ we may then write this $f[g(x)]$, which is a shorthand way of saying $f[g(x)] = f(1 - x^2) = \sin(1 - x^2)$, i.e., it is instructing us to replace x by the function $1 - x^2$, i.e., $g(x)$, in the function $f(x)$.

Addition and multiplication of functions are based on intuitively reasonable definitions. For the above functions f, g

$$f(x) + g(x) = \sin x + (1 - x^2)$$

and

$$g(x)f(x) = (1 - x^2)\sin x.$$

Watch out for ambiguity

Be careful with notation. If we write $f(x)g(x) = \sin x(1 - x^2)$ this is ambiguous. It is not clear whether we mean $\sin[x(1 - x^2)]$ or $(\sin x) \times (1 - x^2)$.

In school trigonometry angles are usually measured in degrees. Many tables use degrees as a basic unit when tabulating values of sines, cosines and tangents. It is well known that 1 degree represents 1/90 of a right angle, or that the number of degrees in a right angle is 90. Most pocket calculators will calculate trigonometric functions like the sine and cosine if you enter an angle in degrees. Many provide another option, letting us express angles using a unit called a *radian.*

An alternative way of measuring angles

The concept of a radian is associated with the circumference of a circle with unit radius, i.e., $r = 1$. The circumference of this circle, measured in the same units as the radius, is 2π. We may think of the circumference as being described by using a pair of compasses to sweep out a complete circular arc on a Cartesian coordinate system with centre at the origin and radius 1 unit. The sweep starts from the point (1,0) on the x-axis and proceeding anticlockwise until we return to the starting point. At that stage the compasses have swung through a total of 360°, or four right angles. Remembering that $r = 1$, the length of the arc, here the complete circle, is 2π. We define the radian by equating 360° to 2π radians. This implies that 1 radian is equivalent to $360/2\pi = 180/\pi \approx 57.296$ degrees. A right angle, which corresponds to 1/4 of the circumference of the complete unit circle, is equivalent to $2\pi/4 = \pi/2$ radians.

In the calculus, and often more generally in a mathematical context, results involving trigonometric functions have simpler expressions if angles are measured in radians. Thus, if we write $f(x) = \sin x$, or $g(x) = (\sin x)/x$, we shall henceforth assume, unless a specific statement is made to the contrary, that x is measured in radians.

The remaining three functions of x given on p. 142. i.e.,

$$f(x) = \frac{x^2 - a^2}{x - a}\,, \quad f(x) = \frac{1}{x}\,, \quad f(x) = \frac{\sin x}{x}\,,$$

illustrate situations where, for some value of x, the function is undefined.

Tackling that 0/0 conundrum

In the first of these, setting $x = a$ gives $f(a) = 0/0$, which is not defined. However, if $x \neq a$ we may factorise the numerator and write

$$f(x) = \frac{(x + a)(x - a)}{x - a}\,.$$

Providing $x \neq a$, we may cancel $x - a$ between numerator and denominator reducing the expression to $f(x) = x + a$. For this 'new' $f(x)$ there is no problem

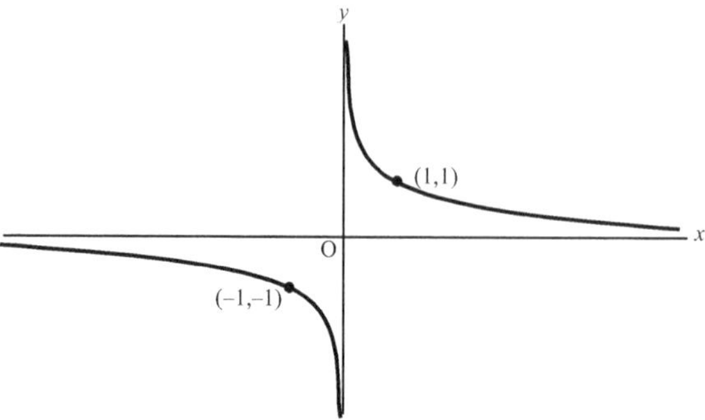

Fig. 10.1. The graph of the function $y = f(x) = 1/x$

when $x = a$, for then $f(a) = 2a$. Intuitively, it then seems reasonable to say that this is the limit as $x \to a$ of the original function $f(x) = (x^2 - a^2)/(x - a)$. To show this is meaningful, we need to be more careful about this intuitive argument. If you are not convinced that there might be a problem look at the trouble we got into by pretending we could divide by zero in equation (1.2).

Consider the even simpler function $f(x) = 1/x$. This gives an undefined $1/0$ when $x = 0$. If we set x at some very small value slightly greater than 0 then $1/x$ becomes very large. It grows ever larger as we bring x closer to 0. For example, if $x = 1/1000$ then $1/x = 1000$, while if $x = 1/10^6$ then $1/x = 10^6$. This tempts us to say that $f(x)$ tends to infinity as x tends to 0. What if we allow x to take values slightly less than zero? If $x = -1/1000$ then $1/x = -1000$, while if $x = -1/10^6$ then $1/x = -10^6$. The behaviour as x approaches zero through negative values is such that we conclude $f(x)$ tends to minus infinity as x tends to zero. This behaviour is illustrated by a sketch of the function $y = 1/x$, which has the form shown in Fig. 10.1. The curve has two branches, and the limiting value as x approaches zero depends upon whether the approach is through negative values, often referred to as 'from the left', or through positive values, often referred to as 'from the right'. Here the terms left, right have an obvious association with the graph of the function. When these two limiting values differ there is no way we can define a unique limiting value when x approaches zero.

This contrasts with the case where $f(x) = (x^2 - a^2)/(x - a)$. If we redefine the function by writing it as $f(x) = (x^2 - a^2)/(x - a)$ if $x \neq a$ and $f(x) = 2a$ when $x = a$, the graph of the function is a straight line passing smoothly through the point $(a, 2a)$. You may think this a somewhat esoteric argument bringing us to a rather obvious conclusion. But we shall see in Chap. 11 that an argument like this is central to the method for determining what is called the derivative of x^2. Do not worry now about what this means if you are not already familiar with the basics of differential calculus.

In this example, a more formal way of looking at what happens to $f(x)$ as $x \to a$, is to set $x = a + h$ and then let h (which may be positive or negative but not zero) become very small, so that x is close to a. For any permissible h we have

$$f(x) = \frac{(a+h)^2 - a^2}{a+h-a} = \frac{2ah+h^2}{h}.$$

Because we have specified that h is not zero, this reduces to $f(x) = 2a + h$. This tells us that by making h sufficiently small we can get the value of $f(x)$ as close to $2a$ as we please. Mathematicians have a formal way of expressing this. They say:

> A function $f(x)$ has limit l as x tends to a if, corresponding to any positive number ϵ, however small, we can find a positive number η such that
>
> $$|f(x) - l| < \epsilon$$
>
> for all $x \neq a$ that satisfy the inequality $|x - a| < \eta$.

Here the expression $|(x - a)|$ means the magnitude of $x - a$, i.e., the value of that quantity without regard to sign. As explained earlier, this is called the *modulus*, and it is always nonnegative. Thus 2 and -2 each have modulus 2.

In the above example it suffices to choose η equal to ϵ to establish that $l = 2a$ for then

$$\left|\frac{x^2 - a^2}{x - a} - 2a\right| < \epsilon$$

if

$$|x + a - 2a| = |x - a| < \epsilon,$$

confirming that we may choose $\eta = \epsilon$. The important point to note in the definition of a limit is that ϵ is chosen and fixed. The challenge is to find a η satisfying the relevant condition.

If this is the first time you have met this idea don't worry if you find it hard to comprehend. Most students do, although it is an important concept in higher mathematics.

The function $f(x) = \sin x/x$ is not defined when $x = 0$. It is not easy to see what the limit, if any, might be simply by looking at the function. A hint may be obtained from tables of values of $\sin x$, remembering that we measure x in radians here, whereas most tables state values for the sine of an angle measured in degrees. However, modern pocket calculators will often give values of $\sin x$ directly if x is entered in radians. To get some idea of how the function behaves as x decreases towards zero, and at the same time to remind you of the relationship between degrees and radians, Table 10.1 gives some relevant values.

Since π radians corresponds to 180°, this means 1 degree corresponds to $\pi/180 \approx 0.0174533$ radians. Remembering this, it is easy to obtain the entries

Table 10.1. Values of $f(x) = (\sin x)/x$ for some angles between $\pi/6$ and zero radians, noting that $\pi/6$ radians is equivalent to 30°

Angle (y^o)	x (radians)	$\sin x$	$\sin x/x$
30	0.523599	0.500000	0.9549
20	0.349066	0.342020	0.9798
10	0.174533	0.173648	0.9949
5	0.087266	0.087156	0.9987
2	0.034907	0.034899	0.9998
1	0.017453	0.017452	0.9999
0.5	0.008727	0.008727	1.0000
0.1	0.001745	0.001745	1.0000

in Table 10.1 using either a calculator or a table of trigonometric functions. For example 10° is equal to $0.0174533 \times 10 = 0.174533$ radians. Tables, or a calculator, give $\sin 10° = 0.173648$. Thus $f(x) = 0.173648/0.174533 = 0.9949$. Table 10.1 suggests that the limit is probably 1. It does not establish this beyond doubt. What happens, for example, if we approach zero through negative values of x? It is well known (see Appendix, Sect. A.3) that $\sin(-x) = -(\sin x)$, so it immediately follows that in this case $f(-x) = f(x)$. This means we get the same values for $(\sin x)/x$ as those in Table 10.1 if we replace each x by $-x$.

The limit may be established by a rigorous geometric argument given in full, for example, in [13] (p.308). Here is an abridged version that appears in a similar form in many elementary trigonometry books. In Fig. 10.2 the circle has unit radius. We consider a situation where the angle AOB is x radians, subject to the restriction $0 < x < \pi/2$.

The area of the triangle OAB (Appendix, Section A2) is

$$\frac{1}{2}\text{OA.OB}\sin x = \frac{1}{2}\sin x\ ,$$

since OA = OB = 1. The area of the sector OAB (i.e., the area bounded by the radius OA, the arc AB and the radius OB) is $x/2$, since the area of the whole circle of unit radius is π, and this is the sector swept out by an angle 2π. It follows by simple proportionality that an angle x sweeps out a sector of area

$$\pi \times \frac{x}{2\pi} = \frac{1}{2}x.$$

Finally, the area of the triangle OAC is

$$\frac{1}{2}\text{OA.CA} = \frac{1}{2}\tan x$$

since $\tan x = \text{CA}/\text{OA}$ and OA is of unit length.

From the geometry of Fig. 10.2 it is obvious that

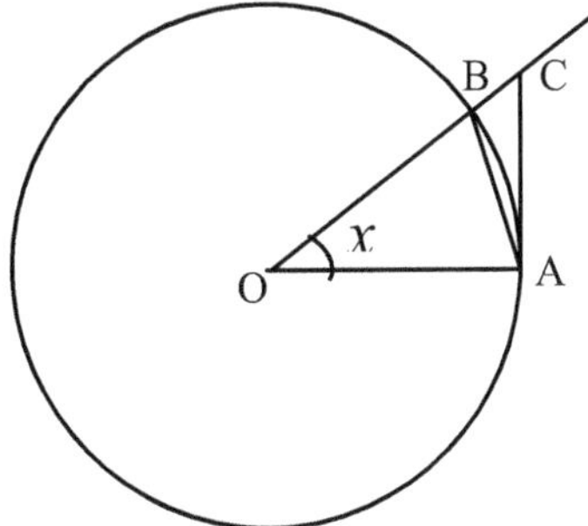

Fig. 10.2. A diagram that is useful to determine the limit of $(\sin x)/x$ as $x \to 0$

$$\text{area triangle OAB} < \text{area sector OAB} < \text{area triangle OAC.}$$

$$\text{i.e., } \frac{1}{2}\sin x < \frac{1}{2}x < \frac{1}{2}\tan x\ ,$$

or dividing through by $(\sin x)/2$ that

$$1 < \frac{x}{\sin x} < \frac{1}{\cos x}\ .$$

Taking reciprocals in an inequality with the same sign on both sides (Appendix, Sect. A.1) reverses the inequality sign, so

$$1 > \frac{\sin x}{x} > \cos x\ .$$

It is well known that when x approaches zero, then $\cos x$ approaches 1. This means that $(\sin x)/x$ is 'sandwiched' between 1 and a value of $\cos x$ which becomes closer and closer to 1 as $x \to 0$. It is not difficult to show that a similar result holds if we allow $x \to 0$ through negative values.

Attention all pilots

You may think the limiting result for $(\sin x)/x$ is interesting, but only a curiosity so far as the real world is concerned. Not so. It is a useful practical result and what makes it particularly useful is that the approach to the limit is quite rapid. For an angle as large as $10°$ approximating to $\sin x$ by x (measured in radians) gives a result (Table 10.1) almost correct to three decimal places. This fact is often used for quick calculation of course corrections in air navigation.

We are often interested in the behaviour of some function $f(x)$ as $x \to \infty$. We may sometimes determine this by using the fact that when $x \to \infty$ it is clear that $1/x \to 0$, although strictly speaking this needs a formal verification based on the definition of a limit. Accepting, however, that this limit is zero, we may use it to find, for example, the limit as $x \to \infty$ of $f(x) = (2x+3)/(x+100)$. To do this we divide numerator and denominator by $1/x$, giving

$$f(x) = \frac{2 + 3/x}{1 + 100/x} .$$

Both $3/x$ and $100/x$ tend to zero as $x \to \infty$, so the limit as $x \to \infty$ of $f(x)$ is $(2+0)/(1+0) = 2$.

A good puzzle about limiting behaviour is provided by the function $\sin(1/x)$ as $x \to 0$. It is not defined when $x = 0$. As $x \to 0$ clearly $1/x$ increases through very large values while $\sin(1/x)$ is confined to values between -1 and 1. The precise values it takes in that range will oscillate quite rapidly for large $1/x$. This is because (Appendix, Sect. A.3), if k is any positive or negative integer, $\sin(1/x) = 0$ if $1/x = k\pi$. This is equivalent to $x = 1/(k\pi)$. We also know (Appendix, Sect. A.3) that $\sin(1/x) = 1$ if $1/x = (4k + 1)(\pi/2)$. This is equivalent to $x = 2/[(4k + 1)\pi]$. Similarly, we may verify that $\sin(1/x) = -1$ when $x = 2/[(4k - 1)\pi$. If we successively put $k = 1, 2, 3, 4, \ldots$, since the denominators of these fractional values of x will increase without limit, the values of x for which $\sin(1/x)$ takes the values $0, 1, -1$ will cluster closer and closer to the y-axis, while the value of the function oscillates between -1 and 1. Similar arguments apply for negative x. This means that all we can say about the function is that it oscillates between -1 and 1 increasingly rapidly as $x \to 0$ through either positive or negative values.

This concludes a very brief introduction to limits — a topic that is sometimes the subject of many chapters when treated rigorously. We have tried only to give a glimpse of where rigour and care are needed, and have pointed out some of the pitfalls.

The notion of limits is vital in the next chapter when we introduce the calculus. There, determining the limit as a function approaches an undefined 0/0 is often the essence of the problem.

10.3 Loose Ends

Continuity. Closely associated with the notion of a limit is that of continuity. Loosely speaking, a function $f(x)$ is continuous at a point $x = a$, if it is defined when $x = a$ and does not change abruptly as x increases from values below a, through a, to values above a. In terms of limits, this means that by making a positive number h sufficiently small, we can make both $|f(a + h) - f(a)|$ and $|f(a) - f(a - h)|$ as close to zero as we please. This is just another way of saying that the function is continuous at $x = a$ if both

$$\lim_{h \to 0} f(a + h) = f(a) \text{ and } \lim_{h \to 0} f(a - h) = f(a)$$

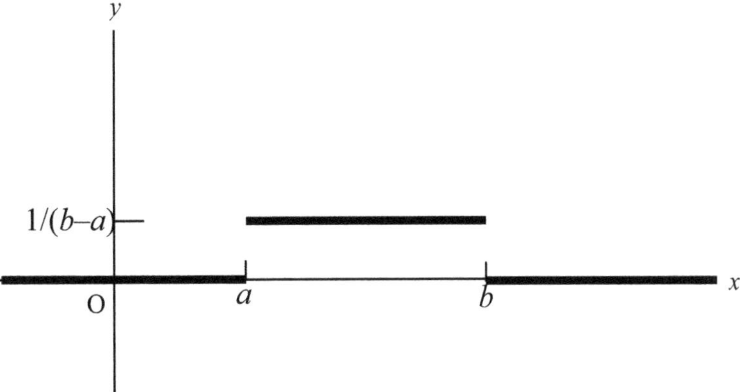

Fig. 10.3. The function $f(x)$ shown by bold segments has discontinuities at $x = a$ and $x = b$

which implies that $f(x)$ is defined when $x = a$.

In many applications of mathematics we meet functions that are not continuous at one or more points. It is clear from Fig. 10.1 that $f(x) = 1/x$ is not continuous at $x = 0$. It is not even defined when $x = 0$.

The function $f(x) = 0$ if $x < a$ and $f(x) = 1/(b - a)$ if $a \leq x < b$, and $f(x) = 0$ if $x \geq b$ arises in applications of mathematics in both statistics and electrical engineering. It is continuous everywhere except at $x = a$ and $x = b$. Near the former it jumps suddenly from the value zero to the value $1/(b-a)$, and at the latter from the value $1/(b - a)$ back to zero. For example, if h is positive, then for this function $f(a - h) = 0$ no matter how small h is. We have defined $f(a) = 1/(b-a)$; thus $f(a-h)$ is not equal to $f(a)$. It is easy to see that for any positive $h < b - a$, that $f(a+h) = f(a) = 1/(b - a)$, whence $f(a + h) - f(a) \to 0$ as $h \to 0$. This property is referred to as *continuity to the right* when $x = a$. Fig. 10.3 is a sketch graph of this function $f(x)$, where the bold line segments represent the function.

If we redefine the function with only slight changes, so that $f(x) = 0$ if $x \leq a$ and $f(x) = 1/(b - a)$ if $a < x \leq b$, and $f(x) = 0$ if $x > b$, there would still be a discontinuity at $x = a$, but the function would be continuous to the left. This follows since it takes the value zero both for values of x at, and for those below, $x = a$.

Physical situations where there are sharp discontinuities in some function are common. In an electric circuit, the current may change from zero at a given time, x_1, to some positive value i at the flick of a switch. At the flick of another switch at some later time, x_2, the flow may be reversed to take the value −i. I am using i here as the symbol for current commonly used by electrical engineers, and not as the imaginary unit $\mathrm{i} = \sqrt{-1}$. As mentioned already, engineers often use the letter j for the square root of −1 to avoid confusion with their traditional symbol i for current.

11

The Calculus

11.1 Dawn of a New Era

A fruitless war of words has raged as to whether Isaac Newton (1642–1727) or Gottfried Leibniz (1646–1716) discovered the calculus. Calculus as we know it was formalized by these two, but it had its origins with the Greeks, and was further developed by Pierre de Fermat (1601–65) and others.

Calculus took a clearly recognisable form only after a formalization of the concept of a limit. We saw in the previous chapter that this concept is especially useful for coping with problems that arise because division by zero is not defined.

It is now widely, but not universally, accepted that Newton and Leibniz independently developed their respective ideas about the subject. Had they not so done someone else would have done this at about the same time or soon after, for there nearly always comes a time when conditions are right for a major break through in some area of mathematics, as in almost any field of science. That is no reason to belittle Newton or Leibniz, two all-time Greats among mathematicians. Newton wrote in a letter to Robert Hooke (1635–1703) 'If I have seen further it is by standing on the shoulders of giants'.

The basic tools for handling, in a unified framework, problems about maxima and minima, curvature, and more generally those of rates of change is the *differential calculus*. General methods for calculating enclosed areas with arbitrary, and not necessarily straight-line boundaries, and a host of more sophisticated but closely related problems, such as the calculation of volumes and moments of inertia, are unified by the *integral calculus.*

Integral calculus has the longer history, dating back at least to the time of Archimedes around 250BC. The Newton-Leibniz approach linked differential and integral calculus in a framework that applied to general situations, whereas previously each new problem had called for a different line of attack.

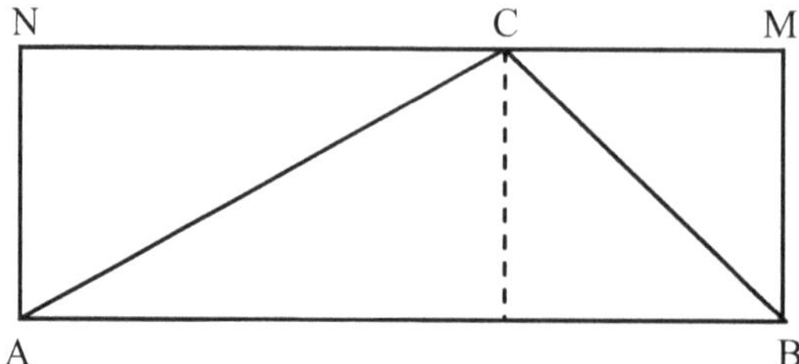

Fig. 11.1. The area of the triangle ABC is half that of the rectangle ABMN

11.2 Calculating Areas

Calculating the area A of a rectangle with adjacent side lengths a, b is a simple exercise in school arithmetic where we learn $A = ab$. Euclidean geometry immediately shows us that the area of a triangle ABC, is one half the area of a rectangle having one side of length AB, and the other side of length equal to that of the perpendicular from the vertex C to the side AB. This is easily seen from Fig. 11.1.

Given these results for the rectangle and triangle, one may find the area of any closed polygon by dividing it into rectangles and/or triangles and adding the areas of these subdivisions. These simple cases do not call on the concept of a limit.

An early encounter with a limit

By considering limits, Archimedes solved the problem of calculating the area of a circle of radius r and known circumference $2\pi r$ in a way that contains the germ of modern integral calculus. He first calculated the area of a regular polygon with n sides inscribed in the circle. He did this by dividing that polygon into n identical isosceles triangles as illustrated in Fig. 11.2. Each triangle has two sides of length equal to the radius r of the circle and a vertex at the centre of the circle. The remaining side of each triangle is one of the n segments forming the perimeter of the polygon.

As we increase the number, n, of sides of the regular polygon, the total perimeter length approaches the known circumference $2\pi r$ of the circle, and its area approaches the unknown area of the circle from below.

Let δ_1 be the length of each side of the inscribed regular polygon. The area of each triangle in the inscribed polygon is approximately $r\delta_1/2$ when δ_1 is small, since the altitude of each triangle is then approximately r. Adding over all triangles, the area of the polygon is $r \times$ (perimeter of polygon)/2 and, as indicated, the perimeter approaches the circumference of the circle from below. Thus the limiting area is $(r/2) \times 2\pi r = \pi r^2$.

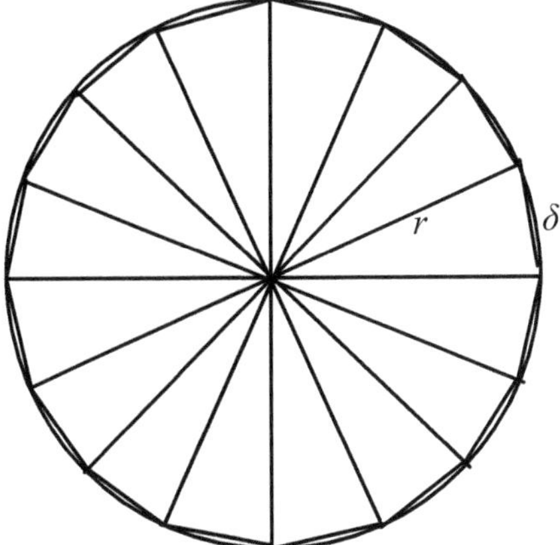

Fig. 11.2. The perimeter of an inscribed polygon approaches the circumference of the circle as the number of sides tend to infinity

By a slight modification of this process we may obtain the same limit using a circumscribed regular polygon (one outside the circle), and considering the limit as we increase the number of sides. The limit is now approached from above.

Archimedes' procedure uses the limiting concept inherent in the integral calculus. Unfortunately, the division of a particular class of regular polygons into triangles to determine the area does not readily generalize from a circle to any closed curve.

The fundamental problem tackled by the integral calculus is essentially:

> Given any reasonably well behaved function $f(x)$, what is the area bounded by that curve between the values $x = a, x = b$, the ordinates at $(a, 0), (b, 0)$, and the segment of the x-axis in the interval between $(a, 0)$ and $(b, 0)$?

We shall not be specific at this stage about what is meant by a 'reasonably well behaved function'. A sufficient, but not a necessary condition, is that the function should be defined and continuous everywhere in (a, b). The situation is shown for an arbitrary function in Fig. 11.3. The hatched region is the area to be calculated.

11.3 An Unconventional Approach

For reasons that will become apparent later, in formal courses the differential calculus is almost always introduced before the integral calculus, despite the fact that the latter has older roots. Courant and Robbins in [12, 13] introduce the integral calculus before differential calculus. As a student I found that approach attractive, so follow it here.

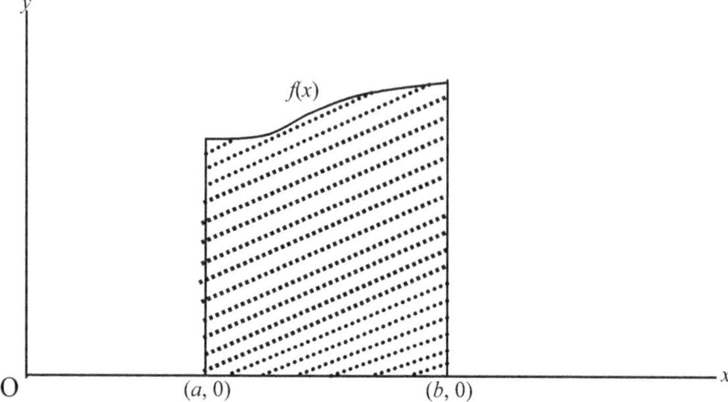

Fig. 11.3. A fundamental application of integral calculus is to calculate the hatched area

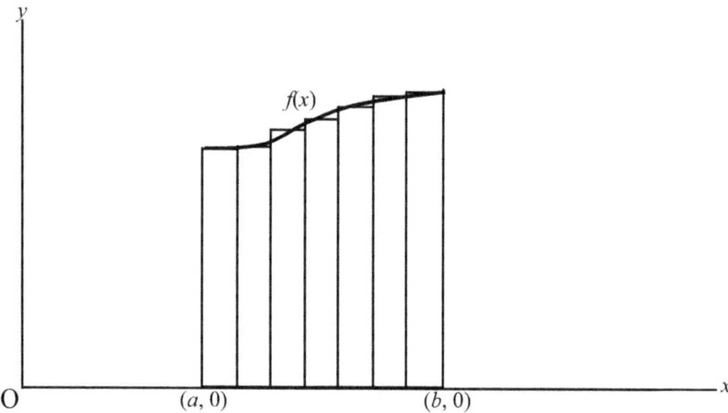

Fig. 11.4. A set of rectangles with total area approaching that under a function $f(x)$ between $x = a$ and $x = b$

To find the hatched area in Fig. 11.3 the first step is to get an approximation based on a series of small non-overlapping rectangles. These need not all be the same width, but there are practical advantages in making them so by a suitable choice of a common width. A set of rectangles that would give a reasonable approximation to this hatched area is shown in Fig. 11.4.

The base of each rectangle is on the x-axis. The height equals a value of $f(x)$ between the maximum and minimum value attained in the subinterval associated with that rectangle. By introducing narrower rectangles, the sum of their areas give improved approximation to the hatched area in Fig. 11.3. This is often referred to as the area 'under the curve'. The integral calculus provides a way to compute this limiting value. We now consider some generalities and then show how this works for a few specific cases.

δx is not $\delta \times x$. It is just a product of the imagination

We approximate to the hatched area using n rectangles, each of width δx, where δx is a composite symbol representing the width. We used it in a similar sense in describing Archimedes' method for determining the area of a circle. Suppose $f(x_i)$ has a value between, or equal to, either the greatest and least value taken by $f(x)$ in the ith interval. We take this as the height of the corresponding rectangle. The sum of the areas of all n rectangles is

$$S_n = f(x_1)\delta x + f(x_2)\delta x + f(x_3)\delta x + \ldots + f(x_n)\delta x. \qquad (11.1)$$

Mathematicians have a convenient shorthand for this sum. They use what they call the *sigma* notation, writing it

$$S_n = \sum_{i=1}^{n} f(x_i)\delta x, \qquad (11.2)$$

where $\sum_{i=1}^{n}$ means *take the sum of the terms that follow*. Here, these terms are $f(x_i)\delta x$ for all integer values of i between $i = 1$ and $i = n$.

The sigma notation is used widely in mathematics. For example,

$$\sum_{r=1}^{k}\left(\frac{1}{r^3}\right)$$

is shorthand for

$$\sum_{r=1}^{k}\left(\frac{1}{r^3}\right) = 1 + \frac{1}{2^3} + \frac{1}{3^3} + \ldots + \frac{1}{k^3}.$$

In (11.1) or (11.2) it is easy to see that $\delta x = (b - a)/n$. As n increases we get ever better approximations to the hatched area in Fig. 11.3. This is a limiting process when $n \to \infty$. We define the area A as that limit, and write

$$A = \lim_{n\to\infty}\sum_{i=1}^{n} f(x_i)\delta x = \int_a^b f(x)\mathrm{d}x. \qquad (11.3)$$

This notation is due originally to Leibniz. It replaces the Greek capital S, i.e., $\sum$, by a symbol $\int$ called an integral sign, and the width, δx, of each rectangle by a symbol $\mathrm{d}x$ to indicate that this interval is becoming arbitrarily small, a consequence of increasing n. The subscript a, and superscript b, on the integral sign indicate the range of x values over which the integration is performed.

This is neat in theory, but what happens in particular cases? Does any general pattern emerge to enable us to find A for a particular $f(x)$? We can

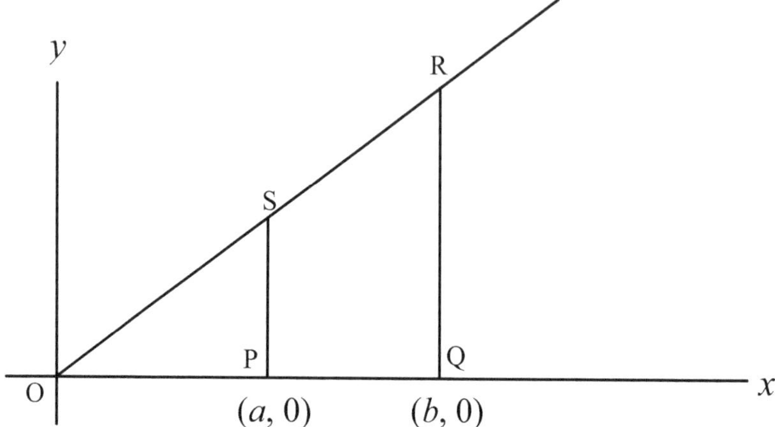

Fig. 11.5. PQRS is area under the curve $y = x$ between $x = a$ and $x = b$

write down (11.2) for a chosen n [this implies $\delta x = (b-a)/n$] and, if we know $f(x)$, we can choose appropriate $f(x_i)$, although such choices in general will not be unique. Indeed, the choice of an $f(x_i)$ will only be unique if $f(x)$ is constant in the ith subinterval. Therefore, if a limit does exist, the choice of a particular permissible $f(x_i)$ when n is large must not be crucial.

Here is a trivial example. Take $f(x) = c$ where c is some positive constant. Then $f(x)$ is a straight line parallel to the x-axis at distance c above it. This means the area enclosed by $f(x)$ and the x-axis between the ordinates at $(a, 0)$ and $(b, 0)$ is a rectangle of width $b-a$ and height c and thus the area is $c(b-a)$. It is easy to show that the limit (11.3) takes this value since, for any n, (11.2) becomes

$$S_n = \sum_{i=1}^{n} c\delta x = c\sum_{i=1}^{n} \delta x = cn\delta x = cn\frac{b-a}{n} = c(b-a) \; .$$

Since it does not involve n, this is also the limiting value when $n \to \infty$. Here, applying the limiting process is using a sledgehammer to crack a nut.

The same is almost true if we set $f(x) = x$, a situation illustrated in Fig. 11.5. The area of interest, PQRS, is a trapezium in which PQ $= b-a$, PS $= a$ and QR $= b$. The standard formula for this area, easily obtained by Euclidean geometry, is

$$A = (b-a) \times \frac{1}{2}(b+a) = \frac{1}{2}(b^2 - a^2) \; .$$

Using (11.2) for a fixed n, since $y = f(x) = x$ we may set $f(x_i) = a + i\delta x$, $i = 1, 2, \ldots, n$. This amounts to choosing the greatest value of $f(x)$ in each interval. Then

$$\begin{aligned} S_n &= \sum_{i=1}^{n}(a+i\delta x)\delta x = (na+\delta x+2\delta x+\ldots n\delta x)\delta x \\ &= na\delta x + (1+2+\ldots+n)(\delta x)^2 = na\delta x + \frac{1}{2}n(n+1)(\delta x)^2 \,, \end{aligned}$$

since we proved in Sect. 4.5 that the sum of the integers form 1 to n inclusive is $n(n+1)/2$. Remembering that $\delta x = (b-a)/n$, we get

$$S_n = a(b-a) + \frac{(b-a)^2}{2} + \frac{(b-a)^2}{2n} \,.$$

When $n \to \infty$ the last term tends to zero. Thus

$$\int_a^b x\mathrm{d}x = a(b-a) + \frac{(b-a)^2}{2} = \frac{1}{2}(b^2-a^2) \,.$$

Again, this is using a sledge-hammer to crack a nut.

11.4 A Less Obvious Case

Things become more interesting when $f(x) = x^2$. The curve is a parabola that passes through the origin and the points $(1,1)$ and $(-1,1)$. It is sketched in Fig. 11.6, We seek the area under the curve specified as PQRS. Following arguments similar to those above, we set $f(x_i) = (a+i\delta x)^2$ whence

$$\begin{aligned} S_n &= \sum_{i=1}^{n}(a+i\delta x)^2\delta x \\ &= na^2\delta x + 2a\delta x(1+2+\ldots+n)\delta x + (1^2+2^2+\ldots+n^2)(\delta x)^3 . \end{aligned}$$

I do not prove it, but the sum of the squares of the first n natural numbers is $n(n+1)(2n+1)/6$. You may (and should) verify this using the method of mathematical induction described in Sect. 4.5. Remembering that $\delta x = (b-a)/n$, it now only needs simple algebra to show that

$$S_n = a^2(b-a) + [(n+1)a(b-a)^2/n] + [n(n+1)(2n+1)(b-a)^3/6n^2] \,,$$

which in turn may be written

$$S_n = a^2(b-a) + \left(1+\frac{1}{n}\right)a(b-a)^2 + \left(1+\frac{1}{n}\right)\left(2+\frac{1}{n}\right)\frac{(b-a)^3}{6} \,.$$

Remembering that $1/n \to 0$ as $n \to \infty$, we find

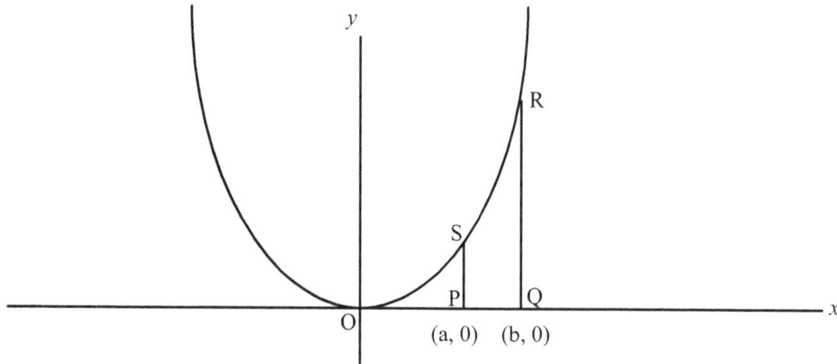

Fig. 11.6. An area under the parabola $y = x^2$

$$\int_a^b x^2 \mathrm{d}x = a^2(b-a) + a(b-a)^2 + (b-a)^3/3 ,$$

which may be shown, after algebraic manipulation, to equal $(b^3 - a^3)/3$.

This is the first result we have obtained using the integral calculus that is not a mere confirmation of one obtainable by easier methods. However, the approach has many practical limitations. The required limit was only obtainable in the last example after simple, but tedious, algebra.

Nevertheless, a pattern is emerging, because the relevant integrals over the interval (a, b) for x and x^2 were respectively $(b^2 - a^2)/2$ and $(b^3 - a^3)/3$. Intuitively, this suggests that for x^3 the area might be $(b^4 - a^4)/4$, and more generally for x^k, where k is a positive integer, that it might be $(b^{k+1} - a^{k+1})/(k+1)$.

We do not establish this here, because we learn in Sect. 11.7 that it is easy to to do so using the differential calculus, a subject we have yet to develop. This is a pragmatic reason why it is conventional to learn about the differential calculus before introducing the integral calculus.

11.5 Removing Some Limitations

So far we have considered only cases where $f(x)$ is positive for all x in the interval (a, b). This is not an essential restriction. Nor is it only possible to calculate areas bounded by a curve, pairs of ordinates, and the x axis.

The extension to calculating any closed area with boundaries defined by one or more 'well behaved' functions is straightforward if we formulate rules for combining integrals, and specify some conventions regarding signs. We do not prove the relevant results, but will show that some of the rules are intuitively reasonable. They provide useful extensions from a few basic results.

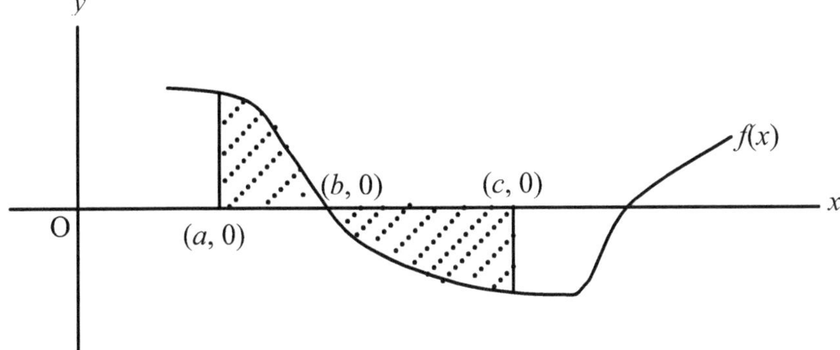

Fig. 11.7. The hatched area above the x-axis is positive and that below is negative

The tacit assumption made so far that $f(x)$ is positive for all x in the interval (a, b) is not essential. For example, if instead of $f(x) = x$, we take $f(x) = -x$, then following arguments used above, we get the same limit except that the sign is reversed.

In general, if the region of interest is entirely below the x-axis, the area computed is negative. What if the area of interest is partly above and partly below the axis? If the total area above the axis is the same magnitude as the total area below, then the integral will take the value zero.

Figure 11.7 shows a function for which there are both positive and negative areas associated with x values in an interval (a, c). Here, we may either want the absolute, or total, area disregarding sign, or else the 'algebraic' area in which the (negative) area below the x-axis is combined with the (positive) area above.

In Fig. 11.7 if $f(x)$ has the curve shown there, the hatched area above the x-axis is given by $\int_a^b f(x)\mathrm{d}x$, which is positive. That below is given by $\int_b^c f(x)\mathrm{d}x$, which is negative. Thus the absolute area is

$$A_{\text{abs}} = \int_a^b f(x)\mathrm{d}x + \left|\int_b^c f(x)\mathrm{d}x\right|$$

and the algebraic area is

$$A = \int_a^b f(x)\mathrm{d}x + \int_b^c f(x)\mathrm{d}x \ .$$

Another rule is that the integral over (a, b) of the sum of two functions is the sum of the integrals. Also the integral of a constant times a function is that constant times the integral of the function. There are obvious extensions to more than two functions. If c, d, e are constants, and $f(x)$, $g(x)$, $h(x)$ are integrable functions over (a, b), then

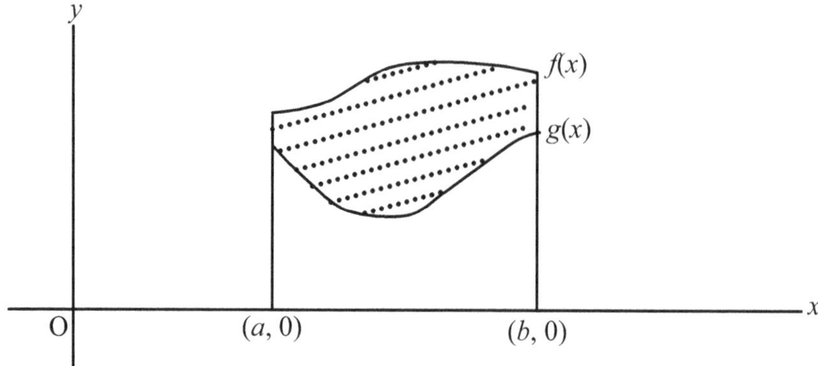

Fig. 11.8. Area bounded by two curves $f(x)$, $g(x)$ and ordinates at $x = a$ and $x = b$

$$\int_a^b [cf(x) + dg(x) + eh(x)]\mathrm{d}x = c\int_a^b f(x)\mathrm{d}x + d\int_a^b g(x)\mathrm{d}x + e\int_a^b h(x)\mathrm{d}x.$$

Using this rule, and the results already obtained, or quoted, for the integration of x^k, it follows that for a polynomial

$$f(x) = a_o + a_1x + a_2x^2 + \ldots a_nx^n$$

we have

$$\int_a^b f(x)\mathrm{d}x = a_0(b-a) + a_1\frac{b^2-a^2}{2} + a_2\frac{b^3-a^3}{3} + \ldots + a_n\frac{b^{n+1}-a^{n+1}}{n+1}\,.$$

Using the summation rule with a choice of constants 1 and -1, the hatched area in Fig. 11.8 is given by

$$\int_a^b f(x)\mathrm{d}x - \int_a^b g(x)\mathrm{d}x\,.$$

11.6 Differential Calculus

Practical reasons for reversing the historic order

Despite its longer history, the practical tools of integration are best developed using results from differential calculus. The latter is concerned primarily with determining slopes, and the associated concept of rates of change.

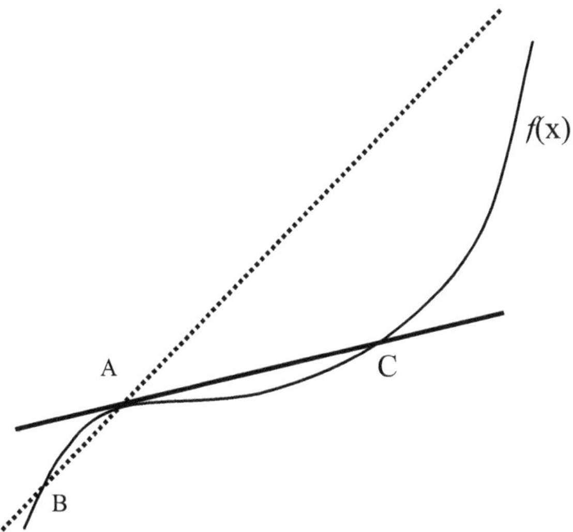

Fig. 11.9. The tangent to a curve $y = f(x)$ at some point A may cut the curve at other points, e.g., C

It surprises most people coming new to calculus to learn that there is a connection between determining the area under a curve and evaluating the slope of a tangent to that curve at any point on it.

One word – two meanings

A tangent to a circle is a line that touches the circle at only one point. Also the slope of a straight line is defined in Cartesian co-ordinate terms (Sect. 7.2) as the tangent of the angle between the line and the positive direction of the x-axis. In the first case *tangent* refers to a specific straight line and in the second to a trigonometric function with that name. (see Appendix, Sect. A.3).

This may seem confusing, but the two concepts are related, and the differential calculus is concerned with determining the slope of a line we call a tangent to a curve. The tangent to any curve is a generalization of the concept of the tangent to a circle.

A tangent at a point, A, on the circumference of a circle may be described by considering the limiting behaviour of a chord AB when we regard A as fixed, and allow another point B, also on the circumference, to tend to (i.e., move towards) A. The tangent is then the 'limiting chord' as B approaches A. We may generalize this idea to any curve by first extending slightly the idea of a chord to that of a *secant.* A secant is a line that cuts a curve at two points A and B. The segment AB between those two points is called a *chord* of the

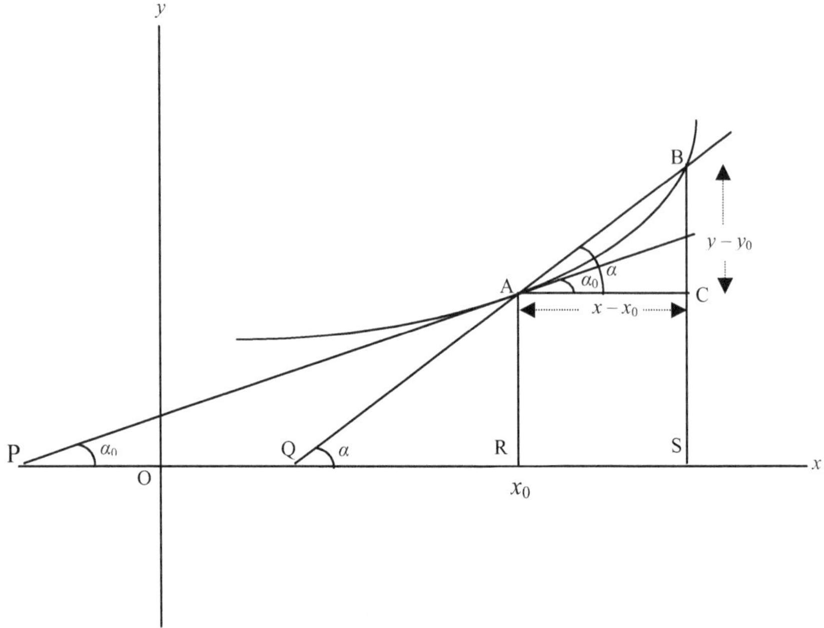

Fig. 11.10. Determination of the slope of a line drawn tangentially to a curve

curve. If we regard A as fixed, and allow B to tend to A, the limiting secant is the tangent at A. Unlike the case of the circle, a tangent to a curve may cut that curve at some other point. Fig. 11.9 shows a case where this happens. The dotted line BA is a secant to the curve $f(x)$ and if we move B towards A, keeping A fixed, the solid line AC shown in that figure is a tangent at A. The tangent cuts the curve again at C, but it is not a tangent at this latter point.

The limiting process above is the key to the differential calculus approach to finding the slope of the tangent to a curve at any given point.

Motivation for studies about tangents to a curve came in the seventeenth century when mathematicians like Newton became interested in problems about rates of change in dynamics, and also in determining the maxima and minima of general functions $f(x)$, but more about that later.

11.7 Derivatives

In the remainder of this chapter we look at some basic calculus and obtain results that we need in later chapters. For the reader meeting the calculus for the first time, the rest of this chapter may prove heavy going. Some of the hints in the notes *How to Read Mathematics* following the Preface to this book, may be helpful if you are new to the calculus.

Fig. 11.10 gives a geometrical picture of the limiting process for determining the slope at a tangent at a fixed point A on a curve $y = f(x)$. AB is a chord of the curve. The coordinates of A are (x_0, y_0) and those of B are (x, y). We want the limiting value of the slope of the chord AB as B approaches A. In terms of the diagram, this is the limiting value, $\tan\alpha_0$, of $\tan\alpha$, where α is the angle between the secant forming the chord AB and the x-axis . From Fig. 11.10 we see that

$$\tan\alpha = \frac{y - y_o}{x - x_o} = \frac{f(x) - f(x_o)}{x - x_o} .$$

We want the limit of this expression as $x \to x_0$. If we denote the difference $x - x_0$ by δx and the difference $y - y_0$ by δy, then $\tan\alpha = \delta y/\delta x$. Taking the limit as $\delta x \to 0$, the slope of the tangent line PA (Fig. 11.10) is given by

$$\tan\alpha_0 = \lim_{\delta x\to 0} \frac{\delta y}{\delta x} = \lim_{\delta x\to 0} \frac{\delta f(x)}{\delta x} , \qquad (11.4)$$

where $\lim_{\delta x\to 0}$ is the usual mathematical shorthand for the limit as δx tends to zero for the expression that follows.

This limit is denoted by

$$\frac{\mathrm{d}y}{\mathrm{d}x}$$

and is called either the *derivative* of $y = f(x)$, or the *differential coefficient* of $f(x)$ with respect to x. The derivative is itself also a function of x. From this viewpoint, an appropriate alternative notation often used is $f'(x)$. Whereas $f(x)$ specifies the distance of the curve above (or below) the x-axis at the point with coordinate x, $f'(x)$ gives the slope of the tangent to the curve at the point with that x-coordinate. This slope is often referred to as the *gradient.*

A further notation one may meet is to write $f'(x) = \mathrm{D}f(x)$. Here D is called the *differential operator.* Do not confuse this with the idea of a constant D multiplying a function $f(x)$. The notation implies that D is an order to carry out the operation of differentiating $f(x)$.

One reason for studying the tangent to a curve is to find any maximum or minimum values of a function, and the x values at which they occur. A *maximum* is a point at which the value of $f(x)$ is greater than that at neighbouring points. A *minimum* is a point at which the value of $f(x)$ is less than that at neighbouring points. At a maximum or minimum the slope of the tangent, if there is one, must be zero, because the tangent will be parallel to the x-axis. However, as Fig. 11.11 indicates by the point C, we may have a point of zero slope that is neither a maximum nor a minimum. Such a point is called a *point of inflection.*

There may also be a maximum or minimum point were the slope is not zero because the derivative is not defined at that point, but changes abruptly as the curve passes through it. Point D in Fig. 11.11 is such a point.

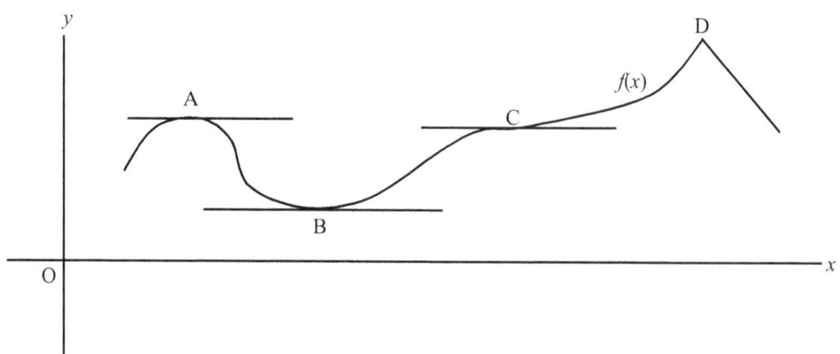

Fig. 11.11. A maximum, A, a minimum, B and a point of inflection, C

Looking at graphs of curves like that in Fig. 11.10, it is easily seen that if the slope of the tangent is changing rapidly at any point of contact,then as x increases through that value, the direction of the curve is changing quickly. If the slope is positive and increasing, the curve is 'rising' ever more steeply. The derivative is a measure of rate of change of the slope of a curve as measured by the tangential direction at any point. We shall see in Sect. 12.8 that this concept is important in dynamics.

As a curve approaches a maximum, as at A in Fig. 11.11, the derivative is positive for x-values slightly less than that at A, since the angle at which the tangent meets the positive direction of the x-axis is less than a right-angle. The slope is zero at A since the tangent is parallel to the x-axis. It is negative for x-values slightly greater than that at A. Thus, as x increases through a maximum of $f(x)$ the sign of $f'(x)$ changes $+, 0, -$. By similar reasoning, when passing through a minimum, it changes $-, 0, +$. When passing through a point of inflection it may change $+.0, +$ (as at the point C). This is referred to as a *point of inflection on a rising curve.* If it changes $-, 0, -$ we have a *point of inflection on a falling curve.* Try sketching an example of this.

We obtain the derivatives of a few well-known functions. Many of these may already be familiar to some readers.

If $y = f(x) = x$ it is easy to see graphically that this represents a straight line through the origin with slope 1. The tangent to a straight line is (trivially) the line itself, so we expect to find by the limiting definition that $\mathrm{d}y/\mathrm{d}x = 1$. Indeed, since $f(x) = x$ we have

$$\frac{\delta y}{\delta x} = \frac{f(x) - f(x_0)}{x - x_0} = \frac{x - x_0}{x - x_0} = 1$$

for all $x \neq x_0$, so the limiting value is 1. Check for yourself the even simpler result that if $y = f(x) = c$, where c is any real constant, then $f'(x) = 0$. This is clear geometrically since $y = c$ represents a straight line parallel to the x-axis.

If we set $y = f(x) = x^2$ we have

$$\frac{\delta y}{\delta x} = \frac{x^2 - x_0^2}{x - x_0} \,.$$

This is not defined when $x = x_0$. If $x \neq x_0$ we may factorize the numerator and divide out to obtain

$$\frac{x^2 - x_0^2}{x - x_0} = \frac{(x + x_0)(x - x_0)}{x - x_0} = x + x_0$$

and this tends to $2x_0$ when $x \to x_0$. Thus for any x, $f'(x) = 2x$.

An alternative way of arguing is to remember that $x - x_0 = \delta x$. This implies $x = x_0 + \delta x$, whence

$$\frac{\delta y}{\delta x} = \frac{x^2 - x_0^2}{x - x_0} = \frac{(x_0 + \delta x)^2 - x_0^2}{x_0 + \delta x - x_0} = \frac{2x_0\delta x + (\delta x)^2}{\delta x} = 2x_0 + \delta x \,.$$

When $\delta x \to 0$ this tends to $2x_0$. Since this result holds for any x_0 we have again that if $f(x) = x^2$ then $f'(x) = 2x$.

A reunion with an old friend

What we have done involves only a notational change from arguments in Sect. 10.3 where we considered the limit as $x \to a$ for the function $(x^2 - a^2)/(x - a)$. We have replaced a by x_0.

Similarly if $f(x) = x^3$ we have

$$\frac{\delta y}{\delta x} = \frac{x^3 - x_0^3}{x - x_0}$$

and the numerator factorizes to

$$x^3 - x_0^3 = (x - x_0)(x^2 + xx_0 + x_0^2)$$

so that $\delta y/\delta x$ has the limiting value $3x_0^2$ when $\delta x \to 0$, i.e., when $x \to x_0$. Thus, if $f(x) = x^3$ then $f'(x) = 3x^2$.

I do not prove it, although it only requires straightforward algebra to do so, that if n is any positive integer and $f(x) = x^n$ then $f'(x) = nx^{n-1}$. Indeed this form, with a few exceptions, holds also when n is any real number. For example, if $f(x) = 1/x = x^{-1}$ then $f'(x) = -1 \times x^{-2} = -1/x^2$ providing $x \neq 0$. We may establish the result by noting that

$$\frac{\delta y}{\delta x} = \left(\frac{1}{x_0 + \delta x} - \frac{1}{x_0}\right)\frac{1}{x_0 + \delta x - x_0} = \frac{-\delta x}{x_0(x_0 + \delta x)\delta x} = \frac{-1}{x_0(x_0 + \delta x)} \,,$$

which tends to $-1/x_0^2$ as $\delta x \to 0$. Thus, for any $x \neq 0$, $f'(x) = -1/x^2$. If $x = 0$, then neither $f(x)$ nor $f'(x)$ are defined. This is clear from the graph

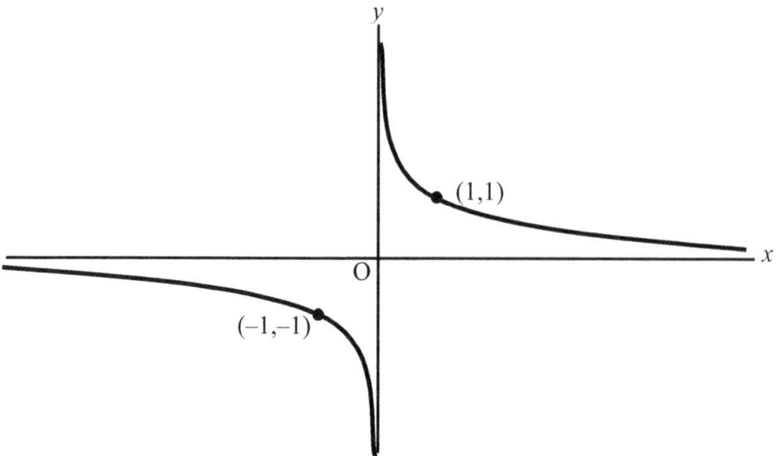

Fig. 11.12. The function $y = 1/x$ has no derivative when $x = 0$

of $f(x) = 1/x$ sketched in Fig. 11.12 (which is the same as Fig. 10.1). You may know this curve is called a *rectangular hyperbola* that passes through the points $(1, 1)$ and $(-1, -1)$.

Differentiation of $\sin x$ or $\cos x$ is straightforward using some results given in the Appendix, Sect. A.3. We obtain the derivative of $\sin x$, where x is expressed in radians.

As in earlier examples, we set $x = x_0 + \delta x$ and consider what happens as $\delta x \to 0$. Then

$$\frac{\delta y}{\delta x} = \frac{\sin x - \sin x_0}{x - x_0} = \frac{\sin(x_0 + \delta x) - \sin x_0}{\delta x} \,. \tag{11.5}$$

In the Appendix, Section A2, we quote the result

$$\sin(\mathrm{A} + \mathrm{B}) = \sin \mathrm{A} \cos \mathrm{B} + \cos \mathrm{A} \sin \mathrm{B} \,.$$

Putting $\mathrm{A} = x_0$ and $\mathrm{B} = \delta x$ we get

$$\sin(x_0 + \delta x) = \sin x_0 \cos \delta x + \cos x_0 \sin \delta x \,.$$

Using this equality in (11.5) we find, after some rearrangement of terms,

$$\frac{\delta y}{\delta x} = \frac{\cos x_0 \sin \delta x + \sin x_0(\cos \delta x - 1)}{\delta x} = \cos x_0 \frac{\sin \delta x}{\delta x} + \sin x_0 \frac{\cos \delta x - 1}{\delta x} \,.$$

We showed in Sect. 10.2 that $(\sin x)/x \to 1$ as $x \to 0$. Thus $(\sin \delta x)/\delta x \to 1$ as $\delta x \to 0$. Also, using standard trigonometric formulae given in the Appendix, Sect. A.3

$$\frac{(\cos\delta x - 1)}{\delta x} = \frac{(\cos\delta x - 1)(\cos\delta x + 1)}{\delta x(\cos\delta x + 1)} = \frac{-\sin^2\delta x}{\delta x(\cos\delta x + 1)}$$
$$= -\left(\frac{\sin\delta x}{\delta x}\right)\left(\frac{\sin\delta x}{\cos\delta x + 1}\right) .$$

This tends to zero as $\delta x \to 0$ since the term in the first bracket tends to 1, and that in the second bracket tends to zero. Thus, for any x, if $f(x) = \sin x$, then $f'(x) = \cos x$. A similar approach establishes that if $f(x) = \cos x$, then $f'(x) = -\sin x$.

A Pandora's box

Knowing the derivatives of x^n, $\sin x$ and $\cos x$ lets us differentiate a wide range of functions once we establish rules for differentiating sums, products, etc. We state, and comment on, some of these rules below and give examples of their use. Proofs are omitted, so please take these rules on trust in illustrations of their use.

11.8 Some Rules for Differentiation

Don't be discouraged too easily

If this is your first meeting with calculus this section may make heavy reading. Be prepared to come back and study the implications of each rule afresh when we use it in later examples. This will help you appreciate their importance and power as tools.

In all these rules $f(x)$, $g(x)$ are differentiable functions of x, i.e., functions whose derivatives exist.

1. *Addition rule.* If a, b are constants and $h(x) = af(x) + bg(x)$ then

$$h'(x) = af'(x) + bg'(x) .$$

This extends in an obvious way to more than two functions. An immediate application is to differentiation of a polynomial once we know the derivative of x^n is nx^{n-1} for any positive integer n. If we write

$$h(x) = a_0 + a_1x + a_2x^2 + \ldots + a_nx^n$$

the rule gives

$$h'(x) = a_1 + 2a_2x + \ldots + na_nx^{n-1} \, .$$

2. *Product rule.* This is more complicated than the addition rule. If

$$p(x) = f(x)g(x) \text{ then } p'(x) = f(x)g'(x) + f'(x)g(x) \, .$$

In words, the derivative of a product of two functions is the sum of the product of the first function with the derivative of the second function, and the product of the second function with the derivative of the first function. For example, if

$$p(x) = (1 + x^2) \sin x \; ,$$

setting $f(x) = 1 + x^2$ and $g(x) = \sin x$, gives $f'(x) = 2x$ and $g'(x) = \cos x$, whence

$$p'(x) = (1 + x^2) \cos x + 2x \sin x \, .$$

3. *Quotient rule* If $q(x) = f(x)/g(x)$ then, providing $g(x) \neq 0$,

$$q'(x) = \frac{g(x)f'(x) - f(x)g'(x)}{[g(x)]^2} \, .$$

Thus, if $f(x) = \sin x$ and $g(x) = \cos x$, then $q(x) = \sin x / \cos x = \tan x$. We have already established that $f'(x) = \cos x$ and $g'(x) = -\sin x$, so the quotient rule gives the derivative of $\tan x$ as

$$q'(x) = \frac{\cos x . \cos x - \sin x(-\sin x)}{\cos^2 x} = \frac{\cos^2 x + \sin^2 x}{\cos^2 x} = \frac{1}{\cos^2 x} = \sec^2 x$$

where $\sec x$ is the reciprocal of $\cos x$, i.e., $\sec x = 1/\cos x$.

4. *Composite functions or functions of functions.* This is a situation that most people take a while to master.

If $u = f(x)$ is a function of x and $y = g(u)$ is a function of u so that y can be expressed as a function of x in the form $y = g(f(x))$, then y is a composite function of x. To differentiate y we use the rule that

$$\frac{\mathrm{d}y}{\mathrm{d}x} = \frac{\mathrm{d}y}{\mathrm{d}u}\frac{\mathrm{d}u}{\mathrm{d}x} \, .$$

For example, if $u = f(x) = 1 - x^2$ and $y = g(u) = \sin u$ then $y = g([f(x)] = g(1 - x^2) = \sin(1 - x^2)$, so we have $\mathrm{d}y/\mathrm{d}u = \cos u = \cos(1 - x^2)$ and $\mathrm{d}u/\mathrm{d}x = -2x$, so that $\mathrm{d}y/\mathrm{d}x = -2x \cos(1 - x^2)$.

5. *Inverse function rule.* This is another rule not easily mastered at first meeting. If $y = f(x)$, and it can be rewritten in the form $x = g(y)$, then $g(y)$ is called the inverse function of $f(x)$. The derivatives of a function and its inverse, are the reciprocals of one another, i.e. $g'(y) = 1/f'(x)$, or $f'(x)g'(y) = 1$.

For example, suppose $y = f(x) = x^{1/5}$. Then $x = g(y) = y^5$, so $g'(y) = \mathrm{d}x/\mathrm{d}y = 5y^4 = 5x^{4/5} = 1/f'(x)$. Thus $f'(x) = (1/5)x^{-4/5}$. This is easily seen to be of the form mx^{m-1}, where $m = 1/5$. Here we have assumed familiarity with the laws for operating with indices. These are given in Sect. A.2. This example, in effect, extends the rule for differentiation of x^m, where m is a positive integer, to m a rational fraction.

Another interesting example of an inverse function is provided by the inverse trigonometric function, $y = \arctan x$. This is the function that specifies an angle, measured in radians, whose tangent is x, i.e., $x = \tan y$. The reason the notation *arctan* is used is to avoid the confusion that would arise with a notation like $\tan^{-1} x$, which might be interpreted as $1/\tan x$. To ensure a unique definition of $\arctan x$ we confine y to the interval $(-\pi/2, \pi/2)$. We saw above that $x = g(y) = \tan y$ has derivative $g'(y) = \sec^2 y = 1/\cos^2 y$, and since $\cos^2 y + \sin^2 y = 1$ this may also be written

$$g'(y) = \frac{\cos^2 y + \sin^2 y}{\cos^2 y} = 1 + \frac{\sin^2 y}{\cos^2 y} = 1 + \tan^2 y = 1 + x^2 \ ,$$

whence

$$f'(x) = \frac{1}{g'(y)} = \frac{1}{1+x^2} \ .$$

At first meeting this looks like a cunning slight of hand, but it is a typical example of how derivatives of inverse functions are obtained.

Readers already familiar with the calculus will know that this sketchy introduction to differentiation does no more than indicate the way we determine the derivatives of many functions using a few relatively simple rules.

11.9 Linking Differentiation and Integration

A small step for mathematicians but a mighty leap for the practical man

The link between differentiation and integration is not intuitively obvious.

Did you notice that in the examples given earlier for evaluating $\int_a^b f(x)\mathrm{d}x$ that x did not appear in the results? Logic demands this be so, since the areas we calculated were all fixed, but x was a variable. For this reason, x in this context is sometimes referred to as a *dummy variable*. We might equally well write $\int_a^b f(u)\mathrm{d}u$. It is the precise nature of the function f and the values assigned to a, b that determine the value of the integral.

We now do something slightly different and consider an integral

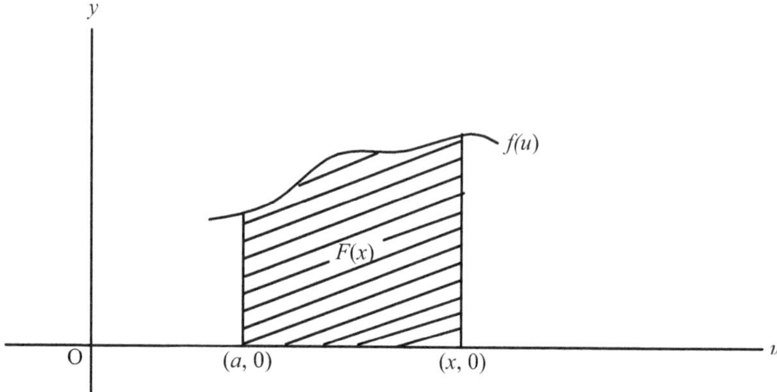

Fig. 11.13. Integration with a variable upper limit x

$$F(x) = \int_a^x f(u)\mathrm{d}u \, , \tag{11.6}$$

where we continue to regard a as fixed, but now regard the upper limit x as variable.

The notation indicates that we are interested in the integral $F(x)$ as a function of this variable upper limit x. A graphical interpretation is given in Fig. 11.13, where $F(x)$ is represented by the hatched area, i.e. that under the curve $y = f(u)$ between $u = a$ and the variable upper limit $u = x$.

Although sometimes expressed in slightly different terms, using this notation the *fundamental theorem of the calculus* states that

$$F'(x) = f(x) \, .$$

The practical import of this theorem is that the process of integration leading from $f(x)$ to $F(x)$ is reversed by applying the process of differentiation to $F(x)$. In other words, integration and differentiation are inverse processes.

Fig. 11.14 helps when outlining an intuitive argument to establish this result when $f(x)$ is a continuous function of x. We consider a small interval from x to $x + \delta x$. It is intuitively reasonable to assume, and can indeed be established subject to certain continuity conditions, that the area under the curve $f(u)$ associated with that interval will be $f(x')\delta x$, where x' is some value of x in the interval $(x, x + \delta x)$.

This area may also be written $F(x + \delta x) - F(x)$, whence

$$f(x') = \frac{F(x + \delta x) - F(x)}{\delta x} \, .$$

If we consider the limit as $\delta x \to 0$, then $x' \to x$ also. The right hand side tends to $F'(x)$ which establishes the result. A rigorous proof requires certain

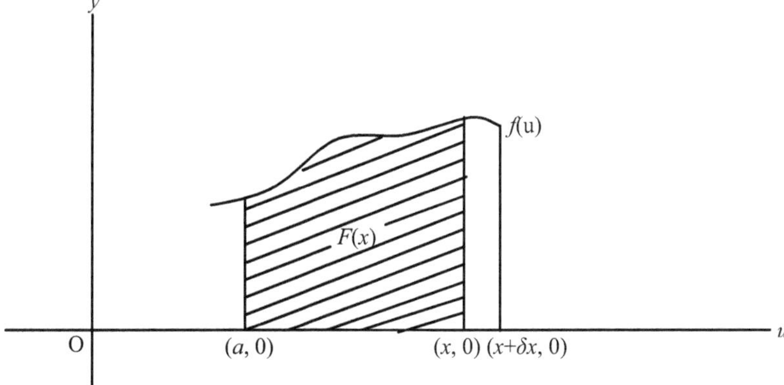

Fig. 11.14. A small increment in area under $f(x)$ between x and $x + \delta x$

continuity conditions and differentiability properties that are not discussed here.

The important practical implication is that if we start with a function $f(x)$, which we know to be the derivative of some function $F(x)$, we are well on the way to evaluating the integral between any limits (a, b).

An unlimited number of functions have the same derivative

A function $F(x)$ having a derivative $f(x)$ is not unique. We may add any constant to $F(x)$ without altering the derivative. This is so because the derivative of a constant is zero. This means that if we know $G(x)$ has derivative $f(x)$, then so has $F(x) = G(x) + c$. By setting $x = a$ it follows from (11.6) that

$$\int_a^a f(u)\mathrm{d}u = 0 = F(a) = G(a) + c$$

whence $c = -G(a)$. This in turn implies

$$\int_a^x f(u)\mathrm{d}u = G(x) - G(a) \text{ or } \int_a^b f(u)\mathrm{d}u = G(b) - G(a).$$

Iin the latter case we have fixed the upper limit at b. Thus, to determine this integral we need only find *any function* $F(x)$ [or $G(x)$] that has derivative $f(x)$. Such a function is often called an *indefinite integral.*

We know, or may easily determine, the derivatives of many functions (many more than those given in this chapter), so this gives a useful way to

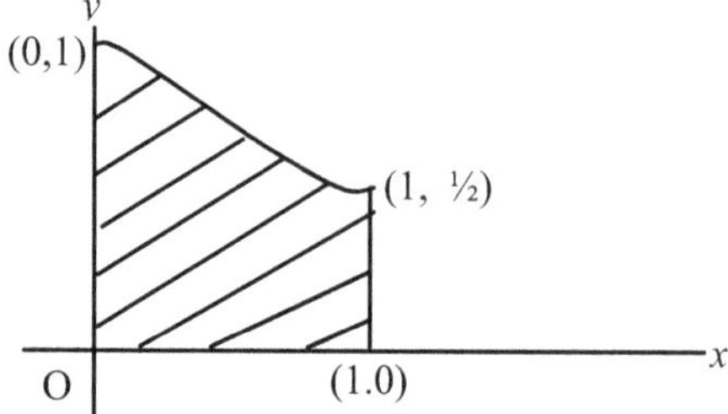

Fig. 11.15. The hatched area under the curve $f(x) = 1/(1+x^2)$ is one-quarter that of a circle of unit radius

obtain many integrals. For example, the derivative of $\sin x$ is $\cos x$. Therefore, we quickly establish that

$$\int_0^{\pi/2} \cos x \mathrm{d}x = \sin\frac{\pi}{2} - 0 = 1 .$$

Also, since nx^{n-1} is the derivative of x^n it is not hard to see that $x^{n+1}/(n+1)$ has derivative x^n whence

$$\int_a^b x^n \mathrm{d}x = \frac{1}{n+1}(b^{n+1} - a^{n+1}) .$$

In Sect. 11.4 before we discussed differentiation, we merely conjectured that this result held for integral n. We have now established that it holds for any two real numbers a, b and for all real n providing $n \neq -1$. We need the restriction $n \neq -1$ because that implies $n + 1 = 0$.

In Sect, 11.8 we showed the derivative of $\arctan x$ was $1/(1 + x^2)$ thus

$$\int_0^1 \frac{1}{1+x^2} \mathrm{d}x = \arctan 1 - \arctan 0 .$$

Considering only x in the interval $(-\pi/2, \pi/2)$ ensures a single value for x. Then the angle with tangent 1 is, in radians, $\pi/4$. The angle with tangent zero is 0. So we get the, perhaps surprising, result that the area under the curve $f(x) = 1/(1+x^2)$ between $x = 0$ and $x = 1$, is $\pi/4$. This is one quarter the area of a circle with unit radius. The area concerned, sketched in Fig. 11.15, bears no geometric resemblance to the shape of a circle. We consider an implication of this result in Sect. 12.6.

11.10 Some Extensions

So far in this chapter we have usually expressed y as a function of x, writing, for example $y = f(x)$, $y = g(x)$. When expressed in this form y is called an *explicit* function of x. We call x the *independent variable* and y the *dependent variable.* Relationships between two variables are sometimes written in an alternative form. For example, in Sect. 7.5 we saw that the equation to a circle with centre at the origin and radius r was

$$x^2 + y^2 = r^2 \ .$$

In this form the relationship between x and y is said to be an *implicit functional relationship.* We can rewrite this as an explicit relationship. This has the form $y = \pm\sqrt{r^2 - x^2}$. The alternative signs $+$ or $-$ reflect the fact that for any value of x in the open interval $(-r, r)$ there are two values of y. These have equal magnitudes but opposite signs. That this is so is evident if we draw a circle of radius r with centre at the origin. Taking the positive sign specifies a semicircle above the x-axis. With the negative sign we obtain a semicircle below the x-axis. If we differentiate the explcit form we get

$$\frac{\mathrm{d}y}{\mathrm{d}x} = \pm\frac{-x}{\sqrt{r^2 - x^2}} \ .$$

This implies $\mathrm{d}y/\mathrm{d}x = 0$ when $x = 0$, and then $y = \pm r$. It is not hard to see that $y = r$ corresponds to a maximum, while $y = -r$ corresponds to a minimum.

We can extend the differential calculus to functions of two or more variables. For illustrative purposes we consider first a function of three independent variables x, y, z giving rise to a dependent variable we shall call u. An example is

$$u = 3x^2yz - \sin y \ .$$

Functions of two or more independent variables no longer have a unique derivative. What we can write down, and attach a sensible meaning to, is the derivative of u with respect to x, if we assume y and z are held constant. This tells us the rate of change of u in the x co-ordinate direction for any specified y and z. Precisely what this means will become clearer in our second example below. We call this derivative *the partial derivative of u with respect to x* and write it $\partial u/\partial x$. It is the derivative obtained by the ordinary rules of differentiation with respect to x when the remaining variables y, z are treated as though they were constants. Similar definitions hold for partial derivatives with respect to the remaining variables.

Thus, if

$$u = 3x^2yz - \sin y \ ,$$

then

$$\frac{\partial u}{\partial x} = 6xyz, \qquad \frac{\partial u}{\partial y} = 3x^2z - \cos y, \qquad \frac{\partial u}{\partial z} = 3x^2y \ .$$

For the second example we consider the function $z = \sqrt{1 - x^2 - y^2}$. Geometrically, this represents the equation to a hemisphere of unit radius with centre at the origin. Here we consider x, y as the independent variables and z as the dependent variable. To verify that the equation represents a hemisphere, a useful starting point is the equation we obtained in Sect. 7.6 for a sphere with radius r and centre at the point with coordinates (a, b, c). The equation was

$$(x - a)^2 + (y - b)^2 + (z - c)^2 = r^2 .$$

Setting $a = b = c = 0$ and $r = 1$, gives the equation to a sphere of unit radius centred at the origin. This is

$$x^2 + y^2 + z^2 = 1 .$$

Expressing this as an explicit function of z gives

$$z = \pm\sqrt{1 - x^2 - y^2} .$$

At the start of this section we obtained an explcit form for the equation of a circle of radius r with centre at the origin. The explicit form of the equation to a sphere extends this concept for the case $r = 1$. It is easy to picture the sphere in a three dimensional Cartesian coordinate system with the x- and y-axes in the horizontal plane and the z axis positive direction vertically above that plane. Then $z = \sqrt{1 - x^2 - y^2}$ corresponds to the upper hemisphere.

It is evident from the geometry of the co-ordinate system that z has a maximum value 1 when $x = y = 0$. Lines parallel to either the x- or the y-axes that pass through this maximum have zero slope (i.e, they will be horizontal. Indeed, any line parallel to the plane containing the x- and y-axes passing through this maximum will have zero slope. Thus, at the maximum we should expect both $\partial z/\partial x$ and $\partial z/\partial y$ to be zero. You should verify that

$$\frac{\partial z}{\partial x} = \frac{-2x}{\sqrt{1 - x^2 - y^2}} .$$

Setting $x = 0$ we get $\partial z/\partial x = 0$. Similarly, in this case $\partial z/\partial y = 0$ when $y = 0$.

That all partial derivatives are zero is a necessary, but not a sufficient, condition for a point to be a maximum or a minimum in situations where there are two or more independent variables. That the condition is not sufficient is hardly surprising, for as we saw in the one independent variable case a zero derivative, although a necessary condition, is not a sufficient condition, for a point to be a maximum or a minimum.

11.11 Loose Ends

This brief introduction to calculus leaves a number of points that we deal with in the next chapter, where we also pick up some loose ends left over from earlier chapters.

Higher derivarives. A loose end it is convenient to tidy up here is an implication of the fact that $f'(x)$ is itself a function of x. We may differentiate this function and form what is called the second derivative of the original function $f(x)$ if the second derivative exists. The second derivative is denoted by $\mathrm{d}^2y/\mathrm{d}x^2$ or by $f''(x)$ or D^2y. While the first derivative is a measure of the rate of change of y with respect to x, the second derivative measures the rate of change of the first derivative with respect to x.

The process of differentiation may be extended by taking the derivative of the second derivative to form the third derivative, and so on. Notations used for the rth derivative are $\mathrm{d}^ry/\mathrm{d}x^r$ or $f^{(r)}(x)$ or D^ry. The notation $f^{(r)}(x)$ is preferred to the simpler $f^r(x)$ because mathematicians use the latter in another context not discussed here.

Velocity. In the dynamics of the motion of a particle, x may be a mcasure of time and y a measure of position specified as distance from some fixed point. If $y = f(x)$ is a function giving the position y at time x, then $\mathrm{d}y/\mathrm{d}x$, or $f'(x)$, is a measure of the 'instantaneous' velocity at time x. Also $\mathrm{d}^2y/\mathrm{d}x^2$, or $f''(x)$, is a measure of the rate of change of velocity, i.e., the acceleration at time x.

The notion of *instantaneous velocity* evolves from the concept that a body moving with constant velocity v for a time t will cover a distance $s = vt$, implying $v = s/t$. Here we use the notation commonly used by physicists (and mathematicians) when studying dynamic systems. If velocity is not constant we define the instantaneous velocity at any instant by considering the limit when a small distance δs is traversed in time δt as $\delta t \to 0$; i.e., the instantaneous velocity at any time t is then

$$v = \lim_{\delta t \to o} \frac{\delta s}{\delta t} = \frac{\mathrm{d}s}{\mathrm{d}t} .$$

The quadratic function. In Sect. 7.6 we found the value of x giving a maximum or a minimum of the function $y = ax^2 + bx + c$. We used the process of completing the square and mentioned that the same result could be got using differential calculus. We saw in Sect. 11.7 that for any function $f(x)$ at a maximum, x_0, $f'(x_0) = 0$. There is a further requirement that as x increases through x_0 the sign of the derivative changes from positive if x is slightly less than x_0 to negative if x is slightly greater than x_0. For a minimum a sign change from negative to positive is required. If the sign of the derivative is the same as we approach x_0 from either side then we have a point of inflection at x_0.

Setting

$$f(x) = ax^2 + bx + c ,$$

and differentiating we get

$$f'(x) = 2ax + b \,,$$

whence $f'(x) = 0$ if $x = -b/2a$. If a is positive we easily see that $f'(x)$ is negative if x has any value less than $-b/2a$, and $f'(x)$ is positive for x greater than $-b/2a$. This follows because the equation $y = 2ax+b$ is a line of positive slope if $a > 0$, implying that $f'(x)$ increases as x increases. Thus, there is a minimum when $x = -b/2a$. Similarly, if $a < 0$ there is a maximum when $x = -b/2a$. This confirms the results we got in Sect. 7.6.

In the light of the comments above about second derivatives measuring the rate of change of the first derivatives since, at a minimum the first derivative changes sign from negative to positive, this implies the first derivative is increasing. Thus if $f''(x_0)$ is positive then there is a minimum at x_0. Similarly if the second derivative is negative there is a maximum. For the parabola considered above $f'(x) = 2ax + b$, so $f''(x) = 2a$, confirming a minimum if a is positive and a maximum if a is negative.

If $f''(x) = 0$ when $f'(x) = 0$ there may be either a maximum, a minimum, or a point of inflection. For example if $y = x^4$ then $\mathrm{d}y/\mathrm{d}x = 4x^3$ which is zero when $x = 0$. Also $\mathrm{d}^2y/\mathrm{d}x^2 = 12x^2$ and this is zero when $x = 0$. However, as x increases through 0 the derivative changes from negative to positive, so we have a minimum. This is easily checked by sketching the graph of $y = x^4$, or by simply noting that y is necessarily positive for any $x \neq 0$. If $y = x^3$ the first and second derivatives are again both zero when $x = 0$. However, in this case there is a point of inflection in the form of a horizontal tangent on a rising curve.

12

Collecting the Pieces

12.1 Tidying up

Some material in this chapter, such as logarithms, will be familiar to many readers. However, we look at some topics in a different way from that adopted in many formal courses.

We also bring together, and expand, several themes developed in earlier chapters — in particular the use of limits to broaden mathematical horizons. We look at simple examples of application of these ideas to practical problems.

The importance of some concepts was recognised, and they were used by many mathematicians, before their formal properties were rigorously examined. In this chapter we quote some results without formal proof. Bear in mind, however, that even if an assertion or result seems intuitively reasonable, it should be treated with caution until it has been backed by a more rigorous analysis. Unless it is straightforward, we shall seldom attempt such analyses here.

12.2 The Binomial Expansion

In elementary mathematics the binomial theorem is met as the expansion of $(x+y)^n$ where n is a positive integer. It expresses the result as a sum of $n+1$ terms of the form $x^{n-r}y^r$, each multiplied by a constant. The summation is over all integer values of r running from 0 to n. The crux of the theorem is to determine each constant. For small values of n we can find these by direct multiplication. Trivially, when $n = 1$ the expansion is just $x+y$ itself, i.e., the terms are xy^0 and x^0y, each associated with a constant 1. If $n = 2$ we know that $(x + y)^2 = x^2 + 2xy + y^2$. We could continue this way, multiplying that result by $(x + y)$ to get $(x + y)^3$. Try this, or perhaps you already know that

$$(x + y)^3 = x^3 + 3x^2y + 3xy^2 + y^3.$$

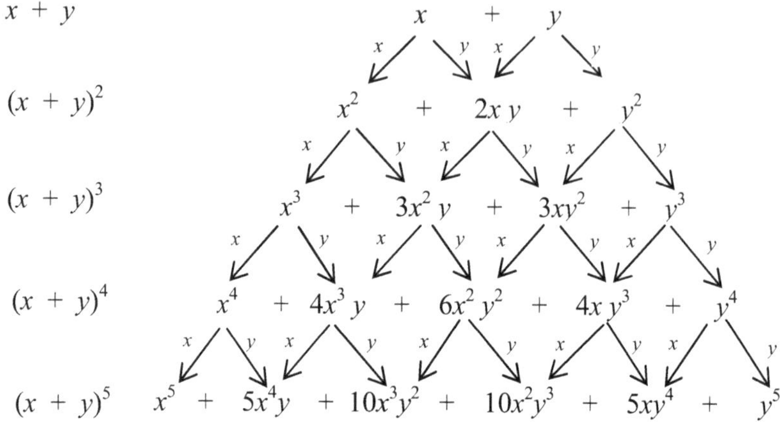

Fig. 12.1. A triangular array of terms in the binomial theorem expansion for $n \leq 5$

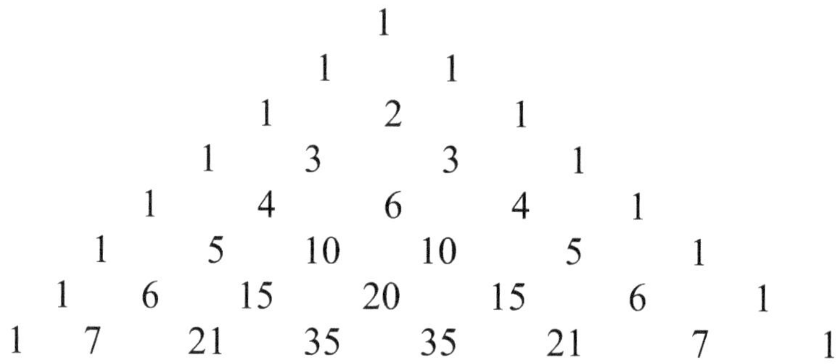

Fig. 12.2. Pascal's triangle of binomial coefficients for $n \leq 7$

Effectively, when we multiply $(x^2 + 2xy + y^2)$ by $(x + y)$, we multiply each term in the former first by x and then by y. We add and group the products so formed. When we have the result for $n = 3$, repeating the process gives the expansion for $(x + y)^4$. This can be extended to higher values of n.

This approach is tedious without a rule that gives us an easy way of determining the coefficients. Blaise Pascal (1623–1662) found such a rule. Fig. 12.1 shows on each line the expansion for the value of n indicated by the $(x + y)^n$ at the left of that line. The arrows labelled x and y indicate the contribution to the expansion in the next line from multiplying a term first by x then by y. The coefficient for each term in the new line is the sum of the coefficients of the contributing terms in the previous line.

In Fig. 12.2 these coefficients are rewritten without the associated $x^{n-r}y^r$ terms. We have added a single 1 as a new first line. Since anything raised to the power 0 is equal to 1 by definition, we may look upon this as representing the expansion $(x + y)^0 = 1$. We have also added a couple of extra lines at the bottom, using rules given below. The triangle of numbers in Fig. 12.2 is called *Pascal's triangle.*

The rules for forming it soon become evident. Each new line has an entry 1 to the left of all entries in the previous line, and also an entry 1 to the right of all entries in the previous line. The intermediate entries lie between the entries in the previous line. Each is the sum of the adjacent entries in the line above.

This is fine if we have such a triangle for all $n \leq 10$, and we are not interested in expansion for higher values of n. If we want the coefficients for $n = 49$ it is a lot of work to extend the table. How do we get around this?

A notation widely used for the entries in the Pascal triangle, i.e., the coefficients corresponding to a term $x^r y^{n-r}$ in the expansion of $(x+y)^n$, is $\binom{n}{r}$, where r runs from 0 to n. For all n, $\binom{n}{0} = \binom{n}{n} = 1$, and the rule for forming the terms in the Pascal triangle implies that

$$\binom{n}{r} = \binom{n-1}{r-1} + \binom{n-1}{r} . \tag{12.1}$$

Using these relationships it is an interesting exercise to prove by induction that

$$\binom{n}{r} = \frac{n(n-1)(n-2)\ldots(n-r+1)}{1 \times 2 \times 3 \times \ldots \times r} = \frac{n!}{r!(n-r)!} . \tag{12.2}$$

The proof proceeds this way. Remembering that we define $0! = 1$, the result is immediately seen to be true for $r = 0$ or $r = n$, for in either case the value is 1. If $r \neq 0$ or $r \neq n$, suppose that (12.2) holds for some given $n = N$, say. Then (12.1) implies that

$$\begin{aligned}\binom{N+1}{r} = \binom{N}{r-1} + \binom{N}{r} &= \frac{N!}{(r-1)!(N-r+1)!} + \frac{N!}{r!(N-r)!} \\ &= \frac{N!}{(r-1)!(N-r)!}\left(\frac{1}{N-r+1} + \frac{1}{r}\right) \\ &= \frac{N!}{(r-1)!(N-r)!}\left(\frac{N+1}{r(N-r+1)}\right) = \frac{(N+1)!}{r!(N-r+1)!} ,\end{aligned}$$

from which we see that if the formula holds for N, it holds for $N+1$. Since

$$\binom{2}{1} = \frac{2!}{1! \times 1!} = 2$$

it holds for $n = 2$. Thus it holds for $n = 3, 4, 5$ etc., and so is universally true.

In Chap. 17 we meet $\binom{n}{r}$ in a different context. There, and as shown in the Appendix, Sect. A.4, it gives the number of ways of selecting sets of r different objects from a set of n objects that are all distinct. After establishing this, it can be used as the basis for an alternative proof of the binomial theorem.

So much for the binomial theorem for n a positive integer. Does it extend to cases where n is not a positive integer? Let us try an intuitive approach and see what happens. We look first at the rth coefficient in the form we found when n is a positive integer, i.e.

$$\binom{n}{r} = \frac{n(n-1)(n-2)\dots(n-r+1)}{1 \times 2 \times 3 \times \dots \times r}.$$

Setting $n = 1/2$ we get

$$\binom{1/2}{r} = \frac{\frac{1}{2} \times (-\frac{1}{2}) \times \dots \times (\frac{3}{2} - r)}{r!}.$$

When n was an integer all terms of the form $\binom{n}{r}$ were zero when $r > n$. If n is not a positive integer we may go on forming terms of this structure indefinitely. Intuitively, this suggests that in these circumstances we might extend the binomial theorem result to form an infinite series.

The following example shows the need for caution. If we set $n = -1$ the coefficients take a simple form because

$$\binom{-1}{r} = \frac{-1 \times (-2) \times \dots \times (-r)}{1 \times 2 \times \dots \times r} = (-1)^r.$$

So the coefficients are alternatively $+1$ when r is even, and -1 when r is odd. Setting $x = 1$ and $y = -x$ and $n = -1$ in the expression $(x + y)^n$, this means that if the binomial expansion is valid then

$$\begin{aligned}(1-x)^{-1} &= 1 + (-1) \times (-x) + 1 \times (-x)^2 + (-1) \times (-x)^3 + \dots \\ &\quad + (-1)^r \times (-x)^r + \dots = 1 + x + x^2 + \dots + x^r + \dots. \end{aligned} \tag{12.3}$$

Does this look familiar? In Sect. 4.3 we considered a geometric progression with first term a and common ratio r. If we set $a = 1$ and $r = x$, we have the situation in (12.3), and the expression on the right is the sum of an infinite geometric progression. The corresponding result in Sect. 4.3 shows the sum of this series is $1/(1-x) = (1-x)^{-1}$. There was also a proviso in Sect. 4.3, where we assumed x to be positive, that x was less than 1. Intuitively, one would expect that the result above is only going to hold with some such restriction on the value of x. This is so. A counter example is given by setting $x = 2$. Then the left hand side becomes $(-1)^{-1} = -1$. Here we use the index rule that that $a^{-r} = 1/a^r$(see Sect. A.2). The right hand side of (12.3) is a divergent series when $x = 2$, so the expansion is not valid.

We do not prove it, but (12.3) is valid if $|x| < 1$, i.e., if $-1 < x < 1$. By this we mean the limit of the series on the right-hand side is given by the value of the expression on the left-hand side. We sometimes express this by saying the series is convergent if x lies in the *open interval* $(-1, 1)$. An open interval is one that does not include the end points, here -1 and 1. If the end points are included the interval is called a *closed interval.* The notation (a, b) is used for the open interval between a and b and $[a, b]$ for the corresponding closed interval.

In general, the binomial expansion of $(1+x)^n$ or $(1-x)^n$ when n is not a positive integer, will only be valid (in the sense that the infinite series is convergent) for certain values of x. There are rigorous methods not described here for testing for convergence.

12.3 A Less Obvious Limit

We use a simple, though not completely rigorous, approach to study one of the most interesting limits in mathematics. We mentioned it briefly in Sect. 10.1. If n is a positive integer, what is the limit as $n \to \infty$ of

$$\left(1+\frac{1}{n}\right)^n? \tag{12.4}$$

When intuition is not even a good servant

As pointed out in Sect. 10.1, if we raise a fixed number a greater than 1 to the power n, then as n increases the value of a^n increases, tending to infinity as $n \to \infty$. For example, $1.1^{100} \approx 13780.6$ and $1.1^{1000} \approx 2.47 \times 10^{41}$. This suggests that we might expect (12.4) to tend to infinity as n increases. On the other hand, as also pointed out in Sect. 10.1, one might argue intuitively that as $n \to \infty$ then $1+1/n \to 1$, and since for all positive integers n we have $1^n = 1$, this suggests the limit of (12.4) might be 1. Neither of these intuitive arguments leads to the correct limit.

For any fixed integer, n, the binomial theorem gives a valid expansion that takes the form

$$\left(1+\frac{1}{n}\right)^n = 1+\binom{n}{1}\left(\frac{1}{n}\right)+\binom{n}{2}\left(\frac{1}{n}\right)^2+\ldots+\binom{n}{r}\left(\frac{1}{n}\right)^r+\ldots+\binom{n}{n}\left(\frac{1}{n}\right)^n$$

where

$$\begin{aligned}\binom{n}{r}\left(\frac{1}{n}\right)^r &= \frac{n(n-1)(n-2)\ldots(n-r+1)}{r!}\left(\frac{1}{n}\right)^r\\ &= \frac{(1-1/n)(1-2/n)\ldots[1-(r-1)/n)]}{r!}\,.\end{aligned}$$

If r is fixed, and we let $n \to \infty$ then $1/n, 2/n, \ldots (r-1)/n$ all tend to zero, so that

$$\binom{n}{r}\left(\frac{1}{n}\right)^r \to \frac{1}{r!}.$$

Thus, providing it is valid to extend this limiting process to the resulting infinite series, we would find

$$\lim_{n\to\infty}\left(1+\frac{1}{n}\right)^n = 1+1+\frac{1}{2!}+\ldots+\frac{1}{r!}+\ldots.$$

Don't forget the dangers of intuition, but here it suggests we have a series we met in Sec. 10.1, where we explained that the sum of this series is the transcendental number denoted by e, and that $\mathrm{e} \approx 2.718$.

The argument above is not rigorous insofar as allowing n to tend to infinity is concerned. But we at least have a series known to be convergent. Indeed, it can be established rigorously that this is the limit.

We may extend the above argument to establish a series expansion for e^x, which is valid for all real x. To do this we consider the limit of $(1+1/n)^{nx}$. Here intuition suggests that since $(1+1/n)^n \to \mathrm{e}$, then a limiting process applied to this raised to the power x might lead to a series for e^x. We cannot proceed directly to a limit as $n\to\infty$, because even when $n\to\infty$ it does not follow that nx will necessarily tend to infinity. To get round this problem we put $nx = m$, giving

$$\left(1+\frac{1}{n}\right)^{nx} = \left(1+\frac{x}{m}\right)^m.$$

We now apply the argument used for expanding $(1+1/n)^n$ to expanding $[1+(x/m)]^m$, only now we let $m\to\infty$ keeping x fixed. You should work through this argument to check that the only essential difference is that now

$$\binom{m}{r}\left(\frac{x}{m}\right)^r \to \frac{x^r}{r!}.$$

If there are no problems with convergence, this implies

$$\mathrm{e}^x = 1+x+\frac{x^2}{2!}+\ldots+\frac{x^r}{r!}+\ldots$$

We do not prove it, but this expansion is valid for all real x.

12.4 An Important Derivative

We indicated in Sect. 11.8 that the derivative of the sum of a finite number of functions, or terms, that are all differentiable is the sum of their derivatives. We also pointed out in Sect. 11.5 that there is a similar result for integrals. Intuitively, one might hope that these results will extend to a limiting infinite series, where each term in the series is some function of a variable x.

It will only do so if certain conditions hold. The first is that the series is convergent for the relevant x. This implies that we might determine the derivative of e^x by term by term differentiation of the series. Remembering that the derivative of a constant is zero, that of x is 1, that of x^2 is $2x$ and so on, i.e., that of x^r is rx^{r-1}, we easily see that if $f(x) = \mathrm{e}^x$ then

$$f'(x) = 1 + \frac{2x}{2!} + \frac{3x^2}{3!} + \ldots + \frac{rx^{r-1}}{r!} + \ldots = 1 + x + \frac{x^2}{2!} + \ldots + \frac{x^{r-1}}{(r-1)!} + \ldots$$

The expression after the second equality sign is $f(x) = \mathrm{e}^x$, the exponential function. Thus, providing term by term differentiation is valid, the exponential function has the property that its derivative is equal to the function itself. In fact, term by term differentiation is valid here. In geometric terms this result implies that the rate of change of the exponential function at any value of x is equal to the value of the function at that x. For positive x, as x increases e^x also increases, and in the limit as $x \to \infty$ the rate of change becomes infinite.

This is the basic idea behind what we call *exponential growth*. More generally, if y represents some quantity such as the weight of an organism, or the number of organisms or individuals in a population such as a bacterial population, at time t, then growth is said to be exponential if it satisfies the relationship

$$y = c\mathrm{e}^{kt}$$

where c, k are positive constants. By the rule for differentiation of a function of a function given in Sect. 11.8, this implies

$$\frac{\mathrm{d}y}{\mathrm{d}t} = ck\mathrm{e}^{kt} = ky. \qquad (12.5)$$

Since $\mathrm{d}y/\mathrm{d}t$ measures the rate of change of y with time t, this implies that the rate of growth is proportional to the size y at time t. Thus the growth rate increases as the size increases, and in theory rapidly approaches infinity.

In practice, in most biological situations growth eventually stops. This may be because animals cannot obtain enough food to sustain growth beyond a certain size. Or perhaps they would collapse under excess body weight if they grew any larger. Similarly, in an epidemic that initially spreads rapidly to individuals in a population, infection eventually dies out because all available individuals become infected, and so on.

If, in (12.5), k is negative the process is one of *exponential decay*. As t increases y approaches zero. This equation is often used to model radio-active decay.

Equation (12.5) is a simple example of a *differential equation*. We say more about how such equations arise in practice in Sect. 12.8.

12.5 Natural Logarithms

How to be natural without being common

If $y = f(x) = \mathrm{e}^x$, the inverse function (Sect. 11.8) is $x = \log y = g(y)$, say, where we have taken e as the 'base' of the logarithm. More fully, if there is any doubt about the base, we may write $\log_{\mathrm{e}} y$. Recalling the rule for differentiation of inverse functions given in Sect. 11.8, it follows that $g'(y) = 1/f'(x) = 1/\mathrm{e}^x = 1/y$, whence interchanging x and y we have, if $y = \log x$ then $\mathrm{d}y/\mathrm{d}x = 1/x$. This result holds only for logarithms to the base e, which are called *natural logarithms*. It depends on the fact that $\log_{\mathrm{e}} \mathrm{e} = 1$.

The rule for differentiation is more complicated for any other base. For example for logarithms to the base 10 — the base for the so-called *common logarithms* — to covert them to the base e we multiply by approximately 2.30259. This follows because $\log_{10} 10 = 1$ while $\log_{\mathrm{e}} 10 = 2.30259$. Thus, if $y = f(x) = \log_{10} x$ then $y = \log_{\mathrm{e}} x/2.30259$ and $f'(x) = 1/(2.30259x)$. It is common practice for pure mathematicians to write $\log x$ when the base is e, but this conflicts with the everyday notation where $\log x$ is also in use for the common logarithm, or logarithm to the base 10. Many mathematicians avoid conflict by writing $\ln x$ for a logarithm to the base e.

12.6 Evaluating π

In Sect. 11.9 we saw that

$$\frac{\pi}{4} = \int_0^1 \frac{1}{1+x^2}\mathrm{d}x\ .$$

If it is legimate to integrate a series term by term, we might perform the integration after expanding $1/(1+x^2) = (1+x^2)^{-1}$ as a binomial expansion. This expansion is easily shown to be

$$(1+x^2)^{-1} = 1 - x^2 + x^4 - x^6 + \ldots + (-1)^{r-1}x^{2r} + \ldots .$$

Can we integrate term by term over the interval $(0, 1)$? It is not difficult to establish convergence of the binomial expansion if $|x| < 1$. But if $x = 1$ the left hand side takes the value $1/2$ while the right hand side oscillates between 0 and 1 as we increase the number of terms. It is certainly an act of faith to believe that the integral will converge if we include this end point, for might it not also oscillate? This is a loose end we do not resolve here. However, it does converge, and term by term integration is straightforward.

For the general term x^{2r}

$$\int_0^1 x^{2r} \mathrm{d}x = \frac{1^{2r+1}}{2r+1} - \frac{0^{2r+1}}{2r+1} = \frac{1}{2r+1} ,$$

whence, if term by term integration is permissible,

$$\int_0^1 \frac{1}{1+x^2} \mathrm{d}x = \frac{\pi}{4} = 1 - \frac{1}{3} + \frac{1}{5} - \frac{1}{7} + \frac{1}{9} - \frac{1}{11} + \frac{1}{13} - \dots .$$

This formula was obtained by James Gregory(1638–1675) a few years before his death. It was also known to Leibniz, who probably derived it independently at a slightly later date.

With a pocket calculator it is not hard to verify that taking terms up to $1/13$ the approximation to π is a not very good 3.26. By taking terms up to $1/31$ it improves only to 3.08. More rapidly convergent series now exist for calculating π. Using these in modern number crunching computers, approximations to π accurate to more than 50 billion (thousand million) digits have been obtained. This is a far cry from the popular approximation $\pi \approx 22/7 \approx 3.142857\dots$, which is only correct to 2 decimal places.

Reasonable rational approximations to π were made by earlier civilizations including the Chinese, Hindu and Babylonian. Archimedes was aware that π lay between $223/71$ and $22/7$. Some 400 years ago π was known accurately to only 6 decimal places. Banks [7] gives an interesting table of the numbers of decimal places to which π was known at various times between 1560 and 1874, ranging from 6 to over 700 decimal places.

A rapidly convergent series for calculating π is given by Bailey et al [4]. It was obtained using a computerised approach with the hexadecimal base 16 and with that base is

$$\pi = \sum_{k=0}^{\infty} \frac{1}{16^k} \left(\frac{4}{8k+1} + \frac{3}{8k+4} + \frac{1}{8k+5} + \frac{1}{8k+6} \right)$$

This and related formulae have some fascinating properties that are discussed by Bailey and Borwein [5].

12.7 Series Expansions

A polynomial of degree n is a versatile function. Writing

$$P_n(x) = a_0 + a_1 x + a_2 x^2 + a_3 x^3 + \dots + a_r x^r + \dots + a_n x^n ,$$

we may regard $P_n(x)$ as a finite series that is the sum of $n+1$ terms of a sequence whose rth term is $a_{r-1}x^{r-1}$. What happens to this series as $n \to \infty$, and why may this be worth knowing?

A rigorous treatment of this topic involves advanced and often subtle mathematics, so we only give an intuitive approach that provides at best a partial answer. An inspiration to explore this approach comes from the fact that when n is finite, by suitable choices of the constants a_r, it is possible to obtain polynomials that give good approximations, at least for some values of x, to a great many functions that are not themselves polynomials.

It is easy to see that if a_n is not zero, then as $x \to \infty$ the polynomial $P_n(x)$ also tends to infinity if a_n is positive. It tends to minus infinity if a_n is negative. Thus, for a function such as $\sin x$, which we know oscillates between -1 and 1, an approximating polynomial of finite degree n will not fit well if $|x|$ is large. A problem also arises if one tries to use a polynomial to approximate to $f(x) = 1/x$ when x takes values near zero. Intuitively it seems clear that the discontinuity in $f(x)$ at $x = 0$ is likely to pose a problem (see Fig. 11.12). It does.

However, for many functions that are continuous everywhere and that have a continuous derivative, polynomial approximations may work well over an appreciable range of, if not all, x values. There are several methods for obtaining polynomials of various degree that give adequate approximations over specified finite intervals. We touch upon this topic again in a different context in Sect. 15.7.

From the mathematical viewpoint, a function behaves particularly well if it is not only differentiable in the sense that $f'(x)$ exists, but if it is one for which all higher order derivatives, $f''(x), f'''(x), \ldots$, also exist. In this situation the behaviour of a polynomial $P_n(x)$ as $n \to \infty$ is of interest. We are then studying the behaviour of an infinite series. Can we use such a series not only to approximate to, but actually to be equivalent to, some given function $f(x)$, which is not itself a polynomial, at least for some values of x? If so, how do we determine appropriate values of each of the coefficients a_r?

This problem was looked at by mathematicians not long after Newton gave us the calculus. One of these was the Scottish mathematician Colin Maclaurin (1698–1746), another the English mathematician Brook Taylor (1685–1731). The latter obtained a fairly general result, whereas Maclaurin considered a special case of great practical importance. The respective results are usually referred to as the *Maclaurin Expansion* and *Taylor's theorem.*

Here we confine ourselves to outlining the way the Maclaurin expansion is obtained, and to indicating how it may be used, sometimes arguing intuitively.

Our objective is to find for a function $f(x)$ an infinite series expansion of the form

$$f(x) = a_0 + a_1 x + a_2 x^2 + \ldots + a_r x^r + \ldots \tag{12.6}$$

given that derivatives of all orders [i.e., $f'(x), f''(x), f'''(x), \ldots$] of $f(x)$ exist. We shall assume, as we did in Sect. 12.4 when seeking the derivative of e^x, that we can differentiate (12.6) term by term. This will not always be a valid assumption, and in particular will almost certainly be invalid unless the series is convergent. However, assuming validity, we have

$$f'(x) = a_1 + 2a_2x + 3a_3x^2 + \ldots + ra_rx^{r-1} + \ldots ,$$
$$f''(x) = 2a_2 + 3.2a_3x + \ldots + r(r-1)a_rx^{r-2} + \ldots ,$$
$$f'''(x) = 3.2a_3 + \ldots + r(r-1)(r-2)a_rx^{r-3} + \ldots ,$$

and continuing this way

$$f^{(r)}(x) = (r!)a_r + (r+1)r(r-1)(r-2)\ldots 3.2a_{r+1}x + \ldots .$$

If, in each of $f(x), f'(x), f''(x), f'''(x), \ldots$ we set $x = 0$, we have

$$\begin{aligned} a_0 &= f(0) \\ a_1 &= f'(0) \\ a_2 &= \frac{1}{2!}f''(0) \\ a_3 &= \frac{1}{3!}f'''(0) \\ \ldots &\ \ldots\ldots \\ a_r &= \frac{1}{r!}f^{(r)}(0) \\ \ldots &\ \ldots\ldots \end{aligned}$$

Substitution in (12.6) gives the Maclaurin expansion

$$f(x) = f(0) + xf'(0) + \frac{x^2}{2!}f''(0) + \ldots + \frac{x^r}{r!}f^{(r)}(0) + \ldots .$$

If $f(x) = \cos x$, it is easy to see that all derivatives exist. Specifically

$$f'(x) = -\sin x,\ f''(x) = -\cos x,\ f'''(x) = \sin x,\ f^{(4)}(x) = \cos x .$$

This pattern is repeated for successive derivatives. Since $\sin 0 = 0$ and $\cos 0 = 1$, it follows that all odd order derivatives are zero when $x = 0$, while the even derivatives are alternatively -1 and 1, giving the following series for $\cos x$:

$$\cos x = 1 - \frac{x^2}{2!} + \frac{x^4}{4!} - \frac{x^6}{6!} + \ldots .$$

As an exercise, establish that if $f(x) = \sin x$, the Maclaurin expansion is

$$\sin x = x - \frac{x^3}{3!} + \frac{x^5}{5!} - \frac{x^7}{7!} + \ldots$$

I do not prove it, but both these series are convergent for all real x.

A remarkable link

Now let intuition run wild. We obtained the series for e^x in Sect. 12.3, and asserted that it held for all real x. Let us assume that it also holds if x is replaced by a complex number z. In particular, if it does, then when $z = \mathrm{i}x$ we could write

$$\mathrm{e}^{\mathrm{i}x} = 1 + \mathrm{i}x - \frac{x^2}{2!} - \mathrm{i}\frac{x^3}{3!} + \frac{x^4}{4!} + \mathrm{i}\frac{x^5}{5!} - \frac{x^6}{6!} - \mathrm{i}\frac{x^7}{7!} + \dots ,$$

where we have replaced i^2 by -1 and i^3 by $-\mathrm{i}$ and i^4 by 1 throughout. Collecting together real and imaginary terms we get

$$\mathrm{e}^{\mathrm{i}x} = 1 - \frac{x^2}{2!} + \frac{x^4}{4!} - \frac{x^6}{6!} \dots + \mathrm{i}\left(x - \frac{x^3}{3!} + \frac{x^5}{5!} - \frac{x^7}{7!} + \dots\right) = \cos x + \mathrm{i}\sin x$$

This expansion can be proved to be valid and the formula

$$\mathrm{e}^{\mathrm{i}x} = \cos x + \mathrm{i}\sin x \tag{12.7}$$

is yet another contribution to mathematics from Leonhard Euler. It has been proved to hold for all real x.

Here is a situation where intuitive arguments have led to a far from obvious relationship between a complex exponential function and the trigonometric ratios. This once again illustrates links between what at first sight appear unrelated concepts. You will also recognise the function on the right of (12.7) as an expression for a complex number with unit modulus and amplitude x. Do not confuse this x with the real part, x, of the complex number $x + y\mathrm{i}$.

Setting $x = \pi$ in (12.7) yields a remarkable relationship between two transcendental numbers contributing to a complex variable that takes a real value, namely

$$\mathrm{e}^{\pi\mathrm{i}} = -1.$$

If we rewrite this as

$$\mathrm{e}^{\pi\mathrm{i}} + 1 = 0,$$

it gives a deceptively simple relationship between 5 basic constants $0, 1, \mathrm{i}, \mathrm{e}, \pi$.

Replacing x by $-x$, it is easy to see that

$$\mathrm{e}^{-\mathrm{i}x} = \cos x - \mathrm{i}\sin x , \tag{12.8}$$

and applying elementary algebra to (12.7) and (12.8) that

$$\cos x = \frac{\mathrm{e}^{\mathrm{i}x} + \mathrm{e}^{-\mathrm{i}x}}{2} \text{ and } \sin x = \frac{\mathrm{e}^{\mathrm{i}x} - \mathrm{e}^{-\mathrm{i}x}}{2\mathrm{i}} ,$$

a far cry from the original definitions of these trigonometric functions as ratios of side lengths in right-angle triangles!

Series expansions using the Maclaurin formula can be obtained for many other functions. Some, however, are only valid for certain values of x. For example, if $f(x) = 1/(1-x)$ then using the rules of differentiation given in Chap. 11, we find

$$f'(x) = 1/(1-x)^2 \text{ whence } f'(0) = 1$$
$$f''(x) = 2/(1-x)^3 \text{ whence } f''(0) = 2$$
$$f'''(x) = 2.3/(1-x)^4 \text{ whence } f'''(0) = 2 \times 3 = 3!$$

and continuing this way that

$$f^{(r)}(x) = 2 \times 3 \times \ldots \times r/(1-x)^{r+1} \text{ whence } f^{(r)}(0) = r!$$

The function and the derivatives are not defined if $x = 1$. In this example it can be proved that the Maclaurin expansion is only valid if $|x| < 1$. It is easily seen to be

$$f(x) = 1/(1-x) = 1 + x + x^2 + \ldots + x^r + \ldots$$

Three routes to the same goal

This is the same result, as it must be to be valid, that we got in (12.3) using the binomial expansion. As we pointed out there it was also the sum of the terms in an infinite geometric progression with first term 1 and common ratio x. Thus we have obtained the same result in three different ways.

The Maclaurin expansion is a particular case of a more general expansion called a Taylor series, which is derived from Taylor's theorem. This again refers to functions for which all derivatives exist and it may be written

$$f(x) = f(a) + f'(a)(x-a) + f''(a)\frac{(x-a)^2}{2!} + \ldots + f^{(r)}(a)\frac{(x-a)^r}{r!} + \ldots .$$

The Maclaurin expansion is obtained by setting $a = 0$.

The basic Taylor series also provides an expression for the 'remainder' if one approximates to $f(x)$ using only a finite number of terms in the above series, but we do not pursue that important topic here.

12.8 Differential Equations

The equation (12.5), i.e.

$$\frac{dy}{dt} = ky ,$$

where k is a known constant is a simple differential equation. It describes exponential growth (k positive) or decay (k negative). We interpret dy/dt as the rate of change of y with respect to t. If y is some size measure such as weight or height of an organism, or the size of a population, this equation tells us the rate at which that size measurement is increasing at time t. A biologist calls this the growth rate of y at time t. In (12.5) we obtained this equation by differentiating the function

$$y = c\mathrm{e}^{kt}.$$

In both the life and the physical sciences a differential equation is often the starting point for exploring the properties of a system. In the above example, if we start from the differential equation (12.5) this implies that the growth rate at any given instant is proportional to the size at that instant. Given this information, the procedure used for finding the relationship between y and t is described as *solving the differential equation*. For this example the solution is almost obvious, for the equation may be written

$$\frac{1}{y}\frac{\mathrm{d}y}{\mathrm{d}t} = k ,$$

and a slight extension of the rules of differentiation and integration from those we have given, but one I won't discuss in detail, justifies the intuitively reasonable idea that the solution may be written

$$\int \frac{1}{y}\mathrm{d}y = \int k\mathrm{d}t$$

giving

$$\log y = kt + C$$

where C is an arbitrary constant. Alternative ways of writing this are

$$y = \mathrm{e}^{kt+C}$$

or

$$y = C'\mathrm{e}^{kt} \tag{12.9}$$

where $C' = \mathrm{e}^{C}$. That this is a solution may be verified by differentiation.

It is not a solution that would satisfy a biologist interested in studying the growth of some organism according to the rule implied by (12.5), because if he or she wanted to know the size when, say, $t = 2$ all (12.9) tells the biologist is that then

$$y = C'\mathrm{e}^{2k}$$

where C' might take any value. The situation here bears similarities to that discussed in Sect. 11.9, where we introduced the indefinite integral as a stepping stone to finding a definite integral. Equation 12.9 is a 'general' solution to the differential equation (12.5). The biologist wants a solution relevant to the entity that he or she is studying. This, called a *particular solution*, may be obtained if we know the value of y at some given time. In practice this is often a time $t = 0$ when the study commences. Suppose that at time $t = 0$ the size is y_0. Then, since (12.9) must hold when $t = 0$, substituting $y = y_0$ and $t = 0$ in (12.9) gives

$$y_0 = C'\mathrm{e}^0$$

and since $\mathrm{e}^0 = 1$, it follows that $C' = y_0$. Thus for the organism, or whatever entity the biologist is studying, the appropriate growth equation is

$$y = y_0\mathrm{e}^{kt}.$$

The condition that $y = y_0$ when $t = 0$ is called an *initial condition*.

Most organisms or populations do not increase in size indefinitely, so biologists often study more realistic equations relating y and t. Many animals, including humans, tend to grow slowly at first, then more rapidly for a time, and finally growth effectively ceases at an adult stage. Similarly, many populations such as those of bacteria or insects, increase in numbers slowly at first, then more rapidly until population growth slows down. The size finally reaches a maximum due to limitations on food supply, population control by predators, or something of that nature.

A number of differential equations have been suggested for these situations. A simple but easily explained one is

$$\frac{\mathrm{d}y}{\mathrm{d}t} = ky(a - y)\,. \tag{12.10}$$

Here we see that $\mathrm{d}y/\mathrm{d}t = 0$, both when $y = 0$ or when $y = a$. The physical interpretation of a (assumed positive) is that it is a supposedly known maximum value that can be attained by y, i.e., there can be no further growth. Equation (12.10) is more realistic than (12.5), although numerous other equations have been proposed that overcome some limitations that still come into play for (12.10). For example, it is not difficult to show that when (12.10) holds the maximum growth rate occurs, i.e., $\mathrm{d}y/\mathrm{d}t$ is a maximum, when $y = a/2$. This is often, but not always, a realistic assumption.

It is not hard to solve (12.10) after one has completed a basic course in solving differential equations, but we simply quote a result. As in the case of exponential growth, to get a particular solution relevant to a real situation, we need an initial condition. This is available if we know, for example, the size y_0 at time $t = 0$. With this initial condition the solution to (12.10) is

$$y = \frac{ay_o}{y_0 + (a - y_0)\mathrm{e}^{-kt}}.$$

I ask you to take this solution on trust, but you can check that when $t = 0$ it does satisfy the initial condition, giving $y = y_0$. Also, as $t \to \infty$, because e^{-kt}

then tends to zero, it follows that $y \to a$, as it should if a is the maximum possible size. The function is called the *logistic growth function.*

Modifications of (12.10) are sometimes used. For example if we replace the y in that equation by a new variable y' that represents the proportion of the final size at any time t, i.e., we make a change of variable and set $y' = y/a$, the equation equivalent to (12.10) takes the form

$$\frac{\mathrm{d}y'}{\mathrm{d}t} = ky'(1 - y') \,, \tag{12.11}$$

where the appropriate initial condition would be some specified proportion of the final size at time $t = 0$.

The examples of differential equations given above have only involved the first derivative of y with respect to t, where we have interpreted y as a size measurement — e.g., a height or weight or number of organisms in some population.

In the physical sciences differential equations play a major role in Newtonian dynamics where y, for example, may be a measure of distance travelled by a body during an elapsed time t. As indicated briefly in Sect. 11.11 in this situation the physical meaning of $\mathrm{d}y/\mathrm{d}t$ is that of rate of change of position at time t. This is the familiar concept of velocity at a given instant.

If we differentiate velocity with respect to time, and form the second derivative $\mathrm{d}^2y/\mathrm{d}t^2$, this represents rate of change of velocity — the physical concept of acceleration. Much of Newtonian dynamics of motion in a straight line is concerned with systems for which the key notion is that of constant acceleration. The classic example is the behaviour of a body falling freely under the influence of gravity. More generally, if a body moving in a straight line is subject to constant acceleration, a, this implies

$$\frac{\mathrm{d}^2y}{\mathrm{d}t^2} = a. \tag{12.12}$$

Integration, being the inverse of differentiation, gives an expression for $\mathrm{d}y/\mathrm{d}t$, i.e., the velocity. The general expression brings in an arbitrary constant which may be 'fixed' for a particular system if we know the initial velocity of the body. If that velocity is u, this implies $\mathrm{d}y/\mathrm{d}t = u$ when $t = 0$.

If we have a particular solution for velocity we may then integrate this new equation, which expresses $\mathrm{d}y/\mathrm{d}t$ in terms to t, u and a, to get an expression for y. The integration involves a further arbitrary constant which can be fixed if we know the value of y when $t = 0$. Very often in practice we start measurement of distance or displacement at time zero, by setting the initial condition on y as $y = 0$ when $t = 0$.

In summary, for the relevant particular solution of (12.12) that gives y as a function of t, we require both a value (initial condition) for $\mathrm{d}y/\mathrm{d}t$, and also one for y, when $t = 0$.

Specifically, solving (12.12) with initial conditions $y = 0$ when $t = 0$ and $\mathrm{d}y/\mathrm{d}t = u$ when $t = 0$, the solution leads to well known formulae in Newtonian dynamics. If we denote the velocity at time t by v, then $v = \mathrm{d}y/\mathrm{d}t$ and differentiating this expression we may write

$$\frac{\mathrm{d}v}{\mathrm{d}t} = \frac{\mathrm{d}}{\mathrm{d}t}\left(\frac{\mathrm{d}y}{\mathrm{d}t}\right) = \frac{\mathrm{d}^2 y}{\mathrm{d}t^2}.$$

Thus (12.12) may be written

$$\frac{\mathrm{d}v}{\mathrm{d}t} = a\,,$$

and on integration this gives the general solution

$$v = at + c\,,$$

where c is an arbitrary constant. The initial condition $v = u$ when $t = 0$ implies $c = u$, whence the appropriate particular solution is

$$v = u + at\,. \tag{12.13}$$

Remembering that $v = \mathrm{d}y/\mathrm{d}t$ and rewriting (12.13) as

$$\frac{\mathrm{d}y}{\mathrm{d}t} = u + at,$$

once again direct integration gives a general solution

$$y = ut + \frac{1}{2}at^2 + c\,,$$

and we obtain the relevant particular solution by determining c to satisfy the initial condition $y = 0$ when $t = 0$. By substitution we immediately find $c = 0$, whence

$$y = ut + \frac{1}{2}at^2\,. \tag{12.14}$$

Equations (12.13) and (12.14) are well known to physics students as the basic equations for motion of a particle moving in a straight line under constant acceleration.

What happens if we throw a stone vertically upwards at an initial velocity of $u = 1962$ cm per second? We shall ignore the effects of air resistance and assume that the stone is subject only to the constant acceleration due to gravity. If we measure velocity in cm per second, the acceleration due to gravity takes the approximate value 981 cm per second per second. This is only true if the direction of motion is vertically downward (i.e. towards the centre of the earth). Because we throw the stone vertically upward, the acceleration upward (sometimes called deceleration) has a negative sign. Thus, in (12.12) to (12.14) we set $a = -981$.

The height, y, of the stone above the projection point ($y = 0$) at time t, measured from zero at the time of projection is, from (12.14), given by

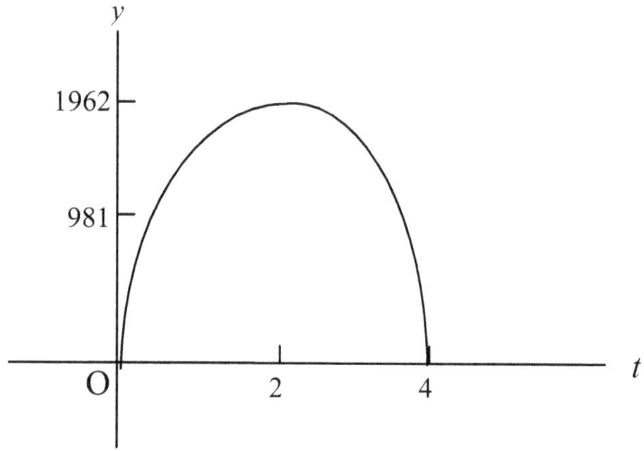

Fig. 12.3. Sketch of parabola in (12.15) for values of t between 0 and 4

$$y = 1962t - \frac{1}{2} \times 981t^2$$

or

$$y = 981t(2 - \frac{1}{2}t). \tag{12.15}$$

Thus, when $t = 2$ the stone will be at a height of 1962 cm. Also, substituting the appropriate values of u and a in (12.13) we see that the equation giving velocity is

$$v = 1962 - 981t\,. \tag{12.16}$$

It follows that when $t = 2$, then $v = 0$. Thus, the stone reaches a maximum height of 1962cm above the ground after 2 seconds. It then starts to drop towards its starting point. When $t = 4$ we again see from (12.15) that $y = 0$, implying the stone will have returned to its starting point just 4 seconds after it was thrown straight up in the air. Unless we want a sore head we should move out of the line of descent within 4 seconds of projecting the stone.

Substituting $t = 4$ in (12.16) gives $v = -1962$, suggesting a downward velocity equal in magnitude to the upward velocity with which we projected the stone. This change in sign reflects an important characteristic of velocity associated with motion in a straight line. Namely, that velocity has both a *magnitude* and a *direction.* Quantities that are specified by both magnitude and direction are called *vectors.* We say more about this important idea in Sect. 13.5, and again under the heading *vectors* in Sect. 25.2.

If we plot a Cartesian graph of (12.15) for values of t between 0 and 4, this takes the form of the familiar parabola associated with a quadratic function of t. One may read from the graph the height above the projection point at any time t in the interval. Fig. 12.3 is a sketch of this graph for our projected stone.

In this example we have used units familiar to all scientists and science students. In the United States, and often also in the United Kingdom, non-scientists frequently measure distances in feet and velocities in feet per second (or in miles per hour). Since there are approximately 2.54 cm in 1 inch, and 12 inches in 1 foot, it follows that 1962 cm is approximately equal to $1962/(2.54\times 12) = 64.37$ feet. Thus a velocity of 1962 cm per second is equivalent to a velocity of 64.37 feet per second. Similarly, an acceleration of 981 cm per second per second is equivalent to an acceleration of $981/(2.54\times 12) = 32.185$ feet per second per second. By inserting these values in (12.13) and (12.14) readers should have little difficulty in confirming that the stone reaches its maximum height in 2 seconds. It returns to its starting point in 4 seconds, with a downward velocity of 64.37 feet per second.

In the above example we have used approximate values both for the acceleration dues to gravity and for the conversion factor 2.54 between inches and centimetres. As these are approximations, one should be careful not to introduce spurious accuracy when making conversions. My pocket calculator gives $981/(2.54\times 12) = 32.18503937$. Since 981 is only an approximation when using metric units, and 2.54 is only an approximate conversion factor, even the approximation 32.185 is of dubious accuracy to more than 1 decimal place.

By the introduction of SI units of measurement as an international standard, scientists are making a serious attempt to standardise units. Their efforts have met with only limited success outside the confines of science. Even in the higher technologies the use of weird mixtures of measurement units is common. Nowhere is this more evident than in aviation, where it is almost universal practice to measure air speed in knots (nautical miles per hour), altitude in feet, airport visibility in metres, distances in statute miles or nautical miles (or sometimes in kilometres), atmospheric pressure in millebars (or inches of mercury in the USA). The confusion caused by any attempts to convert to SI units for aviation would probably increase the risk of accidents.

The assumption in equation (12.12) that the acceleration is constant is not always tenable. It would not be appropriate, for example, to describe the motion of a pendulum. For a simple idealized pendulum, and in many other situations where there is a periodic motion, an appropriate differential equation is that relevant to *simple harmonic motion*. It takes the form

$$\frac{\mathrm{d}^2 y}{\mathrm{d}t^2} = -\omega^2 y \tag{12.17}$$

where ω is a constant. The equation implies that the acceleration is proportional to the displacement, or distance y, from some point where $y = 0$. Given initial conditions for the displacement when $t = 0$, and the velocity when $t = 0$, it is not difficult to show that the solution has the form

$$y = d\cos(\omega t + \alpha) \ ,$$

where the constants d and α are chosen to comply with the initial conditions on y and $\mathrm{d}y/\mathrm{d}t$. We do not discuss the solution further at this stage, except to point out that it represents a periodic motion. We see that the maximum displacement from $y = 0$ is d when $\cos(\omega t + \alpha) = 1$, and $-d$ when $\cos(\omega t + \alpha) = -1$. As t increases the value of y oscillates between $-d$ and $+d$, with the general form of the curve of a scaled cosine function.

Modification to the equation for simple harmonic motion is needed if the motion dies down with time. This happens with a real pendulum which comes to rest after a finite time unless an external force is applied to restore a uniform motion. Equation (12.17) is only an approximation for the motion of any real pendulum. It is only a good approximation when the pendulum swings through a very small arc. The exact equation takes account of some other physical factors.

Simple harmonic motion is discussed further in Sect. 24.6.

12.9 Loose Ends

Solving differential equations. Differential equations are the lifeblood of most mathematical models. We described a few simple ones in the previous section. Many differential equations met in the sciences do not have analytic solutions that take the form of some symbolic expression, usually called a *closed solution*. However, given appropriate initial conditions, or another kind of condition called a *boundary condition* that we have not discussed here, it is often possible to obtain numerical values of the solution at specific points, or over a region. Techniques for doing this belong to a branch of mathematics known as *numerical analysis*, which we introduce in more general terms in Chap. 15.

The most widely studied class of differential equations in conventional courses on the subject are those known as *linear differential equations*. These have the form

$$P_n(x)\frac{\mathrm{d}^n y}{\mathrm{d}x^n} + P_{n-1}(x)\frac{\mathrm{d}^{n-1} y}{\mathrm{d}x^{n-1}} + \ldots + P_1(x)\frac{\mathrm{d}y}{\mathrm{d}x} + P_0(x)y = Q(x)$$

where the coefficients $P_r(x), r = 0, 1, 2, \ldots, n$ of y and its derivatives are all functions of x only. The adjective 'linear' is used to indicate that the equation is linear in y and its derivatives, i.e. there are no non-linear terms in these quantities such as $\sin y$ or $(\mathrm{d}^4y/\mathrm{d}x^4)^3$ in the equation. A special case is the linear differential equation with constant coefficients, where the $P_r(x)$ are all constants, i.e., they do not involve x.

Despite the emphasis on linear differential equations in elementary courses on the subject many, or perhaps most, real world situations — especially those arising in the study of dynamic systems — are nonlinear. As already mentioned they often do not have analytic solutions. During the last 30 or so years a new way of looking at nonlinear systems has led to the development of a new branch of mathematics known as *chaos*. We say more about this in Chaps. 21 and 22.

Equations may involve more than two variables, giving rise to partial differential equations. These, as the name implies, involve partial derivatives. Theoretical physics might almost be said to be built around partial differential equations. One of these is the heat equation described briefly in Sect. 25.2. Another is called the *Navier-Stokes equation*. It is relevant to a range of problems concerned with fluid mechanics, ranging from description of fluid flow, viscosity or turbulence, to what happens at boundaries between different fluids.

Conjuring up integers

Some surprising results. In Sect. 12.7 we met the remarkable relationship $\mathrm{e}^{\pi \mathrm{i}} + 1 = 0$ between 1, 0, the imaginary unit i, and the two transcendental numbers π and e. This is only one of many interesting results that have been thrown up by developments in number theory.

It surprises many to learn that $\mathrm{e}^{\pi\sqrt{163}}$ has an exact value that differs from an integer only in the thirteenth decimal place. To 15 decimal places

$$\mathrm{e}^{\pi\sqrt{163}} = 262\,537\,412\,640\,768\,743\,.\,999\,999\,999\,999\,250.$$

Before explaining why this approximation is no accident, we need a new notation. If we want to multiply together all terms in a sequence we indicate this using a symbol $\prod$ in an analogous way to the use of $\sum$ for addition. Thus

$$\prod_{r=1}^{n} a_r = a_1 \times a_2 \times \ldots \times a_n$$

and

$$\prod_{r=1}^{\infty}(1 + 1/x^r) = (1 + 1/x)(1 + 1/x^2)(1 + 1/x^3)\ldots(1 + 1/x^r)\ldots.$$

The fact that the $\mathrm{e}^{\pi\sqrt{163}}$ is almost an integer follows from what has been described as the 'amazing' fact that

$$x^{1/3}\prod_{r=1}^{\infty}\left(1 + x^{1-2r}\right)^8 - 256x^{-2/3}\prod_{r=1}^{\infty}\left(1 + x^{1-2r}\right)^{-16}$$

is an integer when $x = \mathrm{e}^{\pi\sqrt{163}}$. Indeed, this expression is an integer when $x = \mathrm{e}^{\pi\sqrt{m}}$, if m takes any of the values $3, 4, 7, 11, 19, 43, 67, 163$. This was first proved nearly 100 years ago by Heinrich Weber (1842–1913). A brief account of some of the implications is given by Good [27].

A much older fascinating number with a longer history is the Euler constant. In Sect. 5.4 we established that the harmonic series

$$S_n = 1 + \frac{1}{2} + \frac{1}{3} + \ldots + \frac{1}{n}$$

was divergent if we allow $n \to \infty$. Also, it is easily seen that $\log n \to \infty$ as $n \to \infty$, where e is the base.

What happens to $S_n - \log n$ when $n \to \infty$? It converges to a fixed value, the *Euler constant*, which to 4 decimal places, has the value 0.5772. Surprisingly, it is not known whether this number is rational or irrational, and if the latter whether or not it is transcendental. It is known that if it is rational, then if it were expressed in its lowest terms as a fraction r/s, then s must be a number with more than 244 663 digits.

Number theory, and our understanding of the complexities of infinity, have developed a long way from origins associated with the natural number system.

13

Systems of Linear Equations

13.1 Artificial Simplification

Understanding limits made possible the caculus as we know it today — one of the giant leaps in mathematics. In other areas small steps that may have appeared to be of little importance on their own have had a combined effect as significant as some giant leaps. The study of *linear equations* has developed in a series of steps over several centuries. Many advances were a response to practical problems.

School algebra isn't always the real life brand

At school most of us learn to solve, without the aid of a computer, sets of linear equations that are basically simplifications of 'real-life' ones met in disciplines such as engineering, physics, chemistry or statistics. Simplification takes two forms. The first is by working with only a few variables; the second by choosing numerical values that make calculation easy. This is a sensible approach. Basic ideas are better understood at first meeting if free from technical complexity.

Once we know how to solve a pair of simultaneous linear equations, if asked to solve the pair

$$5x + 2y = 1 \tag{13.1}$$
$$7x - 5y = 17\,, \tag{13.2}$$

most of us would not feel unduly imposed upon. Our response may be less enthusiastic if asked to solve the equations

$$123.75x + 33.97y = 43.29 \tag{13.3}$$
$$72.33x - 11.77y = -6.44\,. \tag{13.4}$$

The number of arithmetic operations (additions, subtractions, multiplications and divisions) needed is the same for solving the pair (13.1) and (13.2) and for solving the pair (13.3) and (13.4). The extra labour to solve the latter comes about because these operations involve more digits and are harder to carry out with only pencil and paper, or even a pocket calculator. Given a good program to solve pairs of simultaneous linear equations, a computer will take about the same time to solve either of the pairs. Computers are not afraid of crunching awkward numbers unless these fall outside extreme limits that do not come into play here.

What may look to be a formidable pair of equations may sometimes prove simple to solve with pencil and paper if they have a particular pattern. For example, in the equations

$$\begin{aligned} 317.4x + 29.2y &= 664.0 \\ 29.2x + 317.4y &= 375.8\ , \end{aligned}$$

the coefficients of x and y are interchanged between equations. Adding and equating the sums on each side we get

$$346.6x + 346.6y = 1039.8\ .$$

Dividing through by 346.6 gives

$$x + y = 3\ .$$

Similarly, subtracting the second equation from the first gives

$$288.2x - 288.2y = 288.2\ ,$$

equivalent to

$$x - y = 1.$$

So the original problem reduces to solving the pair

$$\begin{aligned} x + y &= 3 \\ x - y &= 1. \end{aligned}$$

We easily find the solution $x = 2$ and $y = 1$. This example illustrates a method that may be applied to solving pairs of equations of the form

$$\begin{aligned} ax + by &= c \\ bx + ay &= d\ . \end{aligned}$$

Providing $a \neq b$, these reduce by the above method to the pair

$$\begin{aligned} x + y &= \frac{c+d}{a+b} \\ x - y &= \frac{c-d}{a-b}\ . \end{aligned}$$

Pattern has an important role in mathematics generally, and may often simplify numerical calculations using pencil and paper. However, with modern computer software the tendency is to produce programs that work efficiently whether or not there is a pattern that may make possible 'special' procedures like the one above.

Where computers have the edge

At first sight it may not appear so, but solving, without the aid of a computer, the set of equations

$$x + y + z - w = 13 \tag{13.5}$$

$$3x + 2y - 5z + 2w = -12 \tag{13.6}$$

$$x - 6y - 7z + 3w = 72 \tag{13.7}$$

$$7x - 5y + 2z - 5w = 13 \tag{13.8}$$

will be appreciably more time consuming and need more planning and concentration that solving (13.3) and (13.4). With appropriate software, a computer should solve (13.5) to (13.8) in microseconds, and would take little longer than it did to solve (13.3) and (13.4).

The difficulty with pencil and paper arises because the number of arithmetic operations needed to solve sets of simultaneous linear equations increases dramatically as the number of equations increases. This does not worry a fast number-crunching computer unduly — at least not within certain limits.

The pure mathematics needed to solve sets of simultaneous linear equations was developed in general algebraic terms long before modern computers enabled us to handle large sets of equations in reasonable time. This meant that when fast computers arrived, programmers could use this mathematics to produce software to make best use of a machine's ability to perform calculations at a rate hitherto unattainable.

Tedious routine computations that once took days or weeks — even when using the mechanical or electromechanical calculators widely available in the early- to mid-twentieth century — are now completed by computers in seconds or even microseconds. Calculations once regarded as impossible, because they would have taken large teams several lifetimes to complete, may now be done in minutes or hours.

Before outlining computational aspects of the solution of sets of linear equations, we need some notations that mathematicians have developed for summarizing the processes needed for solution when there are many unknowns. Some of these were developed in the latter part of the nineteenth century, but were explored and exploited fully only during the twentieth century. Key concepts are those of *matrices* and *vectors*. Their development only became possible after the arrival of other versatile general notations.

Simplicity that leaves a false impression

Introductory courses in algebra often leave the impression that we only meet cases where there are two, three, or perhaps four equations in the same numbers of unknowns. However, practising statisticians (and workers in some other disciplines) often meet 30 or more equations in 30 or more unknowns. Without modern algebraic notation it is hard to describe how, even in theory, one might set about solving these.

In the late nineteenth, and even early twentieth, century it was common to see equations in an unspecified number of unknowns written like this:

$$\begin{aligned} ax + by + cz + \ldots + fw + \ldots + gu &= s \\ a'x + b'y + c'z + \ldots + f'w + \ldots + g'u &= s' \\ a''x + b''y + c''z + \ldots + f''w + \ldots + g''u &= s'' \\ \ldots\ldots\ldots\ldots\ldots\ldots\ldots\ldots\ldots\ldots\ldots & \ldots \\ a'''x + b'''y + c'''z + \ldots + f'''w + \ldots + g'''u &= s''' \end{aligned}$$

Such notations are cumbersome and not very informative. Textual explanation is needed to tell us how many equations and unknowns there are. These are important pieces of information. Even if we were to write the equations more fully, and had numerical values for the coefficients, we would still run into trouble if there were more than 26 unknowns.

The most common practical situation where we solve such sets of equations successfully is that where there are a fixed number, say, n equations in n unknowns. We shall see later that there may be solution difficulties even then.

The first step towards a more informative, and simpler, notation is to replace the unknown variables $x, y, z \ldots$ by a single letter with a distinguishing numerical suffix. Different suffixes imply different variables. We might write x_1 in place of x, x_2 in place of y and so on. If there are n unknowns the last is denoted by x_n.

An extension from one to two suffixes, which we used in another context in Sect. 5.4, lets us associate one letter with all the coefficients of the unknowns. If we use the letter a, the general form is to write a_{ij} for the coefficient of x_j in the ith equation. That is, the first suffix tells us in which equation the constant occurs, and the second the unknown to which it is attached in that equation. The constants on the right of each equation may be represented by another letter, say, b with a single suffix representing the number of the equation to which it refers. Thus b_7 refers to the constant after the equals sign in the seventh equation.

Using subscripted a, b for constants complies with the widely used, though not universal, convention of assigning letters near the start of the alphabet to constants and those near the end to variables or unknowns. With these conventions if we have n equations in n unknowns these can be written

$$\begin{aligned} a_{11}x_1 + a_{12}x_2 + \ldots + a_{1n}x_n &= b_1 \\ a_{21}x_1 + a_{22}x_2 + \ldots + a_{2n}x_n &= b_2 \\ a_{31}x_1 + a_{32}x_2 + \ldots + a_{3n}x_n &= b_3 \\ \cdots\cdots\cdots\cdots\cdots\cdots\cdots\cdots\cdots \\ a_{n1}x_1 + a_{n2}x_2 + \ldots + a_{nn}x_n &= b_n \end{aligned} \tag{13.9}$$

It takes time to become at ease with single and double suffix notations, because each symbol gives us up to three things to think about:

- How does the quantity each symbol represents relate to the rest of the notation: e.g., is it a coefficient of an unknown (like a_{23}) or a stand-alone constant (like b_7)?
- What is the significance of the first suffix?
- What is the significance of the second suffix if there is more than one?

Practice soon makes one comfortable with the notation.

13.2 Matrices and Vectors

This new notation still has limitations. Writing down the equations (13.9) takes time, and there is further pattern of which we have not yet taken advantage. For example, the unknowns x_1, x_2, x_3, ..., repeat themselves in the same order in each equation, and the b_1, b_2, b_3, ..., appear in an orderly way in a column on the right-hand side. Also the a_{ij} have a pattern. If we isolate them from their associated x_i, they form a square array of n rows and n columns where the subscripts i, j in a_{ij} indicate it is in the ith row and jth column of that array, as shown below. This array is called a *matrix*.

$$\begin{matrix} a_{11} & a_{12} & a_{13} & \cdots & a_{1n} \\ a_{21} & a_{22} & a_{23} & \cdots & a_{2n} \\ a_{31} & a_{32} & a_{33} & \cdots & a_{3n} \\ \vdots & \vdots & \vdots & \ddots & \vdots \\ a_{n1} & a_{n2} & a_{n3} & \cdots & a_{nn} \end{matrix}$$

A matrix is often enclosed in large square brackets, e.g.,

$$\begin{bmatrix} a_{11} & a_{12} & a_{13} & \cdots & a_{1n} \\ a_{21} & a_{22} & a_{23} & \cdots & a_{2n} \\ a_{31} & a_{32} & a_{33} & \cdots & a_{3n} \\ \vdots & \vdots & \vdots & \ddots & \vdots \\ a_{n1} & a_{n2} & a_{n3} & \cdots & a_{nn} \end{bmatrix}$$

A matrix with this specified pattern may be further abbreviated to a notation $\mathbf{A} = \{a_{ij}\}, i, j = 1, 2, \ldots, n$. Once the matrix is specified this way we need only refer to it by the letter **A** (or in words as 'bold capital A').

A matrix like the one above with the same number of rows and columns, n, is called a *square matrix.* More generally, a matrix may have m rows and n columns, where m and n need not be equal. Such a matrix is referred to as an $m \times n$ matrix. A special case occurs if either $m = 1$ or $n = 1$. An example of the former, a $1 \times n$ matrix, is

$$[x_1 \quad x_2 \quad x_3 \quad \ldots \quad x_n]$$

We may also form an $m \times 1$ matrix which is written as a single column with elements

$$\begin{bmatrix} x_1 \\ x_2 \\ x_3 \\ \vdots \\ x_m \end{bmatrix}$$

Matrices consisting of a single row, or a single column, have a special status. They are referred to respectively as a *row vector* and a *column vector.* Use of lower case bold letters for vectors is a common notation. In the above example we might denote the column vector by **x**.

Interchanging rows and columns is called *transposing* a matrix. The new matrix is called the *transpose* of the original matrix. If we transpose a matrix **A**, a common notation is to write the transpose as $\mathbf{A}^{\mathbf{T}}$. In speech this is usually referred to as '**A** transposed'. A row vector with the same elements as a column vector **x** is, in matrix terms, the transpose of **x** and is written $\mathbf{x}^{\mathbf{T}}$.

We return to vectors in Sect. 13.5, describing how the concept of a vector relates to the idea of a measure such as a length associated with a direction. We met this notion when discussing velocity in Sect. 12.8.

13.3 Matrix Algebra

New rules for addition and multiplication

Without prior motivation the rules for addition and multiplication of matrices may not seem obviously relevant to anything. However, it is easier to give the rules first, and then discuss their practical import using examples.

Two matrices can be added only if each has the same number of rows and columns. If $\mathbf{A} = \{a_{ij}\}$, $\mathbf{B} = \{b_{ij}\}$ are both $m \times n$, then their sum is the $m \times n$

matrix $\mathbf{C} = \mathbf{A}+\mathbf{B} = \{a_{ij}+b_{ij}\}$. That is, the sum of two matrices of the same 'order' (i.e., each with identical numbers of rows and of columns) is obtained by adding the corresponding elements in each.

The definition of the product of two matrices is more complicated. It is only defined if the number of columns in the first equals the number of rows in the second matrix. The product has the same number of rows as the first matrix and the same number of columns as the second. If $\mathbf{A} = \{a_{ij}\}$ is an $m\times r$ matrix and $\mathbf{B} = \{b_{kl}\}$ is an $r\times n$ matrix, then the product $\mathbf{AB} = \mathbf{C} = \{c_{pq}\}$ is an $m\times n$ matrix. The element c_{pq} is the sum of the products of corresponding elements in the pth row of $\mathbf{A}$ and the qth column of $\mathbf{B}$, i.e.

$$c_{pq} = a_{p1}b_{1q} + a_{p2}b_{2q} + \ldots + a_{pr}b_{rq}.$$

Here is a numerical example. If

$$\mathbf{A} = \begin{bmatrix} 2 & 1 & 7 \\ 3 & 2 & 0 \\ 0 & 0 & 1 \end{bmatrix} \quad \text{and} \quad \begin{bmatrix} 1 & 4 \\ 2 & 5 \\ 3 & -1 \end{bmatrix}$$

Then

$$\mathbf{C} = \begin{bmatrix} (2\times 1)+(1\times 2)+(7\times 3) & (2\times 4)+(1\times 5)+[7\times(-1)] \\ (3\times 1)+(2\times 2)+(0\times 3) & (3\times 4)+(2\times 5)+[0\times(-1)] \\ (0\times 1)+(0\times 2)+(1\times 3) & (0\times 4)+(0\times 5)+[1\times(-1)] \end{bmatrix}$$

$$= \begin{bmatrix} 25 & 6 \\ 7 & 22 \\ 3 & -1 \end{bmatrix}$$

If $\mathbf{x}$ is a column vector of n elements, its transpose $\mathbf{x}^{\mathrm{T}}$ is a row vector with the same n elements, then the product $\mathbf{x}^{\mathrm{T}}\mathbf{x} = x_1^2+x_2^2+\ldots+x_n^2$, which is just a number. The product $\mathbf{x}\mathbf{x}^{\mathrm{T}}$ is an $n\times n$ matrix. You should verify that the ith 'diagonal' element (i.e., that with the suffix 'ii') in $\mathbf{y} = \mathbf{x}\mathbf{x}^{\mathrm{T}}$ is $y_{ii} = x_i^2$, while if $i \neq j$ an element with suffix 'ij' has the form $y_{ij} = x_i x_j$. This implies $y_{ij} = y_{ji}$. A square matrix with these properties is called a *symmetric matrix.* A symmetric matrix is equal to its transpose.

How do these rules simplify the notation for simultaneous equations? We have already, for the set of equations (13.9), defined the $n\times n$ matrix $\mathbf{A}$ and the $n\times 1$ column vector $\mathbf{x}$ with elements x_i. We now define $\mathbf{b}$ as the $n\times 1$ column vector with elements b_i occurring to the right of the equality signs in (13.9). Using the rules of multiplication for matrices it is easy to verify that (13.9) may be written in matrix notation as

$$\mathbf{Ax} = \mathbf{b}\,. \tag{13.10}$$

Thus, using matrix notation any set of n simultaneous linear equations in n unknowns takes a form similar to the way we write one equation in one unknown, i.e., $Ax = b$. The latter is a special case of (13.10), because ordinary numbers A, b, x may all be regarded as (1×1) matrices.

Using the new shorthand

At first meeting $\mathbf{Ax} = \mathbf{b}$ may seem little more than an interesting new shorthand that does no more than save some writing. However, it introduces a generality that, among other things, is particularly useful in unifying methods of computation. It allows us to treat a range of problems by giving basically the same instructions to a computer about the operations to be carried out.

In practice, this means that if we want to solve, for example, the set of equations (13.9), then a suitably programmed computer can do this once we feed in numerical values for n and the constants a_{ij}, b_i. This avoids the need to have separate programs for solving sets of 1, of 2, of 3, and so on, equations. The software carries out required matrix operations such as multiplication and addition for any specified value of n. There are technicalities that have to be taken into account to avoid serious rounding errors in large numbers of computerized arithmetic operations, but we concentrate on the basic theory here.

Before we (or the computer) can solve (13.10) we need to define one other matrix operation — the analogue of division for real numbers. In Sect. 2.7 we described division as the inverse of multiplication. For matrices the inverse of multiplication is more complicated. It is basically confined to what is called *inversion of a square* (i.e., $n \times n$) *matrix.*

The relevance of inversion to the solution of simultaneous linear equations is motivated by looking at the solution of the single-variable equation $Ax = b$. This may be written $x = A^{-1}b$. A property inherent in this solution is that $AA^{-1} = 1$, since $A^{-1} = 1/A$.

Suppose we write n equations in n unknowns as in (13.9) in matrix notation as $\mathbf{Ax} = \mathbf{b}$. The solution for the single variable case suggests we might write the solution as $\mathbf{x} = \mathbf{A}^{-1}\mathbf{b}$, providing we can give a meaning to $\mathbf{A}^{-1}$ when we are given $\mathbf{A}$. To provide that meaning is no trivial task.

A first step is to generalize the scalar 1 on the right of the expression $AA^{-1} = 1$ to an $n \times n$ matrix $\mathbf{I}$ which, when multiplied by $\mathbf{A}$, does not alter $\mathbf{A}$, i.e., $\mathbf{AI} = \mathbf{IA} = \mathbf{A}$. It is not difficult to see that this will be so if $\mathbf{I}$ is an $n \times n$ matrix with all main diagonal elements (i.e., those on the diagonal running from top left to bottom right) equal to 1, and every off-diagonal element zero. Here is a numerical example:

$$\begin{bmatrix} 7 & 1 & 2 \\ 4 & 1 & 3 \\ 2 & 0 & 7 \end{bmatrix} \text{ multiplied by } \begin{bmatrix} 1 & 0 & 0 \\ 0 & 1 & 0 \\ 0 & 0 & 1 \end{bmatrix}$$

is easily seen to leave the former unaltered. Carry out the multiplication yourself to check this. To solve the equations (13.9) we need to find a matrix $\mathbf{A}^{-1}$ such that $\mathbf{AA}^{-1} = \mathbf{I}$. The matrix $\mathbf{A}^{-1}$, if it exists, is called the *inverse* of $\mathbf{A}$. This is the hard part. In effect, finding $\mathbf{A}^{-1}$ involves solving the set of equations (13.9) in some systematic way. This is computationally simple only for small values of n.

In the late nineteenth, and the early twentieth, centuries various algorithms for finding an inverse were developed, some more efficient than others. Before the advent of the electronic computer, except in a few special cases, this was laborious. The time taken grew rapidly as n increased. Hand calculation, even when aided by mechanical or electromechanical computers, was something to be avoided if n were much greater than 6 or 7, and virtually out of the question for values of n like 20, 25, 30 or more.

We referred above to '$\mathbf{A}^{-1}$, if it exists'. A hint that it might not always do so is given by the solution of the single-variable equation $Ax = b$. When $A = 0$ the quantity $1/A$ or A^{-1} is not defined.

We shall not describe how $\mathbf{A}^{-1}$ is computed, but merely quote and comment upon a few results.

When $n = 2$ the matrix $\mathbf{A}$ is

$$\begin{bmatrix} a_{11} & a_{12} \\ a_{21} & a_{22} \end{bmatrix}.$$

In this case the inverse $\mathbf{A}^{-1}$ is

$$\begin{bmatrix} (a_{22})/(a_{11}a_{22} - a_{12}a_{21}) & (-a_{12})/(a_{11}a_{22} - a_{12}a_{21}) \\ (-a_{21})/(a_{11}a_{22} - a_{12}a_{21}) & (a_{11})/(a_{11}a_{22} - a_{12}a_{21}) \end{bmatrix}.$$

By forming directly the product $\mathbf{AA}^{-1}$, we may easily verify that this yields the matrix $\mathbf{I}$, i.e.,

$$\begin{bmatrix} 1 & 0 \\ 0 & 1 \end{bmatrix}.$$

For example, the entry 0 in the second row is given by the product rule as

$$\frac{a_{21}a_{22} + a_{22}(-a_{21})}{a_{11}a_{22} - a_{12}a_{21}} = 0 \, .$$

You should check the remaining entries yourself. There is an inbuilt assumption here that $a_{11}a_{22} - a_{12}a_{21} \neq 0$, for if it were zero, the elements in $\mathbf{A}^{-1}$ would not be defined.

In terms of the solution of the pair of simultaneous equations, a nondefined $\mathbf{A}^{-1}$ means either that the equations are inconsistent, and therefore have no solutions, or that they have an infinity of solutions. For each set of paired simultaneous equations below you should verify that $a_{11}a_{22} - a_{12}a_{21} = 0$ implying that $\mathbf{A}^{-1}$ does not exist.

Set I

$$\begin{aligned} 3x + 21y &= 7 \\ 3x + 21y &= -14 \end{aligned}$$

Set II
$$3x + 21y = 3$$
$$x + 7y = 1$$

In *Set* I, there are no values of x and y that satisfy both equations, since for no values of x, y can $3x+21y$ equal both 7 and -14. In terms of a Cartesian graph the lines are parallel. In *Set* II both equations represent the same line. This is easily seen if we divide through by 3 in the first equation, which reduces it to the second. Thus, all points (x, y) lying on the line satisfy both equations. For example, $x = 1$ and $y = 0$ or $x = 8$ and $y = -1$ are just two possible solutions.

For three equations in three unknown the number of possible outcomes grows markedly. Only if the matrix

$$\begin{bmatrix} a_{11} & a_{12} & a_{13} \\ a_{21} & a_{22} & a_{23} \\ a_{31} & a_{32} & a_{33} \end{bmatrix}$$

has an inverse will there be a unique solution. We saw in Sect. 7.6, using a different but equivalent notation, that each equation of the form

$$a_{i1}x_1 + a_{i2}x_2 + a_{i3}x_3 = b_i$$

represents a plane in three dimensions. The existence of an inverse corresponds to the case where the three planes share only one common point. It is not hard to see that any two equations of the form just given represent either (i) planes that are identical or (ii) planes that meet in a line or (iii) planes that have no common points, in which case they are parallel to each other. In case (i) the third equation may represent either (a) a plane identical to that represented by the other two equations, (b) a plane intersecting the plane represented by each of the other two equations or (c) a plane parallel to the plane represented by each of the other two equations. In case (ii) the third equation may represent (a) a plane that intersects the line common to the other two planes at one point only (in which case there is a unique single solution to the three equations) or (b) a plane intersecting the other two planes in lines parallel to the line of intersection to the first two or (c) a plane sharing the line of intersection of the other two planes. In case (iii) the third equation may represent (a) a plane parallel to the first two or (b) a plane intersecting each of the other planes in a different line, the two lines of intersection being parallel to each other or (c) a plane identical with one of the first two.

We shall not go into full detail, but the pattern of the matrix $\mathbf{A}$ and of the vector $\mathbf{b}$ is crucial in deciding which case is relevant. If there is no unique solution, i.e., the inverse of $\mathbf{A}$ does not exist, then a function called the *determinant* of $\mathbf{A}$ takes the value zero. When $n = 2$ the determinant is the quantity $a_{11}a_{22} - a_{12}a_{21}$ we met above. It had to be nonzero if an inverse existed. The determinant of $\mathbf{A}$ when $n = 3$ takes the more complicated form

$$a_{11}(a_{22}a_{33} - a_{32}a_{23}) - a_{12}(a_{21}a_{33} - a_{31}a_{23}) + a_{13}(a_{21}a_{32} - a_{31}a_{22}). \quad (13.11)$$

The determinant (13.11) will take the value zero if the second and third row elements of the matrix are proportional to one another, e.g., if for a non zero constant k, $a_{31} = ka_{21}$ and $a_{32} = ka_{22}$ and $a_{33} = ka_{23}$. This may be verified by substituting these values in (13.11).

We do not prove it, but if the elements of any pair of rows (or columns) in a square $n \times n$ matrix are proportional, then the determinant of that matrix will take the value zero, and the matrix will have no inverse. A matrix that has no inverse is called a *singular* matrix.

13.4 Too Many or Too Few Equations

When there are n unknowns or variables we need n equations that give rise to a non-singular coefficient matrix **A** for there to be a unique solution. There are many situations that arise naturally when we have either fewer or more than n equations.

When there are fewer than n equations we cannot obtain a unique solution. That situation is a generalization of the case of one equation in two unknowns or variables. The equation $3x + 2y = 7$ has no unique solution, In terms of Cartesian coordinates, the coordinates of any of the infinity of points that lie on the line determined by that equation satisfy it. For more than two variables there will in general be more than one solution satisfying the given equations if the number of equations is less than the number of unknowns, or in some cases there may be no solution.

Too many equations?

If there are more than n linear equations in exactly n unknowns the situation is more complicated. When there are $m > n$ equations, the $m \times n$ matrix of coefficients **A** provides some information about possible solutions. It does this through a property known as the *rank* of a matrix. Recalling that determinants are a property of a square matrix, the rank of a matrix **A** is said to be r if the largest non-zero determinant that can be formed by striking out some of the rows and columns is associated with a square matrix having r rows and columns. If $r = n$, the number of variables, the set of equations may have a unique solution, or they may be inconsistent.

For example, if $n = 2$ and we have three equations

$$\begin{aligned} x + 2y &= 7 \\ x + 6y &= 19 \\ 2x - 3y &= -7 \end{aligned}$$

it is easily verified that the rank of the coefficient matrix

$$\begin{bmatrix} 1 & 2 \\ 1 & 6 \\ 2 & -3 \end{bmatrix}$$

is $r = 2$ since, for example, the matrix

$$\begin{bmatrix} 1 & 2 \\ 1 & 6 \end{bmatrix}$$

has a nonzero determinant.

Here the equations are consistent. The solution of any pair is easily found to be $x = 1, y = 3$, and these values satisfy all three equations.

On the other hand, if we replace the third equation by

$$2x - 3y = -8$$

the equations are no longer consistent. The first pair have the solution $x = 1$, $y = 3$ as before, but the first and third now have solutions $x = 5/7$, $y = 22/7$, while the second and third have solutions $x = 3/5$, $y = 46/15$. The solutions differ slightly from pair to pair. If we approximate by rounding each solution to he nearest whole number they would indeed agree, but then not all three equations would hold exactly.

Situations where we have more linear equations than there are unknowns arise in disciplines ranging from astronomy and surveying to statistics, economics and psychology, as well as the life sciences and medicine. In many applications where there are n unknowns and more than n equations, different sets of n equations give nearly, but not quite, the same solutions. How such equations arise, and the practical implications, are discussed in the next chapter.

13.5 A Geometric Interpretation of a Vector

For two variables we may associate a row vector (x, y) with a point P, say, in a Cartesian coordinate system. If O is the origin, these coordinates uniquely specify the length and direction of a line OP. By Pythagoras's theorem the length of OP is the square root of $(x^2 + y^2)$. By convention this is positive. If P has co-ordinates (x, y) and Q has co-ordinates $(-x, -y)$, then OP and OQ represent vectors having the same length but the opposite direction. This is clear from Fig. 13.1.

Fig. 13.2 indicates that the addition rule for vectors is such that the sum of any two vectors OP and OQ is the vector OR. This is the diagonal of a parallelogram with adjacent side lengths OP, OQ . This follows because OQ and PR are equal in length and have the same directions. In vector terms they are simply displacements of one another.

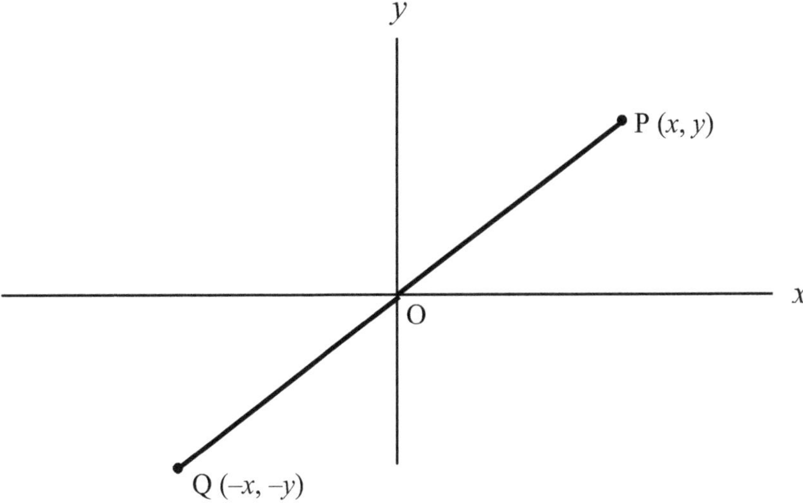

Fig. 13.1. The vectors OP and OQ are equal in length but have opposite directions

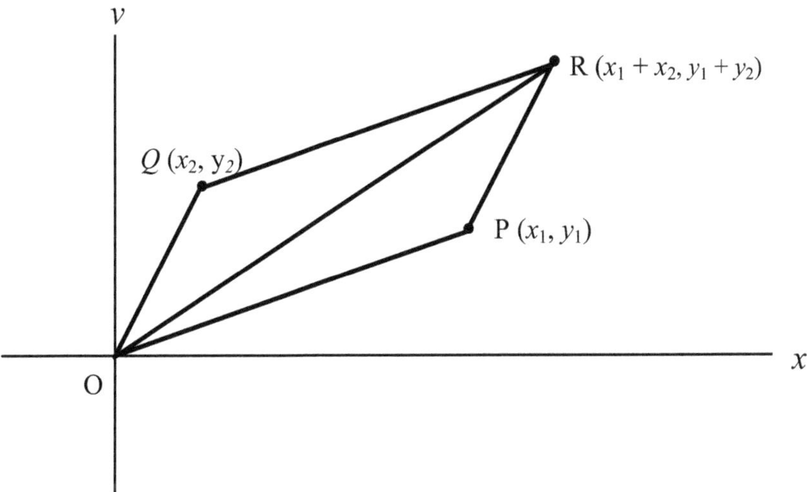

Fig. 13.2. Parallelogram rule of vector addition. The sum of two vectors OP, OQ is the diagonal vector OR of the parallelogram OPRQ

This is called the *parallelogram rule of addition* for vectors. For a two-dimensional vector (x, y), x is referred to as the first component and y as the second component.

There is a fairly obvious extension of the additional rule to three dimensions. Here the parallelogram is replaced by a parallelepiped. This is a three dimensional polyhedron with six faces, opposite faces being parallel planes.

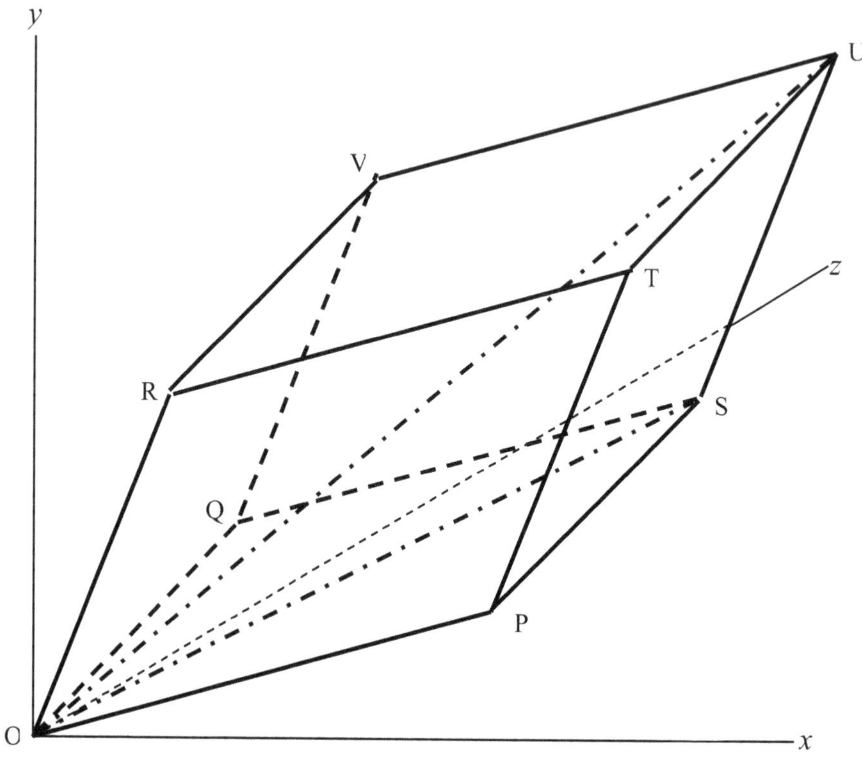

Fig. 13.3. The vector OU is the sum of the three vectors OP, OQ, OR

From Fig. 13.3 we see that the addition rule for vectors implies that in three dimensions the sum of the vectors OP, or (x_1, y_1, z_1), and OQ, or (x_2, y_2, z_2), is given by the vector OS, or $(x_1 + x_2, y_1 + y_2, z_1 + z_2)$. If we add OR, or (x_3, y_3, z_3) to OS this gives the sum of the three vectors OP, OQ, OR as OU (since SU $=$ OR) where OU $= (x_1 + x_2 + x_3, y_1 + y_2 + y_3, z_1 + z_2 + z_3)$. Thus the sum of three vectors is given by a diagonal joining the origin to the opposite vertex of the parallelepiped.

The concept of vectors and the addition of vectors extends to any number of dimensions, although we can no longer give the simple diagrammatic representation that we have in 2 or 3 dimensions.

The association of a length with a direction is important in many physical contexts. We gave velocity as an example in Sect. 12.8. Velocity implies a rate of motion (speed) in a particular direction, as opposed to the same rate in the opposite direction. Some further comments will be found under the heading *vectors* in Sect. 25.2.

14

Observational data

14.1 Data and Yet More Data

We only need a ruler, or some more sophisticated instrument, to measure to some required degree of accuracy the length of a piece of string, or the length or width of a sheet of paper, or a table top. We would only repeat the measurement if we wanted to be sure we had not read or recorded the result incorrectly.

To measure the angle subtended by two stars at some point on the earth's surface is more difficult. Even with a sophisticated instrument there are problems caused by physical phenomena such as diffraction of light as it passes through the earth's atmosphere. If observations are made on board a ship on a rough sea, the vessel's motion must also be taken into account. If several people take observations at or about the same place and time. each using the same or different instruments, the readings are unlikely to be identical even if all observers are competent and careful, and the instruments are the best available.

During the last 250 years there have been steady advances in interpreting and using data that do not fit neatly into a pattern one would regard as ideal. Seamen have long known the value of measuring the relative positions of the sun and stars for determining latitude, and perhaps also longitude (but more about this in Sect. 24.7). Reliable results depend upon measuring angles between heavenly bodies, and using these measurements in appropriate formulae.

In practice, measurements made by astronomers or mariners are usually of limited accuracy for the reasons indicated above. Plugging such data into mathematical formulae based on idealized conditions to calculate a ship's position gives unreliable estimates.

Navigators often want to combine information from repeated or independent observations subject to error in a way that gives a 'best possible' estimate of true position. That is, in a way that is some sense reduces error to a minimum.

Examples of early attempts at combining data like these for astronomic measurements are given by Farebrother [20] (Chaps. 2–4).

It's sometimes sensible to repeat yourself

Errors due to limited accuracy can be allowed for, in part at least, by repeated measurements, but only if it is reasonable to assume that in these repetitions the errors tend to 'cancel' out in some way.

In this chapter we look at several ways of handling measurements subject to error. Sometimes there is a link to the situation mentioned briefly in Sect. 13.4 when we had more linear equations than there were unknowns, and not all those equations were entirely consistent.

Uncontrolled factors causing variations between measurements are commonly met in fields like biology, medical research and the social sciences. These differences are not errors like those in the astronomical example above. For instance, we may visualize and be interested in an ideal or typical height for boys of a given age, but few individuals of that age will have exactly that height. Some will be taller, some will be shorter. A more general example is that where we visualize an idealized relationship between height and age for growing boys. Intuition and prior experience suggests this should be something like a straight line for boys aged between about 6 and 12 years. Such intuitive ideas come from observing and recording heights and ages of many children and plotting these on a graph. The data points often look to lie more or less, but not exactly, on a straight line.

In this context the 'idealized' line suggested by a graphical plot is a hypothetical relationship that might hold if it were not for genetic and environmental factors that produce obervations showing departures from that line.

14.2 Measures that are Typical

We first consider sets of measurements of only one characteristic such as the angle subtended by two stars at some fixed point on the earth's surface, or the weights of a number of men aged 40, or the weekly wages of employees in a particular company.

To measure the angle θ, say, subtended by two stars at a given time at some fixed point we would only need one observation if we had a perfect instrument and could eliminate all sources of error. In practice, with errors present, we might take a sequence of n measurements $x_1, x_2, x_3, \ldots, x_n$. Each of these departs from the unknown true angle θ by some undetermined amount which we may denote by e_i for measurement $x_i, i = 1, 2, \ldots, n$. We express this as

$$x_i = \theta + e_i \ .$$

An astronomer, or navigator, making such observations wants an estimate of θ that is in some sense as close as possible to the true value.

If there are $n = 7$ observations of the angle (measured in degrees) and these are

$$31, 27, 32, 29, 30, 26, 34$$

a common practice is to estimate θ by the arithmetic mean, $\overline{x}$. This is obtained by substituting these values for the x_i, and setting $n = 7$, in the formula

$$\overline{x} = \frac{1}{n} \sum_{i=1}^{n} x_i \, . \qquad (14.1)$$

Here I have used the sigma notation introduced in Sect. 11.3. Thus

$$\overline{x} = (31 + 27 + 32 + 29 + 30 + 26 + 34)/7 = 29.9 \, .$$

The result is rounded to 1 decimal place. This rounding is sensible since readings are only recorded to the nearest degree. To quote the mean (as we are likely to get when using a pocket calculator or basic computer software) as 29.857 142 86 is misleading. It implies spurious accuracy.

Another quantity to estimate the true value which has some appeal is the *median*. If we have an odd number of observations, say $n = 2m + 1$, and arrange these in ascending order, the median is the $(m + 1)$th largest observation. If only one observation takes this value there are m observations that are less than the median, and m that are greater. The median is the 'middle observation' in that sense. Arranging the above data in ascending order i.e. 26, 27, 29, 30, 31, 32, 34 the median, M, is the fourth ordered observation, i.e. $M = 30$.

For an even number of observations, say $n = 2m$, arranged in increasing order, the convention is that the median is half the sum of the mth largest and the $(m + 1)$th largest. This is a number M such that there are m data values less than M and m data values greater than M. Thus, the median of the numbers 1, 2, 7, 21, 25, 31 is $(7 + 21)/2 = 14$. Minor modifications to the above statements are needed if there are 'tied' values for some of the data that correspond to the median, but these are not relevant in this example.

A less widely used estimate of a quantity θ is half the sum of the smallest and largest observation. It is sometimes called the *mid-range* value. For the angle data above this gives an estimate $(26 + 34)/2 = 30$. In this example the mean, median and midrange – 29.9, 30, 30 – are close to one another, but this is not always the situation.

In the angular data, suppose we replace the observation 34 by 42. As an exercise you should confirm that the mean would then be 31, the median would remain at 30, but the mid-range value would be 34. It might be argued that a value 42 is so out of line with the other observations that there must be some unusual error, and therefore that observation should be rejected.

However, there are situations where extreme observations may be correct. For example, the weekly salaries in Euros for seven employees of a small business in Germany might be 510, 530, 554, 670, 681, 693 and 1320. The one high salary might be paid to someone with a special skill. It is easy to check that the mean is now 708.3, the median is 670, and the mid-range is 915. Each differs appreciably from the other two. Here we are not dealing with errors, but seek a measure which gives the best picture of a 'typical' employee salary. Except for the high 1320, most salaries are fairly close to the median. The mean is less typical, in the sense that only salaries of 670, 681 and 693 are near to it. Overall six employees earn less, and only one earns more, than the mean. The mid-range does not give a meaningful picture of a typical salary; as with the mean, only one person earns more than its value, 915. Further, 915 is not close to any individual salary.

14.3 Some Key Properties

Mean or median – which is the better?

What do the alternative choices for a typical value imply mathematically? Let the symbol T represent the 'typicality' measure we are interested in — mean, median, or midrange — for a set of observations $x_1, x_2, x_3, \ldots, x_n$. We may express any observed x_i in a form

$$x_i = T + e_i \ ,$$

where e_i measures the departure of that ith observation from the chosen typical value.

How typical is typical?

How should we decide how ''typical' T is of the observations? One way is to find a value that minimizes some aspect of the departures. For a sensible typical value some departures will be positive, and others negative. We shall not complicate the issue here by considering at this stage statistical considerations involving what are called *bias* or *correlation*. Either, or both, of these may be important in practice and may invalidate the broad assumption that positive and negative departures tend to cancel out.

It seems sensible to choose T so that the sum of the absolute values of the departures (i.e. their magnitudes ignoring signs) is as small as possible. That is, choose T to minimize

$$\sum_{i=1}^{n} |e_i|.$$

A slightly less obvious choice, but one with many practical merits, is to choose T to minimize the sum of the squares of the e_i. i.e., to minimize

$$\sum_{i=1}^{n} e_i^2.$$

Another possibility is to choose T to make the e_i of greatest magnitude as small as possible.

Because it is the most widely used, we consider first the arithmetic mean, denoted by $\overline{x}$. That is, we set $T = \overline{x}$. We may show that this minimizes the sum of the associated e_i^2. Let T' denote *any* typical measure. We now show that $\sum_{i=1}^{n} e_i^2$ is a minimum if $T' = T = \overline{x}$. If you are not used to working with the sigma notation for a sum, you may prefer to check the result by following through the argument setting, say, $n = 3$ and writing out the steps in full with just three observations x_1, x_2, x_3. For any T'

$$\begin{aligned}\sum_{i=1}^{n} e_i^2 &= \sum_{i=1}^{n} (x_i - T')^2 = \sum_{i=1}^{n} (x_i - \overline{x} + \overline{x} - T')^2 \\ &= \sum_{i=1}^{n} (x_i - \overline{x})^2 + \sum_{i=1}^{n} (\overline{x} - T')^2 + 2\sum_{i=1}^{n} (x_i - \overline{x})(\overline{x} - T')\,.\end{aligned}$$

In the final term, i.e., $2\sum_{i=1}^{n}(x_i - \overline{x})(\overline{x} - T')$, the factor $\overline{x} - T'$ is a constant. We may take it outside the summation sign and write that term as

$$2(\overline{x}-T')\sum_{i=1}^{n}(x_i-\overline{x}) = 2(\overline{x}-T')\left[x_1 + x_2 + \ldots + x_n - \frac{n(x_1 + x_2 + \ldots + x_n)}{n}\right].$$

This equals zero, since the term in the square bracket is zero. Thus,

$$\sum_{i=1}^{n} e_i^2 = \sum_{i=1}^{n} (x_i - \overline{x})^2 + \sum_{i=1}^{n} (\overline{x} - T')^2\,.$$

This is a minimum when $T' = \overline{x}$, because the last term is then zero, and since a sum of squares of real numbers must always be non-negative, it is a minimum if it is zero. Thus, the non-zero sum of squares of departures is a minimum when $T' = T = \overline{x}$ is the chosen 'typical' measure.

This result can be obtained more directly using the differential calculus, but one needs to think carefully about notation. In Chap. 11 an independent variable was denoted by x, but here the subscripted x are fixed data values and the quantity that varies is T' . Writing $e_i = x_i - T'$ our problem is to find the value of T' that minimizes the function $f(T')$ given by

$$f(T') = \sum_{i=1}^{n} e_i^2 = \sum_{i=1}^{n} (x_i - T')^2.$$

To find the minimum of this function we differentiate it with respect to T', set the derivative equal to zero, then solve the resulting equation for T'.

If you only met the differential calculus when reading Chap. 11 you may find this difficult, although the rules needed were outlined there. They are those for differentiating a sum and for differentiating a function of a function. If you have difficulty following the argument below, please, at this stage, take it for granted. The derivative $f'(T')$ with respect to T' is

$$f'(T') = -2 \sum_{i=1}^{n} (x_i - T').$$

The value of T' for a minimum of $f(T')$ is obtained by equating this derivative to zero, then solving the resulting equation. Once one is used to the sigma notation, it is easy to see that the solution is $T' = \overline{x}$. It remains only to show this is a minimum. As pointed out in Sect. 11.7 if the derivative changes sign $-, 0, +$ as T' increases through $\overline{x}$ this implies a minimum. Careful consideration should convince you this is what happens here.

If we had another data set for angles in the numerical example in the last section, we would almost certainly get an $\overline{x}$ that differed from that for the original data. Hopefully, if the measurements were all of comparable accuracy, the differences between the two means would be small. One reason for finding the mean of a set of data (called the sample) is to give some idea of what might be the true mean of all measurements of that type that we could conceivably make. This conceivable set might be all possible measurements that have been, or could be, made of the given angle under similar conditions. Statisticians call this set a *population*.

Interpreting how well a characteristic of the sample like the mean (which we can calculate) reflects the corresponding population characteristic (which in general we don't know) is one aim of *statistical inference*, a topic we discuss in Chap. 18.

The sum of the absolute deviations, i.e., $\sum_{i=1}^{n} |e_i|$, is minimized if T is the median. We do not prove this formally, but we indicate how the argument develops for the simplest meaningful case where $n = 3$, when the observations all differ, and are in increasing order x_1, x_2, x_3. Although it is not obvious, the argument can be extended using tedious and careful reasoning to all values of n by mathematical induction or other means. For three observations, the median is the middle value x_2. Suppose any typical measure is T', and for illustrative purposes that T' is not the median, but falls between x_1 and x_2. The situation is shown in Fig. 14.1, where we denote the absolute (i.e. nonnegative) deviation of x_i from the median x_2 by $|e_i|$, and that of x_i from T' by $|f_i|$. In particular $|e_2| = 0$, since x_2 is the median.

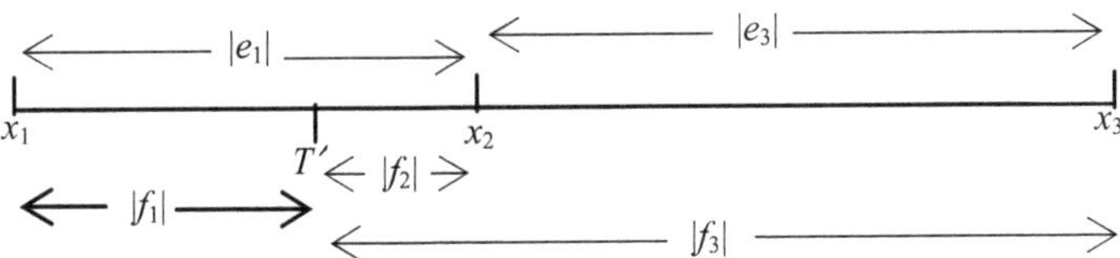

Fig. 14.1. Comparison of deviations from median x_2 with those from a point T'.

Inspecting Fig. 14.1 it is fairly easy to see that

$$|e_1| + |e_2| + |e_3| = |f_1| + |f_3|. \tag{14.2}$$

This must be less than $|f_1| + |f_2| + |f_3|$, since $|f_2| > 0$. It is not difficult to show that a similar argument holds no matter where we position T' other than at the median.

That the mid-range minimises the magnitude of the largest deviation is evident because such deviations must be associated with the most extreme values. The mid-range makes both largest deviations equal in magnitude. Any other choice of T' would give one equivalent deviation smaller in magnitude, and one larger in magnitude, than those associated with the mid-range.

The numerical example for angles indicated that sometimes the three measures — mean, median, midrange — have similar values. The example for salaries showed that they may differ markedly. In particular, the differences may be large if one or two observations depart noticeably from all the others. This need not always be the case. In the data set 21, 67, 68, 69, 71, 72, 73, 74, 119 both 21 and 119 are far removed from the rest of the data, which are confined to the range 67 to 74. The mean, median and midrange are similar, being respectively 70.44, 71 and 70. Can you see what is the essential difference between the situation here and that for the salary data?

The mean is the most widely used measure of a typical or central value. That it minimizes the sum of squares of the deviations of observations from it has important mathematical consequences in the context of statistical inference. However, it has the disadvantage, as in the salary data, that sometimes it is strongly influenced by just one or two observations well away from the rest of the data.

On the other hand the median depends only on the central observation, or on a pair of observations that are central, in the sense that it has the same number of observations above and below these values. This is why it is little influenced by a 'wild' or 'outlying' observation. Because of this property the median is described as a *robust* measure of a typical value. The concept of robustness implies that wild or extreme values have little influence.

The mid-range in most circumstances has the least to commend it. It is sometimes strongly, but as our numerical examples showed, not always, influenced by outlying values. As it ignores 'middle of the road' values, this seems a not very sensible approach, except in the special situation where it is important to minimize the maximum $|e_i|$. A procedure than minimizes a maximum is sometimes called a *minimax* procedure. What do you think is meant if a procedure is described as a *maximin* procedure?

14.4 Points that Almost Lie on a Line

We indicated in Sect. 14.1 that astronomers and surveyors often want to find the best fitting relationship between two or more variables when they are given a set of observations. Similar problems also arise widely in statistics and other branches of twenty-first century mathematics.

The simplest case is when we have data for n observed points with coordinates (x_i, y_i) that lie close to, but not all exactly on, a straight line. In Sect. 14.1 we gave as an example the relation between heights (y) and ages (x) of boys in the age range 6-12 years.

Here is another example. Table 14.1 gives the outside mean air temperature (°C) and electricity consumption (kW h) for a house heated solely by that method on 9 randomly chosen days during one winter.

Table 14.1. Electricity consumption (y kW h) for heating a house at temperatures (x°C)

Temperature x	0.2	1.4	2.3	1.8	4.1	7.3	5.9	7.3	9.4
Electricity consumption y	154	143	135	139	117	83	98	79	66

These data are plotted in Fig. 14.2. The graph differs from those used in Chap. 7 to represent points in a Cartesian coordinate system. First, scales on the x and y axes differ and second, a break has been introduced in the y-axis which effectively implies we omit y values less than about 60. This is a sensible dodge. It makes best use of the area to highlight features of interest. It shows clearly how the points appear in relation to each other, with a strong suggestion that they lie almost, but not quite, on a straight line. A disadvantage is that we may get a false impression of the slope of the line that gives an appropriate fit, or accidently misjudge the point at which such a line would cut the x-axis.

Given the data in Table 14.1, how do we obtain a line that, in some sense, provides a best fit to the data? In a more general context, if we have n observations (x_i, y_i) that lie close to some straight line, and believe that ideally, they would lie on such a line, how do we *estimate* the equation to that line?

In a broader context departures from the ideal situation may reflect measurement error, genetic variation, economic fluctuations, or some other factor relevant to the data at hand. We write the unknown 'ideal' equation as

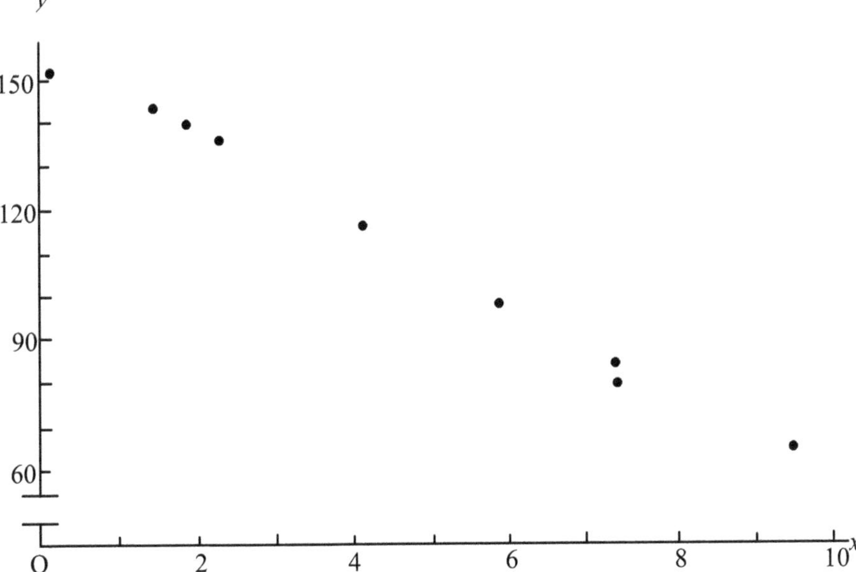

Fig. 14.2. Plot of energy consumption y against temperature x

$$y = a + bx \; . \tag{14.3}$$

The problem is to find appropriate estimates of the unknown coefficients a, b.

If all observations took 'ideal' values, any two separate points, say (x_i, y_i) and (x_j, y_j), would lie exactly on the line and we would have no problem. Substituting these values in (14.3) would give two equations in two unknowns. Here a and b are the unknowns, rather than the conventional x and y. These equations would be

$$y_i = a + bx_i \; , \tag{14.4}$$

$$y_j = a + bx_j \; . \tag{14.5}$$

Solving (do this as an exercise), we find

$$\hat{b} = \frac{y_j - y_i}{x_j - x_i} \; , \tag{14.6}$$

and

$$\hat{a} = \frac{y_i x_j - y_j x_i}{x_j - x_i} \; , \tag{14.7}$$

where we have replaced a, b by $\hat{a}, \hat{b}$. This is to indicate that these are the values of a, b that are solutions to (14.4) and (14.5). as distinct from the unknown a, b in (14.3).

It is not hard to verify that we may also write

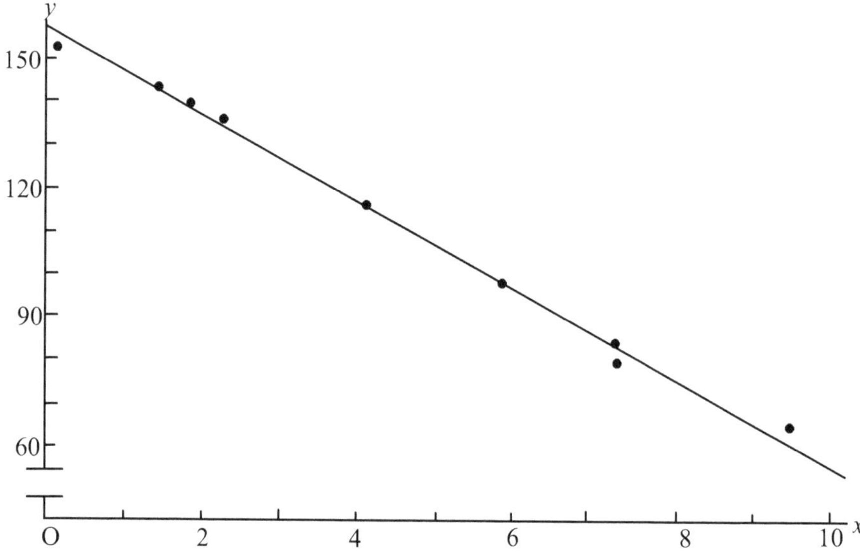

Fig. 14.3. The line passing through the data points (1.4, 143), (4.1, 11.7)

$$\hat{a} = y_i - \hat{b}x_i \text{ or } \hat{a} = y_j - \hat{b}x_j.$$

If we consider the two points (1.4, 143) and (4.1, 117) occurring in Table 14.1, you should check that (14.6) and (14.7) give

$$\hat{b} = -9.630 \text{ and } \hat{a} = 156.48.$$

The equation to the line passing through the points (1.4, 143) and (4.1, 117) is therefore

$$y = 156.48 - 9.630x \qquad (14.8)$$

Fig. 14.3 is a reproduction of Fig. 14.2 with this line superimposed.

Any rational person would agree that (14.8) gives a sensible fit to all data points. It would not seem unreasonable to use this line, for example, to estimate the energy consumption at a temperature of 3°C. Setting $x = 3$ in (14.7) gives $y \approx 127.6$, implying an expected energy consumption of 127.6 kW h. We say 'expected' because it is what we expect to be consumed *if* the line gives an idealized consumption rate.

Why choose the points (1.4, 143), (4.1, 11.7) to estimate our ideal line? We have no reason to assume these are ideal values rather than some other pair. Had we instead chosen from Table 14.1 the points (7.3, 79) and (9.4, 66), the line we would have obtained is that shown in Fig. 14.4. We would be unenthusiastic about this as an adequate 'ideal' line fitted to the data.

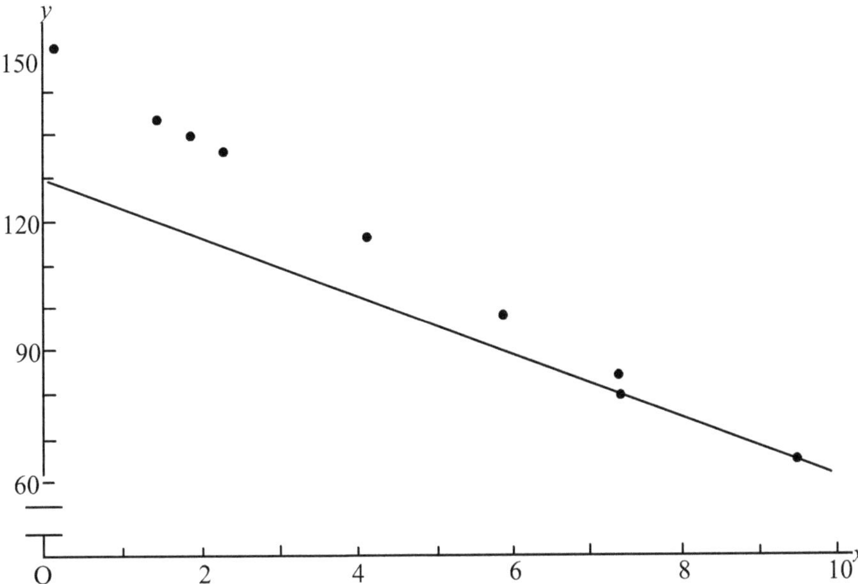

Fig. 14.4. The line passing through the data points (7.3, 79), (9.4, 66)

Some eighteenth and nineteenth century mathematicians thought out a neat but simple solution to the problem that some choices of paired points give a more appropriate line than others. They argued that since we could choose any pair of points and fit a line through them, why not do this for all possible pairs, then calculate the mean of all the resulting $\hat{a}$ and $\hat{b}$ to give an ideal line? This is simply applying the idea of taking a mean, but instead of doing this for the original observations, we do it using numbers derived from those observations.

Birth of an intuitively sound idea

This is neat, but there are snags. For a start, intuition tells us that if all departures from the ideal line due to measurement error, genetic variation, or whatever are of broadly the same magnitude, then we should get better estimates of b, the unknown ideal slope, using pairs of points relatively far apart. Intuitively, we feel this should be taken into account by giving less weight to points close to each other. The need to do something like this is illustrated in our example by the points (7.3, 83) and (7.3, 79). Choosing these, the estimate $\hat{b}$ given by (14.6) is then $-4/0$. Remembering that division by 0 is not defined, we can only note that taking a limit as the denominator $x_j - x_i$ approaches zero, this tends to infinity. The line through these points is easily seen, if we visualise it drawn in Fig. 14.2, to pass vertically through the chosen points. It is a line parallel to the y-axis. Including the slope of this line in our pairs used to compute the mean leads to an 'infinite' estimate for the slope. Any sensible method would leave out such pairs, but even pairs with x values close to each other may give unreliable estimates of b.

We saw in Sect. 14.2 that in such circumstances the mean may be unreliable as a measure of a typical value. Some modern methods of estimating b use the median of the $\hat{b}$ for all possible pairs, but a more widely used procedure is that of least squares.

The rationale is that we specify the ideal relationship between x and y to have the form (14.3), i.e.,

$$y = a + bx$$

where a, b are, of course, unknown. We now assume that each of our datum pairs $(x_i, y_i), i = 1, 2, \ldots, n$ satisfy a relationship

$$y_i = a + bx_i + e_i \ ,$$

where e_i is an unknown value that is in general different for each i. It is reasonable to hope that for the ideal line these differences will tend largely to cancel one another out, some being positive and some negative.

We may look at this as an extension of the model $x_i = m + e_i$ in Sect. 14.3 with x_i replaced by y_i and m replaced by $a + bx_i$. If the ideal line were that shown in Fig. 14.5, where three observed points are designated $(x_1, y_1), (x_2, y_2), (x_3, y_3)$, then the e_i are easily seen to be the departures from the ideal line measured parallel to the y-axis. For points below the line these departures have a negative value. For those above, they are positive.

Writing $e_i = y_i - a - bx_i$, $i = 1, 2, \ldots, n$ and remembering that we do not know a, b, or the e_i, the principle of least squares is to determine values for a, b that minimize the sum of squares of the e_i. That is, that minimize

$$U = \sum_{i=1}^{n} e_i^2 = \sum_{i=1}^{n} (y_i - a - bx_i)^2 \ .$$

Remembering that the unknowns are a and b, we may find the values of these that minimize U by first finding any minima of U considered now as a function of both a and b. This requires calculation of partial derivatives as described in Sect. 11.10. Basically, it calls for differentiation of U with respect to both a when we regard b as fixed, and with respect to b when we regard a as fixed. We then equate each of these partial derivatives to zero and solve the resulting equations for a and b. The rules we gave in Sect. 11.10 suffice for carrying out this differentiation, but until you have some practice at the

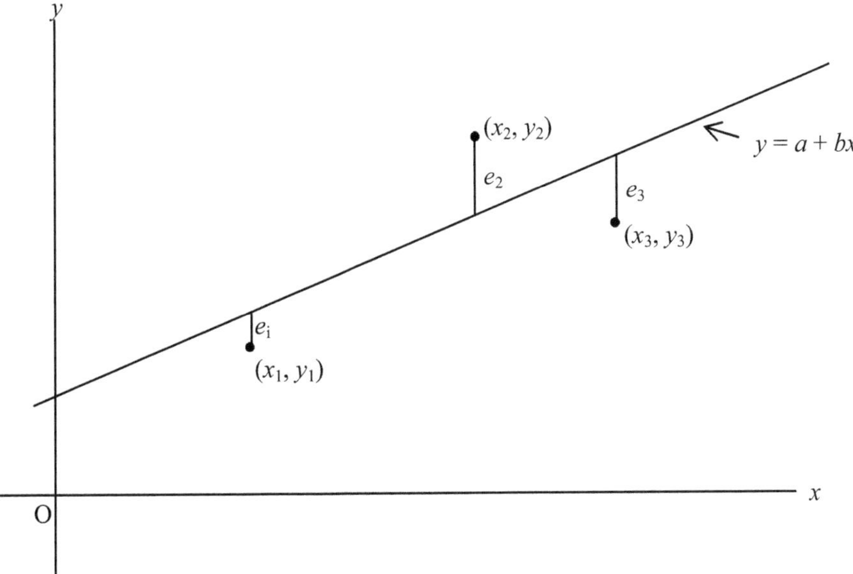

Fig. 14.5. The least squares method minimizes the sum of squares of the e_i

calculus you may find this tricky. If so, please take the following derivation on trust.

We write the partial derivatives as $\partial U/\partial a$ and $\partial U/\partial b$. Remember that for partial differentiation with respective to a, we hold b constant. Similarly, in partial differentiation with respective to b, we hold a constant. Thus

$$\frac{\partial U}{\partial a} = -2\sum_{i=1}^{n}(y_i - a - bx_i) \tag{14.9}$$

and

$$\frac{\partial U}{\partial b} = -2\sum_{i=1}^{n}x_i(y_i - a - bx_i)\ . \tag{14.10}$$

Equating each derivative to zero, and solving for a and b (14.9) gives

$$\sum_{i=1}^{n}y_i - b\sum_{i=1}^{n}x_1 = na\ , \tag{14.11}$$

and remembering that the respective means are

$$\bar{y} = \sum_{i=1}^{n}\frac{y_i}{n} \text{ and } \bar{x} = \sum_{i=1}^{n}\frac{x_i}{n}$$

we may write (14.11) as

$$a = \bar{y} - b\bar{x} . \tag{14.12}$$

Substituting this value for a in (14.10) after equating to zero, followed by some algebraic manipulation not detailed here, we find

$$b = \frac{\sum_{i=1}^{n}(x_i - \bar{x})(y_i - \bar{y})}{\sum_{i=1}^{n}(x_i - \bar{x})^2} . \tag{14.13}$$

You should carry out the algebra to confirm this result.

We have not established that these give a minimum. To do so is a piece of standard differential calculus involving two independent variables which we have not described, so please take this on trust. To indicate that these are the specific values of a, b that minimize the sums of squares of the e_i it is usual to write these minimizing values as $\hat{a}, \hat{b}$.

Any general statistical software package worthy of the name will include a program to calculate $\hat{a}, \hat{b}$ for any set of n values of (x_i, y_i). Even without such software a pocket calculator makes computation reasonably easy. Many calculators having a special facility for calculating both means and sums of squares. For small data sets even a calculator that only adds, subtracts, multiplies and divides makes it not too hard to calculate $\hat{a}, \hat{b}$.

A simple dodge to make calculations easier

The calculations are easier if (14.13) is put in another form which can be obtained by straightforward but tedious algebra. It is

$$\hat{b} = \frac{\sum_{i=1}^{n} x_i y_i - \left(\sum_{i=1}^{n} x_i\right)\left(\sum_{i=1}^{n} y_i\right)/n}{\sum_{i=1}^{n} x_i^2 - \left(\sum_{i=1}^{n} x_i\right)^2/n} . \tag{14.14}$$

This looks more complicated than (14.13), but when expressed in words it shows a pattern that makes computation easy. In words (14.14) is

$$\hat{b} = \frac{\text{sum of products of } x_i \text{ and } y_i - (\text{sum of } x_i) \times (\text{sum of } y_i)/n}{\text{sum of squares of } x_i - (\text{sum of } x_i)^2/n} ,$$

where n is the number of pairs (x_i, y_i).

Using a pocket calculator it is not necessary to record on paper every detail of the calculation of a sum or sum of squares or a sum of products. However, for completeness we show in Table 14.2 each relevant term that has to be calculated for the data in Table 14.1. Try using a pocket calculator to check the relevant computation.

Rembering that $n = 9$, substituting the relevant values shown in Table 14.2 in (14.14) gives

Table 14.2. Computation by least squares of a line of best fit to the data in Table 14.1

x	y	xy	x^2
0.2	154	30.8	0.04
1.4	143	200.2	1.96
2.3	135	310.5	5.29
1.8	139	250.2	3.24
4.1	117	479.7	16.81
7.3	83	605.9	53.29
5.9	98	578.2	34.81
7.3	79	576.7	53.29
9.4	66	620.4	88.36
$\sum x = 39.7$	$\sum y = 1014$	$\sum xy = 3652.6$	$\sum x^2 = 257.09$

$$\hat{b} = \frac{3652.6 - (39.7 \times 1014)/9}{257.09 - (39.7)^2/9} = -10.0070 \, ,$$

and from (14.12) we easily confirm that

$$\hat{a} = 1014/9 - (-10.0070) \times 39.7/9 = 156.809.$$

Thus the equation to the least squares line of best fit is

$$y = 156.809 - 10.007x \tag{14.15}$$

Any standard statistical software regression program, where one merely has to read in the data and click the mouse a couple of times, should confirm this. Most such *regression programs* also give a wealth of other information valid for making statistical inferences (Chap. 18). Plot the data and the line (14.15) on a graph to see for yourself that it gives a sensible fit.

We may view the situation here this way. We are given $n > 2$ data points (x_i, y_i), which, if they all lay on one straight line, would each satisfy one of n relationships of the form

$$y_i = a + bx_i \, ,$$

where the unknowns are a, and b. However, in practice the points don't all lie exactly on the same line, so we have the situation described in Sect. 13.4 where we have more equations than we have unknowns, and the equations are not entirely consistent. Least squares gives a solution that tries to smooth out these inconsistencies.

14.5 A Duality

There is an interesting duality between lines and points for exact relationships between variables.

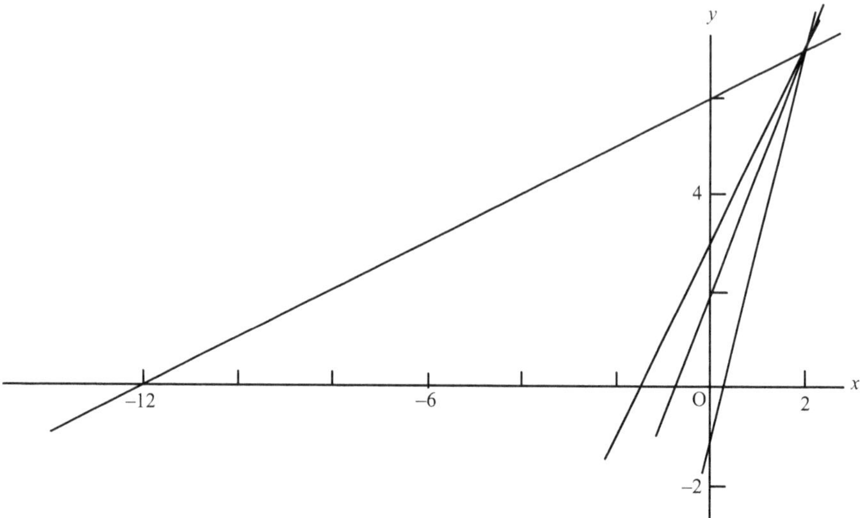

Fig. 14.6. Four consistent equations with common solution $x = 2$, $y = 7$

Consider the 4 equations in 2 unknowns

$$y = 3 + 2x \tag{14.16}$$

$$y = 6 + 0.5x \tag{14.17}$$

$$y = -1 + 4x \tag{14.18}$$

$$y = 2 + 2.5x \tag{14.19}$$

By substitution, we easily verify that $x = 2$ and $y = 7$ satisfies all four equations. They have a common solution and are therefore all consistent. Graphically this solution is represented in Fig. 14.6 as the point with Cartesian coordinates $(2, 7)$ where all lines meet.

The general form of each equation in (14.16) to (14.19) is

$$y = a + bx$$

where the values (a, b) associated with the equations are $(3, 2)$, $(6, 0.5)$, $(-1, 4)$, $(2, 2.5)$. These points are plotted in Fig. 14.7and all lie on a straight line.

Figs. 14.6 and 14.7 suggest that if we write the equations in the form $y = a + bx$ then, if they are all consistent, there is a linear relationship between the coefficients (a, b).

The difficulty facing earlier astronomers and navigators was that due to errors in measurement the situations in Figs. 14.6 and 14.7 did not pertain exactly. In the notation used by the early workers in this field, their erroneous measurements were denoted by (a, b). They sought a common solution (x, y) satisfying more than two equations. A unique solution did not materialize

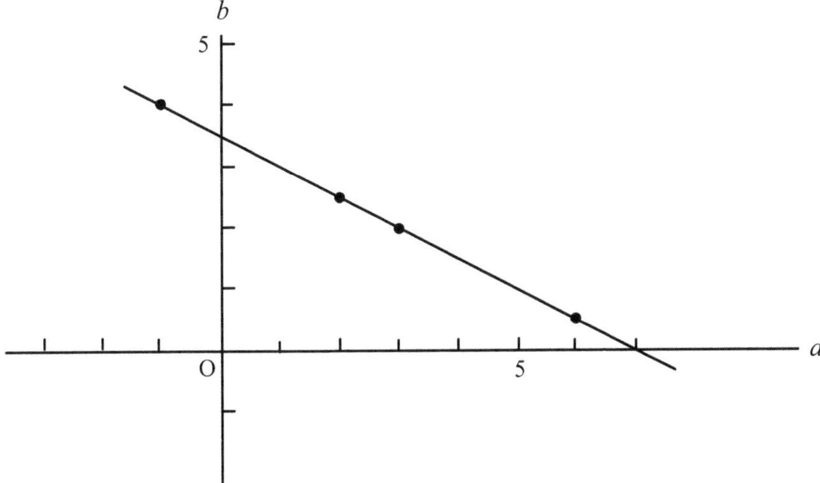

Fig. 14.7. The coefficients a, b of x, y in (14.16) to (14.19) lie on a straight line

because lines representing different possible pairs of equations met in different points. The situation is illustrated by the set of equations

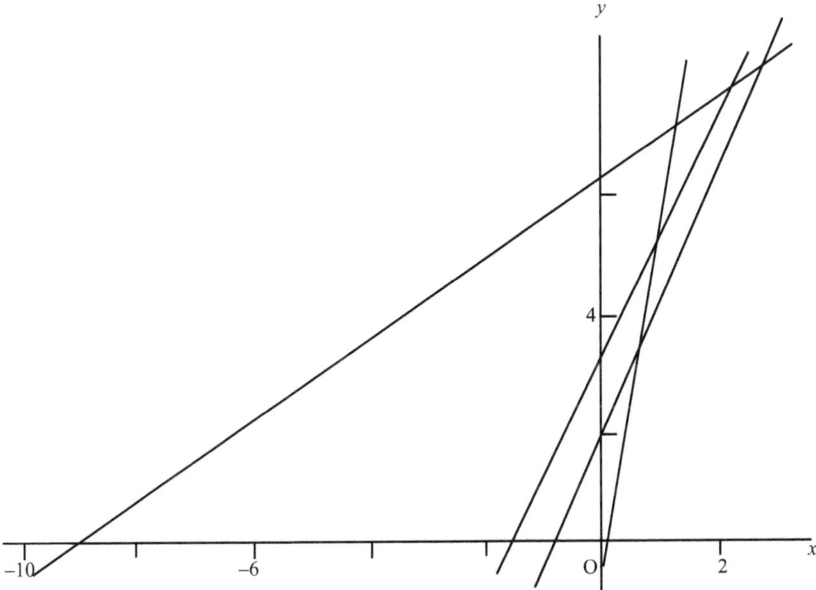

Fig. 14.8. The approximate solution of four equations in two unknowns that are not quite mutually consistent

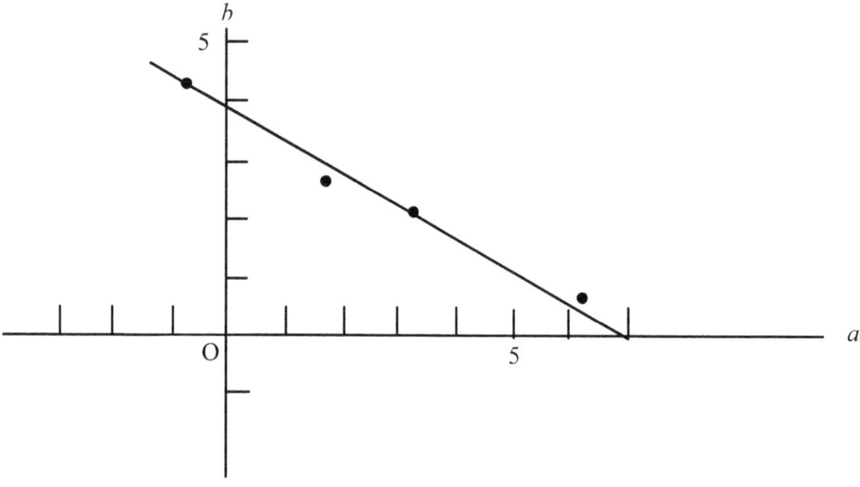

Fig. 14.9. Coefficients a, b of x, y in four relationships that are not quite consistent with each other

$$
\begin{aligned}
y &= 3.2 + 2.1x \\
y &= 6.3 + 0.7x \\
y &= -0.7 + 4.2x \\
y &= 1.8 + 2.7x
\end{aligned}
$$

which have slightly different values of (a, b) from those in the set (14.16) to (14.19). The corresponding lines are shown in Fig. 14.8.

The corresponding (a, b) values are plotted in Fig. 14.9. You can see that they almost, but not quite, lie on a straight line.

The problem of obtaining a best value of some ideal point of intersection (x, y) in this case is exactly equivalent to that considered in the line fitting example in the previous section, except that we have interchanged x, y and a, b. We now seek in some sense the 'best possible' fixed values of x, y given a set of observed paired values of a and b which we may denote by $(a_i, b_i), i = 1, 2, \ldots, n$. Either notation is valid. It is little more than a historical accident that the early astronomers and navigators tended to label their observations a, b, etc., while their present-day counterparts and statisticians tend to label these x, y, etc.

14.6 Loose Ends

Choice of sample data. On p. 220 we remarked that the data for temperature and electricity consumption were recorded for randomly chosen days. This is a statistical requirement. It does not necessarily ensure that least squares

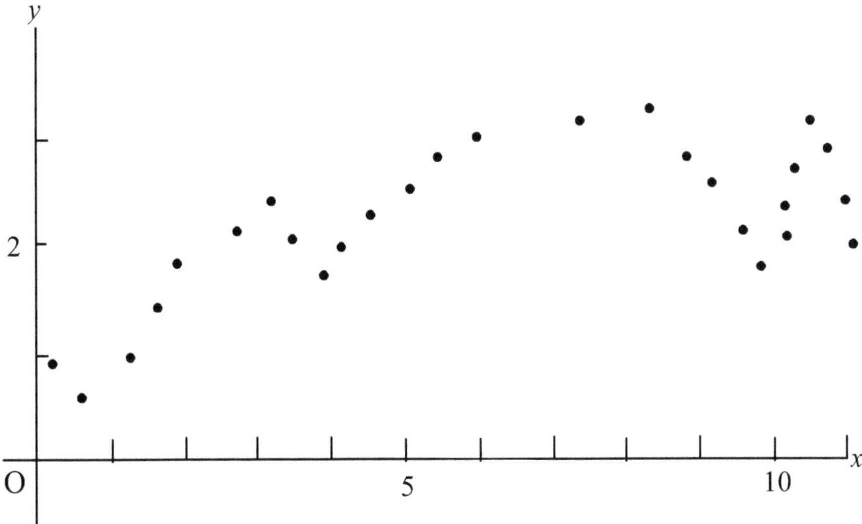

Fig. 14.10. A data set that is clearly not adequately described by a straight line or parabolic fit

is the most appropriate tool for obtaining estimates of a, b from which valid statistical inferences can be made. What it does is make the data more likely to be relevant to what is happening over the winter as a whole than would be the case if we had used data for consecutive days. Questions of statistical independence, a basic but quite sophisticated concept, are involved here. Independence is discussed informally in the context of probability in Chap. 17.

Extensions beyond straight lines. The method of least squares developed for determining an idealized straight line fit to n data points (x_i, y_i) extends to more than two variables. Given n data points corresponding to $p+1$ variables $(x_1, x_2, \ldots, x_p, y)$ statisticians and others often use least squares to estimate coefficients α, β_1, β_2, $\ldots$, β_p in an idealized relationship

$$y = \alpha + \beta_1 x_1 + \beta_2 x_2 + \ldots + \beta_p x_p.$$

Obtaining these estimates is the starting point for the statistical method called *regression analysis*. The straight line fit we have discussed in this chapter is the special case where $p = 1$.

Least squares is also used for fitting curves other than straight lines to data when that is appropriate. The equations that have to be solved to estimate the parameters are often quite complicated.

Research workers often encounter data sets that have only a vaguely defined pattern. Fig. 14.10 represents a graph of such a set.

The ideal curve to reflect the pattern in these data is certainly not a straight line. Least squares may be extended to fit polynomials of varying degree to such data. These polynomials may, with appropriate care, be used to estimate y corresponding to new values of x within the range of observations. There are dangers when using this method if one wants to extrapolate beyond the range of observed x values. This is partly because, unless a_n tends to zero, for values of x of large magnitude, a polynomial rapidly shoots off either towards ∞ or towards $-\infty$. There are often physical reasons connected to the source of the data that indicate this is inappropriate.

How well a polynomial of the form

$$y = a_0 + a_1x + a_2x^2 + \ldots + a_mx^m$$

will fit data like those in Fig. 14.10 is highly dependent on the choice of m. If $m = 1$ we get a straight line, and if $m = 2$ a parabola, neither of which would be adequate for the data in Fig. 14.10.

Splines. No polynomial of degree less than 6 would give a satisfactory fit to the data in Fig. 14.10. In many situations it is not possible to attach any reasonable physical meaning to polynomial coefficients. An alternative way to obtain curves that fit reasonably to such data is to fit curves of lower degree locally. One procedure uses what are called *splines.*

There are several variants of splines, and we shall not attempt a detailed explanation. In essence the range of x values is broken down into a number of smaller intervals. The end point of each interval is called a *knot.* Between any two neighbouring knots a polynomial of some chosen low degree (often a cubic, setting $m = 3$) is fitted locally (i.e. ignoring points outside the interval between the selected neighbouring knots). There is also a restriction that at each knot the polynomials in the adjacent intervals meeting at that knot shall have equal derivatives at that knot.

Regression diagnostics. Another topic in regression that has been a growth area in the last thirty years is that of *regression diagnostics.* In many data sets there may be one or two observations among many that do not comply with the general data pattern. In the one-variable case in Sect. 14.2 we showed that the mean may be strongly influenced by one observation well away from the bulk of the data. In regression analysis the slope and intercept of a fitted line may be influenced strongly by one point that does not follow the general data pattern. Regression diagnostics aim to detect such points, and to make allowance for them in a way that is appropriate to the given situation.

15

Numerical Analysis or Sophisticated Arithmetic

15.1 When Analysis Fails

We have seen how the challenge of a real world problem often leads to dramatic advances.

I have mentioned that users of mathematics often complain that mathematicians are good at telling them whether there is a solution to a problem, but sometimes less good at telling them how to find that solution even when one is known to exist.

For many real-world problems mathematical analysis cannot provide a solution, yet it is obvious that one must exist. We showed in Sect. 11.3 how the integral calculus enables us to find the area bounded by a curve $f(x)$, the x-axis and the ordinates at $x = a$ and $x = b$. To do this we only need to find a function $F(x)$ whose derivative with respect to x is $f(x)$, then the area is $F(b) - F(a)$. This simple statement hides two potential practical difficulties. The first is that if $f(x)$ is a complicated function it may be difficult to find $F(x)$. The second is that there are many, often simple, but important cases for which no elementary or explicit function $F(x)$ exists that has the given $f(x)$ as its derivative.

Statisticians, engineers, astronomers, physicists and others often meet the function

$$y = f(x) = \mathrm{e}^{-x^2/2} .$$

No explicit function of x has this as its derivative. Nevertheless, it is easy to see that there is a finite area beneath this curve between, for example, $x = 0$ and $x = 1$. In situations like this numerical analysis comes to our aid. We return to this problem, with a small modification, in Sect. 15.8.

The term *numerical analysis* was seldom used before World War II, although many techniques used in this branch of mathematics go back to the seventeenth century. Forerunners of the subject went under names like the *calculus of finite differences*, the *calculus of observations*, or *numerical calculus*.

Numerical analysts may accuse me of slighting their subject by describing it as sophisticated arithmetic, but broadly speaking that is what it is, and no slight is intended. The importance of these methods cannot be over-estimated. Space travel would be impossible without contributions from numerical analysts at the design and planning stages.

Tasks tackled by numerical analysts include solving differential equations, especially ones that may not have analytical solutions. Another useful one is approximating to a complicated function by a simpler one that reflects the properties of the complicated one closely enough for practical purposes, and where the latter may be handled more easily, either analytically or numerically.

The subject also is useful to fill gaps in our information. For example, we may not know the algebraic form of a function $y = f(x)$, for which we are given the values of x and y at each of n points. These values may have been obtained by observation or experiment. On the basis of such information we might compute good approximations to values of the function for values of x between those at which we are given a value of y, or find good approximations to the value of the derivative at specified points, or to the integral of the function over a certain range. These are examples of techniques classed as *interpolation*.

15.2 Solving Equations

An early problem tackled by a numerical method was solving equations of the form $f(x) = 0$. Even if $f(x)$ is a polynomial of degree greater than 2, analytic methods are either cumbersome or apply only for special cases. For the equation

$$f(x) = x - 2\sin x = 0\,, \tag{15.1}$$

where x is measured in radians, there is no obvious analytic approach that helps us find a positive value of x that is a solution. A graphical method is likely to be of limited accuracy. When x is measured in radians, it is easy to verify that

$$f(1) = 1 - 2\sin 1 \approx 1 - 2 \times 0.84 = -0.68$$

and that

$$f(2) = 2 - 2\sin 2 \approx 2 - 2 \times 0.91 = 0.18.$$

The function is continuous, so the change in sign implies that there is a solution between $x = 1$ and $x = 2$. Since $f(x)$ is closer to zero when $x = 2$ this suggests a solution is nearer to 2 than it is to 1. This assertion is based on a subconscious use of a principle once used commonly at school for getting values between those recorded in tables of logarithms, trigonometric functions, square roots, etc. Now that such tables have largely given way to calculators the method is less widely used. The principle is that of *linear interpolation*, which we discuss further in Sect. 15.6.

Effectively, when applied to solving (15.1), linear interpolation finds where the straight line joining the points $(1, -0.68)$ and $(2, 0.18)$ cuts the x-axis. This tells us what the solution would be if the graph of $f(x)$ were a straight line for values of x between 1 and 2. This is only a crude approximation to reality. If you find the equation of the straight line and determine the point where it cuts the x-axis, you should find $x \approx 1.79$. If we take this as an approximate solution there are two sources of error. The first is that, as pointed out, a straight line approximation to $f(x)$ between $x = 1$ and $x = 2$ is a crude one. The second is that we have rounded values of $\sin 1$ and $\sin 2$ to 2 decimal places. In numerical analysis we often meet both *approximation errors* (the first kind) and *rounding errors* (the second kind). Assessing the contribution of each is an important part of the subject — some might even say the most important part — but it is a topic with many ramifications, and we only cover it incidentally in this chapter.

Another bright idea from Newton

A long standing method of getting a better solution to equations of the form $f(x) = 0$ when we know the derivative $f'(x)$ was proposed independently by Newton and a contemporary Joseph Raphson (1648–1715). It is known either as *Newton's method* or the *Newton–Raphson method.*

15.3 Newton's Method

In this, as in many numerical methods, we start with an approxmate solution, x_0, to the equation $f(x) = 0$. Fig. 15.1 shows such an approximation represented by the point M for a general function $f(x)$. For (15.1) $x_0 = 2$ would be a possible approximation.

In Fig. 15.1 the value $f(x_0)$ at $x = x_0$ is represented by the point P, and PQ is the tangent to the curve at that point, i.e. PQ is a line with slope $f'(x_0)$. Q is the point of intersection of this tangent with the x-axis. This provides a new approximation, x_1, for the solution.

From the definition of slope, it follows that the tangent of the angle PQM is $f'(x_0)$, the value of the derivative of $f(x)$ at $x = x_0$. Thus, by elementary trigonometry, $\mathrm{QM} = \mathrm{PM}/f'(x_0)$, and since $\mathrm{PM} = f(x_0)$, we have

$$x_1 = x_0 - \frac{f(x_0)}{f'(x_0)} . \qquad (15.2)$$

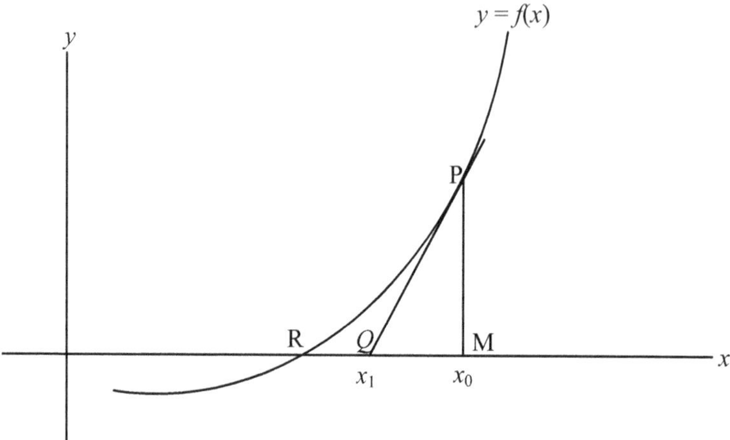

Fig. 15.1. The first step in Newton's method. Q is a better approximation to the root at R than is M

The process is repeated, starting now from x_1, to give (hopefully) a better approximation

$$x_2 = x_1 - \frac{f(x_1)}{f'(x_1)} .$$

It is continued to give further approximations, until the required degree of accuracy is attained. A general step for this 'algorithm' may be written

$$x_{n+1} = x_n - \frac{f(x_n)}{f'(x_n)} . \tag{15.3}$$

The word *algorithm* is used to describe a rule or set of rules in the form of one or more procedures that may be applied automatically without further intervention for a finite number of steps to obtain a solution, or to comply with some stopping rule.

I inserted 'hopefully' beneath (15.2) because the method sometimes runs amok. This may happen if there is a multiple root, or if the slope is close to zero near a root.

We use (15.3) to solve (15.1). At a general step

$$f(x_n) = x_n - 2\sin x_n \text{ and } f'(x_n) = 1 - 2\cos x_n$$

whence

$$x_{n+1} = x_n - \frac{x_n - 2\sin x_n}{1 - 2\cos x_n} = \frac{2\sin x_n - 2x_n \cos x_n}{1 - 2\cos x_n} . \tag{15.4}$$

Setting $n = 0$ and $x_0 = 2$ gives

$$x_1 = (2\sin 2 - 4\cos 2)/(1 - 2\cos 2) = 1.90\ .$$

a result easily confirmed with a calculator, remembering that angles are expressed in radians.

You should verify that if we now set $x_1 = 1.90$ in (15.4) the next approximation is $x_2 = 1.8955$. Then, starting with this value, the next approximation is $x_3 = 1.8954943$. One further iteration gives $x_4 = 1.89549427$. The iterations are converging rapidly, so a solution correct to 8 significant figures is 1.8954943. This is a substantial improvement on the $x = 1.79$ given by linear interpolation between 1 and 2. The accuracy one requires may depend on the nature of a practical application.

The formula (15.3) is a *recursive relation.* A recursive relation requires a starting value x_0 and a relationship of a form like (15.3) that expresses x_{n+1} as a function of x_n for $n = 0, 1, 2, \ldots, r, \ldots$ continuing as long as we please. Relationships of this type are especially important in the computer age, being easily incorporated as algorithms into programmes as subroutines or macros that may be repeated as often as required. Processes using recursive relations are also referred to as *iterative processes* or as *mappings.*

15.4 The Calculus of Finite Differences

We often know only the values of some unspecified function (which may or may not have an analytical form) at a set of equally spaced values of x. In this situation a calculus, usually called the *calculus of finite differences,* evolved more or less in parallel with the differential calculus. The latter is sometimes referred to as the *infinitesimal calculus,* since it is in essence a calculus of infinitesimal differences. Many similarities exist between the two.

The calculus of finite differences can be extended to cover the situation where we are given values of a function at unequally spaced points, but we confine our attention to equally spaced x values.

A table of squares of integers gives values of the function $f(x) = x^2$ for $x = 1, 2, 3, \ldots$. A section of such a table would include the entries in Table 15.1.

Table 15.1. Squares of integers between 870 and 876 inclusive

x	x^2
870	756900
871	758641
872	760384
873	762129
874	763876
875	765625
876	767376

The *first difference* of the squares for the values $x+1$ and x is denoted by Δx (pronounced 'delta x'). For any given integer x it is the difference $\Delta x = (x+1)^2 - x^2$. More generally, for any function $y = f(x)$, if we are given its values $y_0, y_1, y_2, \dots y_n$ at equally spaced values of x, i.e.

$$x_0,\ x_1 = x_0 + h,\ x_2 = x_0 + 2h,\ \dots,\ x_n = x + nh$$

the first differences are

$$\begin{aligned}
\Delta y_0 &= y_1 - y_0\ ,\\
\Delta y_1 &= y_2 - y_1\ ,\\
\dots\dots &\dots\dots\dots\\
\Delta y_r &= y_{r+1} - y_r\ ,\\
\dots\dots &\dots\dots\dots\\
\Delta y_{n-1} &= y_n - y_{n-1}\ .
\end{aligned}$$

Just as we take the derivative of a derivative to form a second derivative, and repeat the process to form higher derivatives, so too we may take differences of first differences to form second differences and then proceed to higher order differences. Second differences are denoted by Δ^2 and defined as

$$\begin{aligned}
\Delta^2 y_0 &= \Delta y_1 - \Delta y_0\ ,\\
\Delta^2 y_1 &= \Delta y_2 - \Delta y_1\ ,
\end{aligned}$$

and so on. More generally, a $(k+1)$th difference is

$$\Delta^{k+1} y_r = \Delta^k y_{r+1} - \Delta^k y_r\ .$$

There is an interesting relationship between these differences and the binomial coefficients. We easily see that

$$\Delta^2 y_0 = \Delta y_1 - \Delta y_0 = (y_2 - y_1) - (y_1 - y_0) = y_2 - 2y_1 + y_0\ .$$

Similarly

$$\Delta^3 y_0 = \Delta^2 y_1 - \Delta^2 y_0 = (y_3 - 2y_2 + y_1) - (y_2 - 2y_1 + y_0) = y_3 - 3y_2 + 3y_1 - y_0\ .$$

Intuition suggests the general result

$$\Delta^k y_0 = \sum_{r=0}^{k} (-1)^r \binom{k}{r} y_{k-r}\ ,$$

where

$$\binom{k}{r} = \frac{k(k-1)(k-2)\dots(k-r+1)}{r!}$$

Table 15.2. A table of finite differences

y_0						
	Δy_0					
y_1		$\Delta^2 y_0$				
	Δy_1		$\Delta^3 y_0$			
y_2		$\Delta^2 y_1$		$\Delta^4 y_0$		
	Δy_2		$\Delta^3 y_1$		$\Delta^5 y_0$	
y_3		$\Delta^2 y_2$		$\Delta^4 y_1$		$\Delta^6 y_0$
	Δy_3		$\Delta^3 y_2$		$\Delta^5 y_1$	
y_4		$\Delta^2 y_3$		$\Delta^4 y_2$		
	Δy_4		$\Delta^3 y_3$			
y_5		$\Delta^2 y_4$				
	Δy_5					
y_6						

Table 15.3. Finite difference table for squares of integers between 870 and 876 given in Table 15.1

x	x^2	*1st difference*	*2nd difference*	*3rd difference*
870	756900			
		1741		
871	758641		2	
		1743		0
872	760384		2	
		1745		0
873	762129		2	
		1747		0
874	763876		2	
		1749		0
875	765625		2	
		1751		
876	767376			

is a binomial coefficient. Intuition here leads to a correct conclusion.

Given a set of values $y_0, y_1, y_2, \ldots$ for equally spaced x, the differences may be set out as in Table 15.2. Each difference in any column is the difference between the entry in the previous column immediately below it, and the entry immediately above it. For example, $\Delta^2 y_2 = \Delta y_3 - \Delta y_2$.

If $f(x)$ is a polynomial of degree n, it is well known, and easy to show, that the rth derivative $f^r(x)$ is a polynomial of degree $n - r$. This means the nth derivative is a constant, and all higher derivatives are zero. For example, for the function $y = f(x) = x^2$ we find $f'(x) = 2x$ and $f''(x) = 2$ and $f'''(x) = 0$. Table 15.1 gave, in the second column, the value of this function, x^2, for integral values of x between 870 and 876. Table 15.3 is a *difference table* for those data.

Table 15.4. Finite difference table for reciprocals of integer between 20 and 29

x	$y = 1/x$	Δy	$\Delta^2 y$	$\Delta^3 y$	$\Delta^4 y$	$\Delta^5 y$
20	0.0500					
		−0.0024				
21	0.0476		0.0003			
		−0.0021		−0.0002		
22	0.0455		0.0001		0.0003	
		−0.0020		0.0001		−0.0005
23	0.0435		0.0002		−0.0002	
		−0.0018		−0.0001		0.0004
24	0.0417		0.0001		0.0002	
		−0.0017		0.0001		−0.0005
25	0.0400		0.0002		−0.0003	
		−0.0015		−0.0002		0.0007
26	0.0385		0.0000		0.0004	
		−0.0015		0.0002		−0.0007
27	0.0370		0.0002		−0.0003	
		−0.0013		−0.0001		
28	0.0357		0.0001			
		−0.0012				
29	0.0345					

Table 15.3 shows close parallels between derivatives and finite differences. All second order differences take the same constant value 2, as does the second derivative. All third order differences are zero, as is the third derivative. While the first derivative takes the value $2x$, the first difference corresponding to integer values x and $x + 1$ is $2x + 1$.

Often tabulated values, unlike those in Table 15.1, refer to a function that is not a polynomial, and they may also involve some rounding. Table 15.4 gives finite differences for the reciprocals of integers between 20 and 29, in all cases rounded to 4 decimal places. Remember the reciprocal of x is $1/x$.

A feature of Table 15.4 is that the second differences are nearly constant and close to zero, as also are the third differences, suggesting that a quadratic almost fits the data. However the fourth, and more especially the fifth differences, are tending to increase in magnitude, a trend that becomes even more evident if we form sixth differences.

This behaviour is typical when there is not an exact polynomial fit. It becomes more marked if the data have been rounded. In situations like this, it is usually appropriate to stop taking differences when they approach zero, and to ignore higher order differences. The behaviour here is in sharp contrast to the situation in Table 15.3, where we had exact tabulated values for a particular second degree polynomial at equally spaced points.

Table 15.5. Finite difference table for squares of integers between 870 and 876 with an error in one entry

x	x^2	*1st difference*	*2nd difference*	*3rd difference*
870	756900			
		1741		
871	758641		2	
		1743		90
872	760384		92	
		1835		−270
873	762219		−178	
		1657		270
874	763876		92	
		1749		−90
875	765625		2	
		1751		
876	767376			

Table 15.6. A table of finite differences with an error e in y_3

y_0				
	Δy_0			
y_1		$\Delta^2 y_0$		
	Δy_1		$\Delta^3 y_0 + e$	
y_2		$\Delta^2 y_1 + e$		$\Delta^4 y_0 - 4e$
	$\Delta y_2 + e$		$\Delta^3 y_1 - 3e$	
$y_3 + e$		$\Delta^2 y_2 - 2e$		$\Delta^4 y_1 + 6e$
	$\Delta y_3 - e$		$\Delta^3 y_2 + 3e$	
y_4		$\Delta^2 y_3 + e$		$\Delta^4 y_2 - 4e$
	Δy_4		$\Delta^3 y_3 - e$	
y_5		$\Delta^2 y_4$		
	Δy_5			
y_6				

15.5 Tabulation Errors

It is sometimes possible to use differences to spot and correct tabulation errors. In Table 15.1 the square of 873 might have been incorrectly recorded as 762 219 instead of 762 129. Transposition of adjacent digits is a common error in recording a number. The difference table is now that in Table 15.5.

Pardon me, your slip is showing

It is easy to find out what the error is when we know the form the differences would take if there were no error. To illustrate this, we trace the effect of an error e in some entry in a general table like Table 15.2. This is shown in Table 15.6.

Take a moment to look at the coefficients of e in successive columns. Do you recognise a pattern? Comparing Tables 15.5 and 15.6 you should find it easy to pin down the rogue entry in the former. For a correct unrounded table of integer squares the third differences should all be zero. It immediately follows by comparing third differences in in the two tables that $e = 90$, and that the error occurs in the entry 762 219. It should be 762 219$-$90 $=$ 762 129.

15.6 Interpolation

Interpolation is useful if we are given values y_i of a function, the precise analytic form of which may not be known, at fixed points with x values $x_i, i = 1, 2, ..., n$ and wish to estimate the value of that function at one or more other values of x. Strictly, if the x_i are arranged in ascending order, any point x should lie in the interval x_1 to x_n if the term *interpolation* is used. For values of x outside the interval (x_1, x_n) the appropriate term is *extrapolation*.

If we had a table of values of a function e.g., $y = \sin x^\circ$, at interval of 1 degree between 0 and 10 degrees we might interpolate to find the value of $\sin x^\circ$ when $x = 7^\circ 21'$ (or, if you prefer, $x = 7.35^\circ$).

There are many different forms of interpolation. We look at only the simplest, and probably the most widely used, namely, *polynomial interpolation*. The appeal of the method is that for any reasonably smooth function a polynomial that passes through all the given points is likely to lie close to the function at other points over the interval covered by these points.

There are exceptions to this. For example, a periodic function that repeats its form may well be better approximated by a trigonometric function, or by combinations of trigonometric functions, unless we are only interpolating over a small interval. A function that becomes infinite for some finite value of x may be better approximated by a rational function. This is, in essence, the quotient of two polynomials.

In general, given n pairs of values (x_i, y_i) where the x_i are all different we may find a polynomial of degree $n-1$ that passes through all the points. Here we may drop the assumption that the x_i are equally spaced.

The simplest case of polynomial interpolation is that between two neighbouring points, ignoring values of the function at other points. This is the familiar idea of linear interpolation used, as mentioned in Sect. 15.2, in reading mathematical tables. It is based on fitting a straight line, i.e., a polynomial of degree 1, through the two points.

Fig. 15.2 illustrates a situation where the coordinates are determined by the values of $y = 2\mathrm{e}^x$ at $x = 0.5, 1, 1.5, 2.0$. i.e.,

x	0.5	1.0	1.5	2.0
y	3.2974	5.4366	8.9634	14.7781

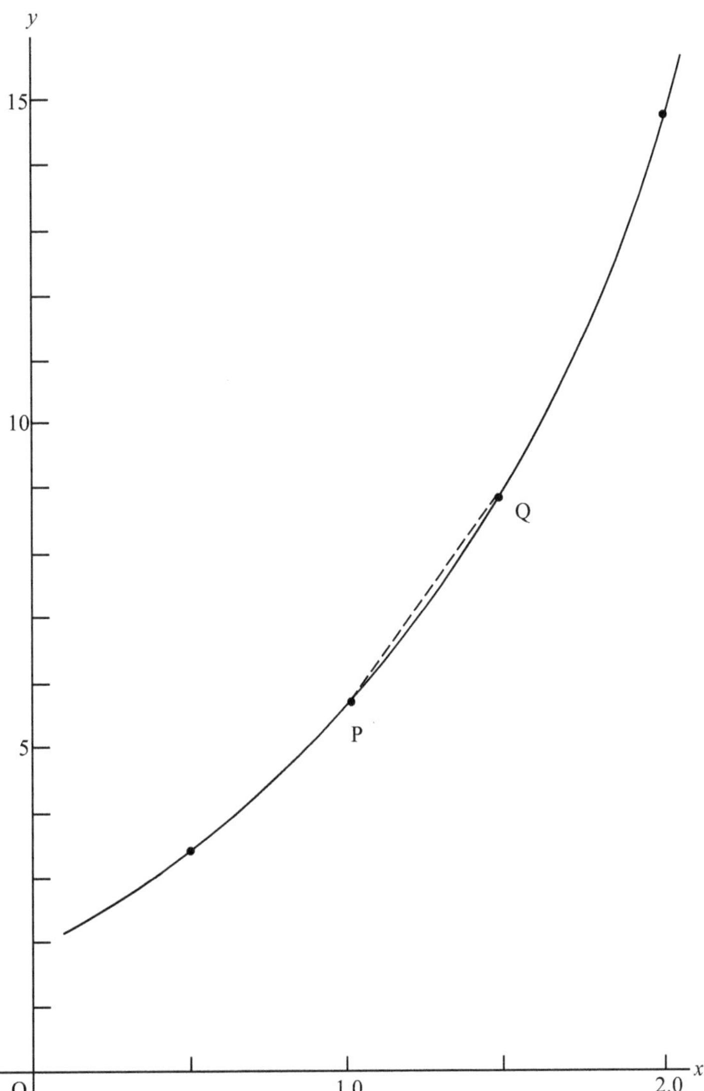

Fig. 15.2. A polynomial fit to four points satisfying the relationship $y = 2e^x$ and also the straight line for linear interpolation between P and Q

To estimate y when $x = 1.2$ we might interpolate linearly between $x = 1$ and $x = 1.5$. To do this we first find the equation for the line PQ joining points with ordinates $x = 1$ and $x = 1.5$. The estimate is the point on this line with x co-ordinate 1.2. More generally, for linear interpolation between and two points (x_1, y_1) and (x_2, y_2) with known coordinates, the required line, as shown in (7.1), has the equation

$$\frac{x - x_1}{x_2 - x_1} = \frac{y - y_1}{y_2 - y_1} .$$

By simple algebraic manipulation, the value of y corresponding to linear interpolation at a value x is

$$y = y_1 + \frac{(x - x_1)(y_2 - y_1)}{x_2 - x_1} .$$

Substituting the values (1, 5.4366), (1.5, 8.9634), and setting $x = 1.2$, we find for our numerical example that

$$y = 5.4366 + \frac{(1.2 - 1)(8.9634 - 5.4366)}{1.5 - 1} = 6.8473.$$

Tables, or a pocket calculator, quickly establish that the correct value (rounded to 4 decimal places) is $2\mathrm{e}^{1.2} = 6.6402$. Linear interpolation appreciably overestimates the value in this case.

15.7 Lagrange's Formula for Polynomial Interpolation

We illustrate a method proposed by Joseph Lagrange (1736–1813) for interpolation. This uses a polynomial of degree $n - 1$ that passes through n given points (x_i, y_i). When $n = 4$ the general formula may be written

$$f(x) = y_1 L_1(x) + y_2 L_2(x) + y_3 L_3(x) + y_4 L_4(x) ,$$

where

$$L_1(x) = \frac{(x - x_2)(x - x_3)(x - x_4)}{(x_1 - x_2)(x_1 - x_3)(x_1 - x_4)} ,$$
$$L_2(x) = \frac{(x - x_1)(x - x_3)(x - x_4)}{(x_2 - x_1)(x_2 - x_3)(x_2 - x_4)} ,$$
$$L_3(x) = \frac{(x - x_1)(x - x_2)(x - x_4)}{(x_3 - x_2)(x_3 - x_2)(x_3 - x_4)} ,$$
$$L_4(x) = \frac{(x - x_1)(x - x_2)(x - x_3)}{(x_4 - x_1)(x_4 - x_2)(x_4 - x_3)} .$$

The pattern of each $L_i(x), i = 1, 2, 3, 4$ becomes clear on inspection. The numerator in $L_i(x)$ is the product of all factors of the form $(x - x_j)$ with the omission of the factor with $j = i$. Each factor in the denominator corresponds to the one in the numerator immediately above it with x replaced by x_i.

It is easy to verify that $f(x)$ is a polynomial of degree three. Also, when $x = x_i$ we have $L_i(x_i) = 1$ and all $L_j(x_i) = 0$ if $j \neq i$. It follows that $f(x_i) = y_i$ at each of the four data points. We could rewrite $f(x)$ in a form

$a + bx + cx^2 + dx^3$ using straightforward but tedious algebra. Some simplification is also possible if the x_i are equally spaced. Generalization to fitting a polynomial of degree $n - 1$ to n data points is straightforward.

We now apply the Lagrange formula to estimating the value of $y = 2e^x$ when $x = 1.2$ using the four data values given in Sect. 15.6. Setting $x = 1.2$ we easily find that

$$L_1(x) = \frac{(1.2 - 1)(1.2 - 1.5)(1.2 - 2)}{(0.5 - 1)(0.5 - 1.5)(0.5 - 2)} = -0.064\,.$$

Similarly,

$$L_2 = 0.672, L_3 = 0.448, \text{ and } L_4 = -0.056,$$

whence

$$\begin{aligned} f(1.2) &= 3.2974 \times (-0.064) + 5.4366 \times 0.672 \\ &\quad + 8.9634 \times 0.448 + 14.7781 \times (-0.056) = 6.6304. \end{aligned}$$

The error is thus slightly less than 0.01, a considerable improvement over linear interpolation. When approximating to a known function it is possible to carry out an analysis of the error, and to set an upper limit to the error resulting from interpolation. We do not pursue this, but in this, and other approximations in numerical analysis, an investigation of error is often of critical importance. One reason for this is that in applications in engineering, space exploration, etc., or in many laboratory experiments, one often needs high precision. Near enough may not be good enough!

There are many other interpolation formulae, and in the case of equally spaced points these are often based on finite differences. One is the *Gregory-Newton formula*. This is especially useful for, and has high precision for, interpolation for values of x between x_1 andx_2 where values of a function are given at ordered equally spaced x-values $x_1, x_2, \ldots, x_n$. A simple modification of the formula is appropriate for interpolation between x_{n-1} and x_n.

15.8 Numerical Integration

For any reasonably well behaved function, if the derivative exists it is usually possible to determine this analytically. The same is not true for integration. For example, there is no known explicit function $y = F(x)$ that has a derivative $f(x) = F'(x)$ where

$$f(x) = \frac{1}{\sqrt{2\pi}} e^{-x^2/2}\,. \tag{15.5}$$

Statisticians and others often want to evaluate the integral of (15.5) over some interval (a, b). In precomputer days one resorted to a table of values of

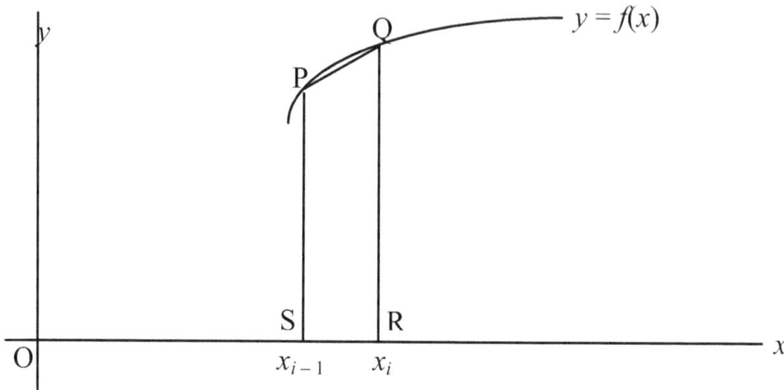

Fig. 15.3. The area of the trapezium PQRS approximates the area under $f(x)$ between x_{i-1} and x_i

this integral that was included in almost any elementary statistics textbook, or collection of standard mathematical or statistical tables.

These tables were all based on numerical approximations that, in effect, replaced the function $f(x)$ by polynomials, or more sophisticated functions, each of which fitted the function extremely well over some small range of x values. Computer software has largely replaced the use of tables, but the programs themselves often use similar computations to evaluate the required integrals.

The process boils down to fitting something like an interpolating function such as a polynomial in a local area, then integrating that polynomial, also locally. The results are added for a sequence of localized sub-intervals that covers the whole required interval.

That sounds a mouthful, but we show how it works for two of the best known simple rules for numerical integration. These are the *trapezoidal rule* and *Simpson's rule*. The former uses straight lines, and the latter quadratic (or second degree) polynomials, as local approximating functions.

To use the trapezoidal rule to integrate a function $f(x)$ over a finite interval (a, b), we divide the interval into n sub-intervals each of length h. It is convenient to make a small notational change and to set $a = x_0$. The first sub-interval extends from x_0 to $x_1 = x_0 + h$, the second from x_1 to $x_2 = x_1 + h$, and proceeding in this way the nth such sub-interval extends from x_{n-1} to $x_n = b = x_{n-1} + h$.

Corresponding to each x_i, we denote the value of the function to be integrated by y_i. The interpolating function in the ith interval $(i = 1, 2, \ldots, n)$ is simply the line PQ joining the points with ordinates y_{i-1} and y_1 as illustrated in Fig. 15.3.

The area under the curve between P and Q in Fig. 15.3 may be approximated by the area of the trapezium PQRS which is well-known, and indeed easily shown using Euclidean geometry, to be $h(y_{i-1}+y_i)/2$. The trapezoidal rule is to add these neighbouring approximations over all n intervals, and this leads to an estimated area

$$A = \frac{1}{2}h(y_0 + 2y_1 + 2y_2 + \ldots + 2y_{n-1} + y_n) \, . \qquad (15.6)$$

We apply this rule to integrate (15.5) between 0 and 1, setting $h = 0.1$. Since $1/\sqrt{2\pi}$ is a constant, we need only integrate the exponential part of the function, then multiply the result by that constant at the end. From tables, or with a pocket calculator, we obtain the following values of

$$y_i = \mathrm{e}^{-x_i^2/2}.$$

x_i	0.0	0.1	0.2	0.3	0.4	0.5	0.6	0.7	0.8	0.9	1.0
y_i	1.000	0.995	0.980	0.956	0.923	0.882	0.835	0.783	0.726	0.667	0.607

whence (15.6) gives

$$A = \frac{1}{2} \times 0.1[1.000 + 2(0.995 + 0.980 + 0.956 + 0.923 + 0.882 + 0.835 \\ + 0.783 + 0.726 + 0.667) + 0.607] = 0.8551.$$

Multiplying by $1/\sqrt{2\pi} = 0.3989$ gives the required integral over the interval (0, 1) to be 0.3411. The true value, to 4 decimal places, is 0.3413.

Simpson's rule was proposed by Thomas Simpson(1710–1761). It uses a second degree polynomial fitted locally to sets of three points and adds neighbouring approximations over the entire interval. The whole interval is broken into an even number of sub-intervals, each of length h. The area under the second degree polynomial fitted to the relevant three points is computed for successive triplets, and these are added together. We omit details of the relatively straight-forward algebra, but the formula for the estimated area A is

$$A = \frac{h}{3}(y_0 + 4y_1 + 2y_2 + 4y_3 + 2y_4 + \ldots + 2y_{n-2} + 4y_{n-1} + y_n) \, ,$$

where $y_0 = f(a)$, $y_n = f(b)$ and n is even. Applying this to (15.5) over the interval (0, 1) with $h = 0.1$, ignoring the constant $1/\sqrt{2\pi}$ as we did for the trapezoidal rule, and using the y_i values given above, we get

$$A = \frac{0.1}{3}[1.000 + 4(0.995 + 0.956 + 0.882 + 0.783 + 0.667) \\ + 2(0.980 + 0.923 + 0.835 + 0.726) + 0.607] = 0.8556.$$

Finally, multiplication by $1/\sqrt{2\pi} = 0.3989$ gives the approximation 0.3413, which is correct to 4 decimal places, slightly better than the trapezoidal rule estimate.

As in other branches of numerical analysis, an error analysis plays an important role in many practical applications.

Computers have made possible other methods of computing integrals based on statistical principles. One of these is described in Sect. 24.3.

15.9 Difference Equations

Differential calculus leads naturally to differential equations as a tool for modelling real-world situations in ways we described in an elementary fashion in Sect. 12.8. We mentioned there that numerical analysis has an important role in providing methods of solution when analytic solutions do not exist, or are difficult to obtain. The details are fairly technical so we do not discuss them.

The calculus of finite differences gives rise to equations called *difference equations*. These have many similarities to differential equations, and also are often difficult to solve analytically. In essence, difference equations involve an independent variable, often denoted by x, and a dependent variable y, and the differences $\Delta y, \Delta^2 y, \ldots$. However, to bring this discussion into line with that in Sect. 12.8 we use t rather than x as the independent variable. This is simply a convenience introduced because in many real-world examples the independent variable is *time*.

We shall assume, if necessary by making an appropriate change of variable, that y is recorded at times $t = 0, 1, 2, \ldots, r, \ldots$. The value of y at a time t is denoted by y_t. We write

$$\Delta y_t = y_{t+1} - y_t, \ \Delta^2 y_t = y_{t+2} - 2y_{t+1} + y_t$$

and so on. A common practice is to express difference equations in terms of $t, y_t, y_{t+1} \ldots$ rather than in terms of $t, \Delta y_t, \Delta^2 y_t \ldots$. In this notation a typical difference equation is

$$y_{t+1} = 2y_t + t^2 - 2t - 1. \tag{15.7}$$

This is called a difference equation of the *first order* because it involves only first differences. The equation

$$y_{t+2} - 3ty_{t+1} + t^2 y_t = 0 \tag{15.8}$$

is of the *second order*, because it involves second differences. A third example, a real-world one, is

$$y_{t+1} = ky_t(1 - y_t) \tag{15.9}$$

Before explaining the real-world interpretation of (15.9) we look at the solution of (15.7). A possible solution is

$$y_t = c2^t - t^2 \tag{15.10}$$

where c is an arbitrary constant. This may be verified by direct substitution. On the left hand side

$$y_{t+1} = c2^{t+1} - (t+1)^2 ,$$

and on the right hand side

$$2y_t + t^2 - 2t - 1 = 2(c2^t - t^2) + t^2 - 2t - 1 = c2^{t+1} - (t+1)^2 .$$

Equality of the two sides establishes the result. If we have an initial condition that $y_0 = 1$ then from (15.10) we get $c = 1$, giving the particular solution

$$y_t = 2^t - t^2 . \tag{15.11}$$

This enables us to write down y_t for any t. In practice, the solution for integer values of t are often of main interest. For a given initial condition, these are obtainable whether or not an analytic solution exists. Even if an analytic solution exists it may not be unique. Indeed, a more general solution of (15.7) than (15.10) takes the form

$$u_t = 2^t(a + b\cos 2\pi t + c\sin 2\pi t) - t^2 ,$$

where a, b, c are constants. We may verify this in a similar way to that in which we verfied that (15.10) was a solution. Here we need to bear in mind the periodicity properties of sines and cosines.

Even if we have no analytic solution, with the initial condition $y_0 = 1$, we immediately deduce, by setting $t = 0$ in (15.7), that

$$y_1 = 2y_0 - 1 = 1 ,$$

and setting $t = 1$, that

$$y_2 = 2y_1 - 2 = 2 \times 1 - 2 = 0 ,$$

Proceeding in this way we find

$$y_3 = -1, y_4 = 0, y_5 = 7, y_6 = 28, y_7 = 79,$$

and so on.

These, and values for $t > 7$,, are easily verified using (15.11), and from that solution it can be shown that $y_t \to \infty$ as $t \to \infty$. This is because 2^t increases more rapidly than t^2. Of course, if we have a solution such as (15.11) there is no need to use iterative solutions based on (15.7). Had we not produced the solution (15.10) 'rabbit out of hat' fashion, we would have to use the numerical approach unless we knew how to get (15.10).

We say a lot about equation (15.9) in Chap. 21. It suffices to remark here that it is the difference equation analogue to (12.11) in Sect. 12.8. It provides a simple model for discrete population dynamics for simple groups of organisms such as an insect population where each generation has a lifespan of 1 day. On each day a new generation hatches, eggs are laid, then the entire population dies at the end of the day, leaving only the eggs to hatch on the following day for a new generation. Here t represents time on an integer scale of 1 day, and y_t is a proportional measure of population size on day t. This is expressed as the ratio of the actual population size to a theoretical maximum. The constant k is a known constant. Given an initial value y_0 at time $t = 0$, where $0 < y_0 < 1$, we may iterate in (15.9) to determine the population size on successive days. I keep you in suspense until Chapter 20, but a study by Robert May and others some 30 years ago of iterative solutions for these and many other difference and differential equations contributed to a revolution in the way mathematicians (and other scientists) now look at dynamic systems.

16

Modern Algebra

16.1 An Escape from Numbers

In Chap. 15 we described numerical analysis as *sophisticated arithmetic*. We might call the content of this chapter *sophisticated algebra*. Just as the preferred name for the former is *numerical analysis*, the latter is better known as *modern* or *abstract algebra*.

Classic algebra evolved from, and reflects properties of, the real and complex number systems. The rules for operating with numbers given in Sect. 2.5 were established first for the natural numbers. As we have seen, their scope was later extended to real and then complex numbers. Without algebra the links between numbers and geometry, implicit in both the use of Cartesian coordinates and the Argand diagram, would not have been possible.

In Sect. 4.2 we met the concept of a set as a collection of items determined by rules which allow us to specify uniquely whether given items (these need not be numbers) are or are not members of that set.

While classic algebra has numbers as its foundation stones, modern algebras are based largely on sets and operations with sets. There are clear links between classic and modern algebras. Indeed, the concept of sets was used by Georg Cantor in 1874 in a 'classic algebra' context to unravel the mysteries of infinity in connection with the real number system. Specifically, he used the concept in establishing the distinction between denumerable infinite sets and those that are nondenumerable. We discussed that distinction in Sect. 5.4 when looking at the extension from rational numbers to a real number system that included irrational numbers.

2 + 2 need not equal 4

About 150 years ago mathematicians realised that some, though not all, rules needed to handle numbers might also be relevant in other contexts.

The algebra of sets, known as *set theory*, applies to well-defined collections of objects. It has similarities to classic algebra as applied to numbers, but the basic axioms and the allowable operations are different. We may have 2 sets, each with 2 items, but between them the sets might only contain 3, or even only 2, items.

Another algebra, *Boolean algebra*, has a much closer relationship to the algebra of sets than it has to classic algebra. Boolean algebra was developed initially as an algebra for logical systems. Today it has important applications in computer science.

A whole family of abstract algebras have been developed largely during the twentieth century. Some, such as *group algebra* have many real-life applications. A few abstruse ones have less obvious connections with real life. All algebras have as their foundation a set of axioms. Rules of operation must be consistent with those axioms.

The notion of, and the notation used for, operations on sets are important in probability and statistics, topics we meet in Chap. 17. In mathematics generally, certain recognisable basic patterns occur in several different contexts, and these often involve operations with sets. One such collection of patterns is the basis of the classification of sets into *groups. Group algebra* is about permissible and meaningful operations with these entities. Before discussing groups we need to say more about sets themselves.

16.2 Set Theory

Classifying objects according to whether they do, or do not, possess well-defined *attributes* or properties is a notion almost as fundamental as that of counting using natural numbers. In Sect. 4.2 we explained that 'a well-defined attribute' meant there is a rule that enables us to decide unambiguously whether or not an object has that attribute. If the objects are integers, then the property of being an even positive integer, is well defined. The property of being a large integer is not, but the property of being an integer greater than one million is. Properties that are not well-defined lead us into the realms of *fuzzy logic*, a topic we mention briefly in Sect. 25.2.

Items or objects that all possess the appropriate defining property, or properties, form a set which may be denoted by some letter, e.g., A. There is virtually no restriction on what these defining properties may be. Thus all current members of the British House of Commons form a set. All members of the United States Senate also form a set. So do all books containing more than 400 pages in the room in which I am writing this sentence.

Each object that is a member of a set is called an *element* of that set. The set consisting of all current members of the British House of Commons has a finite number of elements. You may or may not know what that number is, but by making appropriate enquiries one could find out what it is. The position is similar for the set consisting of all members of the United States Senate.

The set consisting of all prime numbers has an infinite (i.e. unlimited) number of elements (see Sect. 2.9), and such sets are, for brevity, often referred to as 'infinite' sets.

Two sets with the same finite number of elements are said to be of the same size or to be *equivalent*. Thus the set consisting of the numbers 1, 2, 3, and that consisting of the letters a, b, c as well as the set consisting of one banana, one pineapple and one grapefruit are all of the same size.

Sets may be combined in accord with certain rules to form other sets. These rules have analogies to, but differ in some respects from, the rules of addition and multiplication of numbers. The application of these rules is the basis for the algebra of sets.

A basic concept in set theory is that of a *universal set.* I speak of 'a' (not 'the') universal set, for a universal set is simply a set that includes all objects considered in some current context. We may be interested in a set consisting of *all countries that are member states of the European Union*, and also in another set consisting of *all countries in Europe that have a coastline on the Baltic Sea.* An appropriate universal set covering all elements in either of these sets would be one having as its elements *all countries in Europe.* An alternative universal set might be that of *all countries in the northern hemisphere.*

The set consisting of *all countries that are member states of the European Union* forms part of the set *all countries in Europe.* If we take the latter as the universal set, then the former is called a *subset* of that universal set. More generally, we denote any relevant universal set by U. Some writers prefer S, and others I, for a universal set. Sets in which we are interested, and which are subsets of U, may be written A, B, C, ...

If A is a subset of U, we write

$$\mathrm{A} \subset \mathrm{U}$$

The symbol $\subset$ is read as 'is a subset of'. It is the set algebra equivalent of the classic algebra symbol $<$ for 'is less than'. More generally, any set B is said to be a subset of A if there is no element in B that is not also in A. This definition does not exclude the possibility that all elements in A are included in B, and we write this either as

$$\mathrm{B} \subseteq \mathrm{A} \text{ or } \mathrm{A} \supseteq \mathrm{B}$$

where the symbols $\subseteq$, $\supseteq$ are the set theory equivalents of the algebraic symbol $\leq$, $\geq$. If we want to exclude the case where all elements of A are included in B, we refer to B as a *proper subset* of A, and write

$$\mathrm{B} \subset \mathrm{A} \text{ or } \mathrm{A} \supset \mathrm{B}.$$

The notational distinction between $B \subseteq A$ and $B \subset A$ is not observed by all mathematicians. Some use the notation $B \subset A$ to include the case where all elements of A are included in B. In practice this causes little difficulty, for in most circumstances statements made about proper subsets will also hold for all subsets.

If all elements of A are included in B, and all elements of B are included in A, i.e. if both $B \subseteq A$ and $A \subseteq B$ this implies A and B are identical sets, and we may use the obvious notation $B = A$.

The individual objects or elements of a set are often denoted by lower case letters such as x, with subscripts if appropriate. We denote any element, x, of A by

$$x \in A.$$

A useful shorthand for the set A, having elements x with some defining property P, is to write

$$A = \{x : P\}.$$

For example, if A is the set of all prime numbers less than 200, we could write

$$A = \{x : x < 200 \text{ and prime}\}.$$

If the property is obvious the notation may be modified. We might write, for example, $A = \{1, 2, 3, 4\}$ or $B = \{\text{all positive integers } \leq 99\}$ to indicate the set consisting of the numbers $1, 2, 3, 4$ in the first case, and of the numbers $1, 2, 3, \ldots, 99$ in the second.

Don't take analogies too far

Here is an illustration of the danger of carrying analogies between relations for numbers and those for sets too far. It is easy to see that if $A \subseteq B$ and $B \subseteq C$ then $A \subseteq C$. The relation '$\subseteq$' is called an *order relation* because of this property which it shares with the relation '$\leq$' for numbers. However, while for any two real numbers, at least one of the relationships $a \leq b$ or $b \leq a$ must hold this is not true for sets. For example, if $A = \{1, 2, 3\}$ and $B = \{1, 2, 4, 8\}$, then neither is a subset of the other, so neither $A \subseteq B$ nor $B \subseteq A$ is true.

Just as a symbol 0 for zero is important for numbers, making possible our Arabic notation, it is useful to have a symbol for a set with no elements — called an *empty set*. The empty set is sometimes written 0, but a preferred notation is $\emptyset$. The empty set is a subset of any set.

If an element y is not a member of a set A, we write $y \notin A$ (in speech 'y is not an element of A').

Two operations with sets are those of forming the *union* and of forming the *intersection*.

If A and B are any two sets their union, (sometimes also called the logical sum) is written $\mathrm{A} \cup \mathrm{B}$ and is the set of all elements that are in either A or B or both. Thus if

$$\mathrm{A} = \{1, 2, 4, 8, 16, 32\} \text{ and } \mathrm{B} = \{2, 3, 5, 7, 11, 13, 17, 19\}$$

then

$$\mathrm{A} \cup \mathrm{B} = \{1, 2, 3, 4, 5, 7, 8, 11, 13, 16, 17, 19, 32\}.$$

A specific set may often be written in different ways. For example, instead of writing

$$\mathrm{B} = \{2, 3, 5, 7, 11, 13, 17, 19\},$$

we could write

$$\mathrm{B} = \{x : x \text{ any prime number less than } 20\}.$$

The intersection of two sets, written $\mathrm{A} \cap \mathrm{B}$, is the set consisting of all elements common to A and B. Thus if A is the set of all prime numbers less than 20 and B the set of all odd integers, $\mathrm{A} \cap \mathrm{B} = \{3, 5, 7, 11, 13, 17, 19\}$. Because of their shape, the symbols $\cup, \cap$, are sometimes called *cup* and *cap* respectively.

It is not hard, just tedious, to verify the following rules for operations involving intersections and unions. You should compare these with operations for addition and multiplication of numbers to see which are exactly equivalent if we interchange $+$ and $\cup$, $\times$ and $\cap$, and which differ.

$$\begin{aligned}
&\text{(i) } \mathrm{A} \cup \mathrm{B} = \mathrm{B} \cup \mathrm{A},\\
&\text{(ii) } \mathrm{A} \cap \mathrm{B} = \mathrm{B} \cap \mathrm{A},\\
&\text{(iii) } \mathrm{A} \cup (\mathrm{B} \cup \mathrm{C}) = (\mathrm{A} \cup \mathrm{B}) \cup \mathrm{C},\\
&\text{(iv) } \mathrm{A} \cap (\mathrm{B} \cap \mathrm{C}) = (\mathrm{A} \cap \mathrm{B}) \cap \mathrm{C},\\
&\text{(v) } \mathrm{A} \cup \mathrm{A} = \mathrm{A},\\
&\text{(vi) } \mathrm{A} \cap \mathrm{A} = \mathrm{A},\\
&\text{(vii) } \mathrm{A} \cap (\mathrm{B} \cup \mathrm{C}) = (\mathrm{A} \cap \mathrm{B}) \cup (\mathrm{A} \cap \mathrm{C}),\\
&\text{(viii) } \mathrm{A} \cup (\mathrm{B} \cap \mathrm{C}) = (\mathrm{A} \cup \mathrm{B}) \cap (\mathrm{A} \cup \mathrm{C}),\\
&\text{(ix) } \mathrm{A} \cup \emptyset = \mathrm{A},\\
&\text{(x) } \mathrm{A} \cap \emptyset = \emptyset,\\
&\text{(xi) } \mathrm{A} \cup \mathrm{U} = \mathrm{U},\\
&\text{(xii) } \mathrm{A} \cap \mathrm{U} = \mathrm{A}.
\end{aligned}$$

Remember U is the universal set.

16.3 Venn Diagrams

John Venn (1834–1923) devised a useful geometric representation of the relationship between sets. These are called *Venn diagrams*. In these, the universal

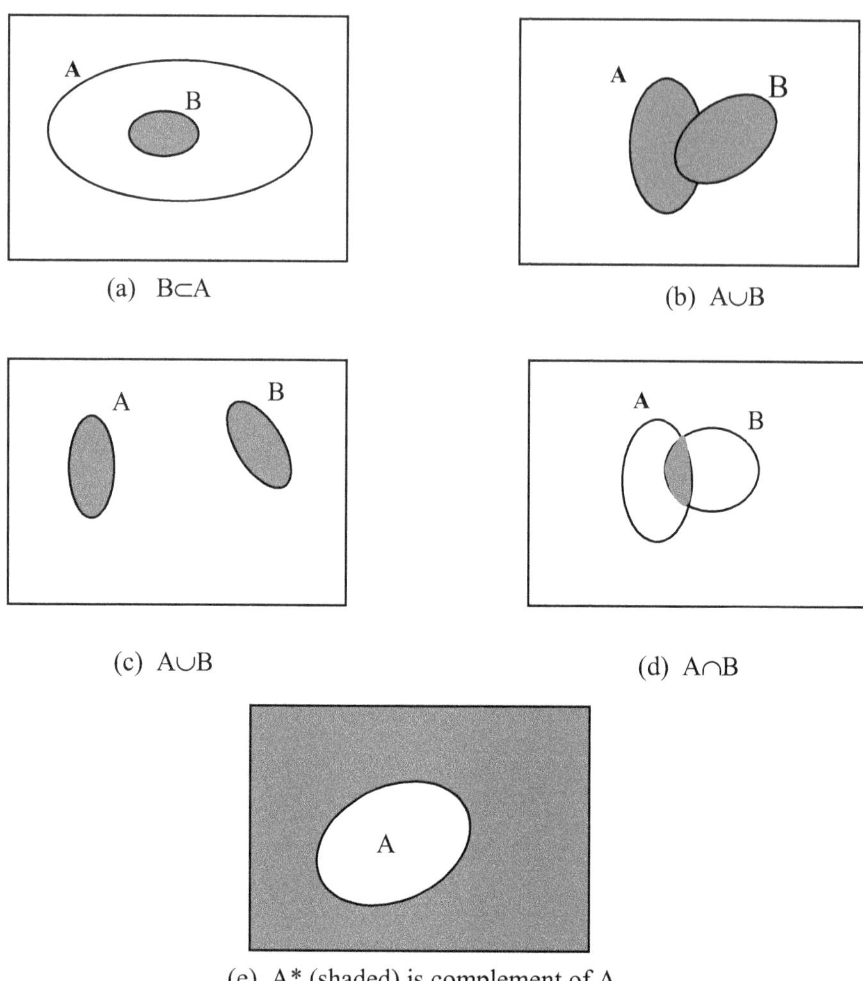

(a) B⊂A (b) A∪B (c) A∪B (d) A∩B (e) A* (shaded) is complement of A

Fig. 16.1. Some Venn diagrams. The shaded areas represent the relevant sets

set is conveniently represented by a rectangle. Subsets of the universal set may be represented by circles, ellipses or less regular closed surfaces lying entirely within that rectangle. If B is a subset of A then the shape representing B lies entirely within that for A as in Fig. 16.1(**a**), where the subset B is shaded.

If A, B are any two sets the union is the combined area covered by the two closed surfaces representing these sets. Examples are given in Fig. 16.1(**b**) and (**c**), where the union is the shaded area.

An example of an intersection is given by the shaded area in Fig. 16.1(**d**). If A is a set in U we may be interested in the set of all elements of U that are not in A. This set is called the *complement* of A, and is often denoted by

A^* or by A' or by $\bar{A}$. An illustration is given in Fig. 16.1(**e**) where A^* is the shaded area.

A major difference between classic algebra and the algebra of sets is illustrated by the set analogue of adding a number to itself, i.e. forming the union of a set with itself. For numbers $a + a = 2a$, but rule (v) above states $A \cup A = A$

Some relationships in set theory have no analogues in classic algebra. One such is the easily verified fact that if $B \subseteq A$ this implies both that $A \cup B = A$ and $A \cap B = B$.

The complement A^* of A with respect to the universal set U has the following properties:

$$\begin{gathered}
\text{(xiii) } A \cup A^* = U, \\
\text{(xiv) } A \cap A^* = \emptyset, \\
\text{(xv) } \emptyset^* = U, \\
\text{(xvi) } U^* = \emptyset, \\
\text{(xvii) If } B \subseteq A \text{ then } A^* \subseteq B^*, \\
\text{(xviii) } (A \cup B)^* = A^* \cap B^*, \\
\text{(xix) } (A \cap B)^* = A^* \cup B^*, \\
\text{(xx) } (A^*)^* = A.
\end{gathered}$$

There is an important duality exhibited in the properties of sets labelled (i) to (xii) and in those labelled (xiii) to (xx) above which relate to complementary sets. If we interchange, wherever they appear

$$\cup \text{ and } \cap$$

and

$$\emptyset \text{ and } U$$

this produces one of the other numbered rules. For example, interchanging $\cup$ and $\cap$ in (vii) gives (viii), and a similar interchange in (xviii) gives (xix). Check for yourself that similar pairings occur in other cases.

We do not prove it, but all the rules (i)–(xx) above for operations on sets can be deduced from three axioms. These are:

$$\begin{aligned}
A \cup B &= B \cup A \\
A \cup (B \cup C) &= (A \cup B) \cup C \\
(A^* \cup B^*)^* \cup (A^* \cup B)^* &= A \qquad (16.1)
\end{aligned}$$

The first two axioms state that forming unions is, like addition in classic algebra, both commutative and associative and indeed corresponds to rules (i) and (iii) given above.

The third axiom, (16.1) looks complicated. It may help you to see the implications if you draw a Venn diagram with sets A and B that intersect. We see that $A^* \cup B^* = (A \cap B)^*$ by (xix) above and that either of these expressions consists of all elements not common to both A and B. The complement of either of these with respect to the universal set is thus $A \cap B$, or the set of elements common to both A and B. The term $A^* \cup B$ consists of all elements that are not in A together with all elements in B. Any such elements that are additional to those already included in A^* must necessarily lie in A. Thus the complement of $A^* \cup B$, i.e., $(A^* \cup B)^*$ consists of all elements in A except those that are also in B. Thus the first term on the left of (16.1) consists of all elements common to A and B while the second term contains all elements in A that are not common to B. The union of these sets thus consists of the elements of A. If you have to read this a couple of times (or even three or four times) to understand it don't be disheartened. New notations and ideas take time to digest.

16.4 Mappings

Consider the function $y = \sin x$ for any real x. While x is an element of the set of real numbers, y can only be an element of the subset of real numbers in the interval $-1 \leq y \leq 1$. In terms of sets the sine function is said to map the set

$$X = \{x : x \text{ any real number}\}$$

into the set

$$Y = \{y : y \text{ a real number in the interval } -1 \leq y \leq 1\}.$$

There is more than one value of x that 'maps' into each value of y. We are measuring x in radians so, for example, for all integers n the points $x = n\pi$ all map to $y = 0$. Such a mapping is called a *many to one* mapping.

In Sect. 4.2 we considered what was essentially a mapping of the positive integers onto the even positive integers, pairing any integer $x = n$ in the set

$$X = \{x : x \text{ a positive integer}\}$$

with $y = 2n$ in the set

$$Y = \{y : y \text{ is an even positive integer}\}$$

In this mapping each element x corresponds to a unique element y and vice versa. Such a mapping is called a *one to one mapping* of X *into* Y.

The concept of mappings can be extended to any set. Here are two examples. Although not in themselves particularly interesting, they indicate the generality of the mapping concept.

Suppose $A = \{a, b, c, d\}$ and $N = \{1, 3, 5, 7, 9\}$ then a possible mapping, f, of A into N would be $f(a) = 3, f(b) = 3, f(c) = 5, f(d) = 9$. A mapping must be defined for every element of the set (here A) which is to be mapped onto another set (here N), but not all the elements in N need be used in the mapping and/or some may be used more than once. In this example only the subset $\{3, 5, 9\}$ of N is used in the mapping and both a and b map into 3.

For a second simple example consider a mapping g of the set $\mathrm{A} = \{b, e, a, n\}$ into $\mathrm{B} = \{c, o, u, n, t, e, r\}$, defined by $g(b) = r, g(e) = u, g(a) = n$ and $g(n) = t$. Thus the mapping produces the set $\{r, u, n, t\}$

Several technical terms and notations are used to cover the ideas inherent in these examples, but we omit details as we do not need them in this book.

An important class of mappings are *one to one* mappings. An example already given is the mapping of the positive integers into the even positive integers. More formally, a mapping f of X into Y is *one to one* if, for all $x_i, x_j \in \mathrm{X}$, $x_i \neq x_j$ implies $f(x_i) \neq f(x_j)$. The mapping $g : \mathrm{A} \to \mathrm{B}$ in the second example above is one-to-one.

Many algebraic functions defined for certain sub-sets of the real or complex numbers are one-to-one mappings. For example if X is the set of all real numbers $x > 1$ then $f(x) = 2/x$ is a one-to-one mapping into the subset of the real numbers, Y, consisting of the real numbers $0 < y < 2$.

16.5 Group Algebra

We have mentioned that abstract algebras are about systems of operations built up from specified axioms. A simple axiomatic system is that for *group algebra*.

The motivation for groups and group algebra, like so many ideas in mathematics, can be traced back to the natural numbers. If we consider the operation of addition in the natural number system adding two natural numbers gives another natural number. So the natural numbers form a closed system with respect to addition. However, if we define subtraction as the inverse of addition, we need to extend the system by introducing zero and negative integers. Thus the natural numbers do not form a closed system with respect to subtraction, i.e., subtraction of two natural numbers may lead us outside the natural number system.

If we now confine ourselves to addition and subtraction among the set of all positive and negative integers (including zero), and regard these as elements that form our universal set, then this set is closed with respect to addition and subtraction.

However, as we discovered in Chap. 2 — and no doubt you knew long before you read this book — if we start with the universal set of positive and negative integers, and consider multiplication and its inverse process division for any two integers, we have to extend that universal set from the integers to the rational numbers to include all possible outcomes of division. There is a further complication that we cannot include division by zero, as this is undefined. However, if we take as our universal set all positive and negative rational numbers, but exclude zero, then the operations of multiplication and division of two rational numbers p, q always leads to another rational number. There is a loose end here, nowhere have we established that the quotient of two rational numbers is a rational number, so please take this on trust though it is not difficult to prove. Can you see how to prove it yourself? Note that for addition and subtraction zero cannot be excluded from the universal set, because for any non-zero a we have $a + (-a) = 0$ and $a - a = 0$.

Group algebra is concerned in essence with a study of sets that are closed in the sense that, when we perform some defined operation on elements in that set, the result is also an element of that set. The class of operations involved are called *binary* operations because they are applied to any two elements in the set. In group algebra we consider only binary operations that give rise to elements in that same set.

In line with the discussion above, if our set is *all positive integers* addition is such a binary operation, since the result of addition of two natural numbers is to produce a natural number. For the same set multiplication is such a binary operation. The binary operations of subtraction or division take us outside the set of positive integers, so they are not the type of binary operation considered in group algebra. Subtraction would, however, be such a binary operation for the set of all integers (positive, negative and zero). But division is not such a binary operation with respect to that set, because it takes us outside the set of all integers.

If we consider the set of all 2×2 nonsingular matrices with elements that are real numbers, both matrix addition and multiplication are binary operations with respect to that set that keep us within the set.

Formally, a closed set G forms a group with respect to a binary operation, which is conventionally denoted by $\circ$, if it satisfies the following conditions:

- 1. The operation $\circ$ is associative, i.e. if a, b, c are elements of G then $a \circ (b \circ c) = (a \circ b) \circ c$;
- 2. There exists an element I of G such that for any a in G, $a \circ \mathrm{I} = \mathrm{I} \circ a = a$;
- 3. Corresponding to each element a there is an element a' such that $a \circ a' = a' \circ a = \mathrm{I}$.

It is important to remember that $\circ$ is a binary operation such that $a \circ b$ is a member of G for all a, b in G.

Nonempty sets for which only the first axiom holds are said to form *semigroups*. The first axiom is true for any set S that is closed with respect to the binary operation $\circ$ providing that if s_1, s_2, s_3 are an ordered triplet of elements that we get the same result if we perform first the operation $s_1 \circ s_2$ then $(s_1 \circ s_2) \circ s_3$, or if we first perform the operation $s_2 \circ s_3$ then $s_1 \circ (s_2 \circ s_3)$, where the operation within brackets takes precedence.

The set of all real numbers is a semigroup with respect to the operations of either addition or multiplication, since it is closed with respect to either of these operations, and the associative law holds for both.

Does the set S = {1, i, −1, −i} form a semigroup under addition and multiplication by the usual rules? It is immediately obvious that it does not do so under addition, since $1 + 1 = 2$ and thus S is not closed under this operation. It is easily seen to be closed under the operation of multiplication, and since multiplication in the complex number system is associative it follows that this is a semigroup.

Subsets of a set S, where S forms a semigroup, need not themselves be semigroups. The set $S_1 = \{i, -i\}$ does not form a semigroup under multiplication since $i \times (-i) = 1$, and 1 is not an element of S_1, so S_1 is not closed with respect to multiplication.

You may recall that in Chap. 3 we formed addition and multiplication tables for the digits 0 to 9. In considering binary operations it is often convenient to form tables showing the results of any relevant binary operations for a particular set. For example, for the binary operation multiplication for the set S = {1, i, −1, −i} the appropriate table is

	1	i	**−1**	−i
1	1	i	−1	−i
i	i	−1	−i	1
−1	−1	−i	1	i
−i	−i	1	i	−1

This table immediately shows that S is closed under multiplication, since it contains no element that is not an element of S.

The above table pertains to a specific operation (multiplication) for a very special subset of the complex number system. We may set out tables of this kind for sets with arbitrary elements with respect to any operation $\circ$.

The following table shows that the set S = {*banana, plum, apple*} is closed with respect to an operation $\circ$ that accords with that table

	banana	**plum**	**apple**
banana	*banana*	*banana*	*apple*
plum	*plum*	*plum*	*apple*
apple	*apple*	*apple*	*apple*

Does the set from a semigroup with respect to $\circ$? To answer this question we must consider all operations $\circ$ on ordered triplets. If at least one of these is not associative, this is not a semigroup. To save a lot of writing we abbreviate the fruit names to their initial letters b, p, a. Also, without loss of generality, we use the 'multiplication' notation expressing $a \circ b$ as ab as we do when we write ab for $a \times b$ when dealing with real numbers. With this notation we find, rather tediously, that for the 27 possible ordered triplets in every case the operation is associative. We give a few examples in Table 16.1. You should verify for at least a few more cases, that association is established. The complete set of 27 triplets is given in Table A.1 in Sect. A.6 of the appendix

Table 16.1. A check that an arbitrary operation $\circ$ on a set S $= \{b, p, a\}$ is sssociative

$(bb)b = bb = b$ and $b(bb) = bb = b$
$(bb)p = bp = b$ and $b(bp) = bb = b$
$(bb)a = ba = a$ and $b(ba) = ba = a$
$(bp)b = bb = b$ and $b(pb) = bp = b$
$(bp)p = bp = b$ and $b(pp) = bp = b$
$(bp)a = ba = a$ and $b(pa) = ba = a$
$(ba)b = ab = a$ and $b(ab) = ba = a$
$(ba)p = ap = a$ and $b(ap) = ba = a$

Just as in ordinary multiplication of numbers, once association is established brackets may be omitted in the relevant operation.

Small changes can be destructive

Had the operator table for this example been that given above, except that the operation $ab = a$ was replaced by $ab = p$, the set would no longer form a semigroup with respect to the operation. It would still be closed, but the association law would not hold. One counter instance shows this. We now find

$$(aa)b = ab = p \text{ but } a(ab) = ap = a \ ,$$

whence $(aa)b \neq a(ab)$.

The group of positive integers, P, say, obviously forms a semigroup with respect to the operation multiplication, since multiplication of two elements of that group gives another element of the group, and multiplication is associative. It also satisfies the second requirement defining a group, because the element 1 has the property $a \times 1 = 1 \times a = a$ for all $a \in$ P. However, It does not satisfy the third condition, since only if $a = 1$ is there an element a' such that $a \times a' = 1$.

If we extend the set to all positive rational numbers, S, say, this forms a group with respect to the operation multiplication, since corresponding to any element of S, i.e., to any rational number a, there exists another element of S, namely $1/a$, such that $a \times (1/a) = (1/a) \times a = 1$. A small alteration to S destroys this property. Had we defined S as the set of all nonnegative rationals (i.e. added zero to the set), the inverse would not exist for the rational number 0, and so S would no longer be a group.

However, the semigroup consisting of all nonzero real numbers (positive or negative) is easily shown to form a group with respect to multiplication. On the other hand the set of all real numbers, R, (including zero) does not form a group since zero has no inverse. This same set R does form a group with respect to addition. It is easily verified that the three conditions are satisfied, since addition is associative and there is an identity element, zero, such that $a + 0 = 0 + a = a$. The inverse of any element a is $-a$.

Addition of integers has the commutative property that $a + b = b + a$. Groups that have the corresponding property $a \circ b = b \circ a$ are called *commutative*, or Abelian, groups.

We do not prove it, but it is not difficult to show that if an identity element exists it is unique. As the above examples indicate, it is possible for some but not all, elements of a semigroup to have an inverse.

The examples concerning the group properties of numbers might seem little more than a different way of looking at the obvious, but they highlight important oddities that are decisive in determining whether certain closed sets do or do not form a group. Clearly the golden rule that thou shalt not divide by zero is reflected in the fact that a set containing zero does not form a group under multiplication. We see later that singularity of certain matrices leads to similar constraints and, as we have shown, such singularities are important in deciding whether or not sets of linear equations have unique solutions.

Although it is not a technique used in this book, in many branches of both pure and applied mathematics one often wants to change variables x, y to variables u, v, say, where there are known functional relationships between the variables. An important transformation of this type is that where the functional relationships are:

$$x = u \cos\theta + v \sin\theta \tag{16.2}$$

$$y = -u \sin\theta + v \cos\theta. \tag{16.3}$$

This transformation implies $x^2 + y^2 = u^2 + v^2$. This follows by straight-forward algebra, remembering that for all θ, $\cos^2\theta + \sin^2\theta = 1$. We showed in Sect. 7.4 that for Cartesian coordinates $x^2 + y^2$ is the square of the distance of a point with coordinates (x, y) from the origin. Thus, for this transformation, the point (u, v) is at the same distance from the origin in the new co-ordinate system as was (x, y) in the old. The transformation is sometimes described as one that *preserves distance*. Using matrix notation we may write (16.2) and (16.3)

$$\begin{bmatrix} x \\ y \end{bmatrix} = \begin{bmatrix} \cos\theta & \sin\theta \\ -\sin\theta & \cos\theta \end{bmatrix} \times \begin{bmatrix} u \\ v \end{bmatrix}$$

The 2×2 matrix

$$\begin{bmatrix} \cos\theta & \sin\theta \\ -\sin\theta & \cos\theta \end{bmatrix}$$

defines the transformation from x, y to u, v. We now show that for all real θ the set M of 2×2 matrices of the above form constitute a group with respect to matrix multiplication. We first show that M is closed under multiplication. This follows from elementary trigonometry, since for any two real values α, β of θ we have, by the multiplication rule for matrices

$$\begin{bmatrix} \cos\alpha & \sin\alpha \\ -\sin\alpha & \cos\alpha \end{bmatrix} \times \begin{bmatrix} \cos\beta & \sin\beta \\ -\sin\beta & \cos\beta \end{bmatrix}$$

$$= \begin{bmatrix} \cos\alpha\cos\beta - \sin\alpha\sin\beta & \cos\alpha\sin\beta + \sin\alpha\cos\beta \\ -\sin\alpha\cos\beta - \cos\alpha\sin\beta & -\sin\alpha\sin\beta + \cos\alpha\cos\beta \end{bmatrix} .$$

Using the addition formulae given in the Appendix, Sect. A.3, this may be written

$$\begin{bmatrix} \cos(\alpha+\beta) & \sin(\alpha+\beta) \\ -\sin(\alpha+\beta) & \cos(\alpha+\beta) \end{bmatrix}$$

which is a member of the set M. We did not prove it in the introduction to matrix multiplication in Sect. 13.3, but it may be establish that matrix multiplication is associative for square matrices. Thus the set M forms a semigroup with respect to multiplication. To establish that it forms a group, we need to find an identity element included in M, and show also that each element of M has an inverse. It follows from arguments in Sect. 13.3 that the identity matrix must be

$$\begin{bmatrix} 1 & 0 \\ 0 & 1 \end{bmatrix}$$

and it is easily seen by considering the element of M where $\theta = 0$ that

$$\begin{bmatrix} \cos 0 & \sin 0 \\ -\sin 0 & \cos 0 \end{bmatrix} = \begin{bmatrix} 1 & 0 \\ 0 & 1 \end{bmatrix} .$$

We leave it as an exercise to show that every element of M has an inverse and that M is commutative with respect to multiplication. Here is a hint of how to go about this. In the argument used above to show that the group is closed with respect to multiplication set $\beta = -\alpha$, then it is easily shown that the inverse of

$$\begin{bmatrix} \cos\theta & \sin\theta \\ -\sin\theta & \cos\theta \end{bmatrix}$$

is obtained by replacing θ by $-\theta$. Also, by interchanging α and β, it is easily established that M is commutative with respect to multiplication.

It is not difficult to show that all nonsingular 2×2 matrices form a group with respect to matrix multiplication. We leave this as an exercise. However, unlike the special case considered above, the group is in general not a commutative group. This is easily shown by a counter example. Suppose

$$\mathbf{A} = \begin{bmatrix} 1 & 3 \\ 2 & 0 \end{bmatrix} \text{ and } \mathbf{B} = \begin{bmatrix} 1 & 1 \\ 1 & 1 \end{bmatrix}$$

then

$$\mathbf{AB} = \begin{bmatrix} 4 & 4 \\ 2 & 2 \end{bmatrix} \text{ but } \mathbf{BA} = \begin{bmatrix} 3 & 3 \\ 3 & 3 \end{bmatrix} .$$

The non-singularity condition is needed to ensure that any 2×2 matrix that is an element of the group has an inverse.

We showed earlier that the set $\mathrm{S} = \{1, \mathrm{i}, -1, -\mathrm{i}\}$ was closed with respect to multiplication. It forms a semigroup because, by the usual rules of multiplication of complex numbers, multiplication is associative. It is in fact a commutative, or Abelian, group. It is easily verified that 1 is the identity element, and that each element has an inverse. That of 1 is 1, that of i is –i, that of –1 is –1 and that of –i is i. The group is Abelian because multiplication is commutative.

Our final example of a group concerns the set $\mathrm{T} = \{0, 1, 2, 3\}$. This forms a group with respect to the operation addition modulo (4), a concept we met in Sect. 3.3. Fairly obviously T is closed under this form of addition since the only possible remainders on division of the sum by 4 are 0, 1, 2, or 3; e.g., $(2 + 3)(\mathrm{mod}\ 4) = 5(\mathrm{mod}\ 4) = 1(\mathrm{mod}\ 4) = 1$. The law of association holds for this operation as a consequence of the associative law for addition. The identity element is 0. Each element has an inverse. That of 0 is 0, that of 1 is 3 since $(1+3)(\mathrm{mod}\ 4) = 4(\mathrm{mod}\ 4) = 0$. Similarly that of 2 is 2, and that of 3 is 1.

At first glance the group $\mathrm{S} = \{1, \mathrm{i}, -1, -\mathrm{i}\}$ with respect to multiplication and $\mathrm{T} = \{0, 1, 2, 3\}$ with respect to addition modulo (4) seem to have little in common, apart from both involving sets of size 4. In the next section we see there is a closer connection.

16.6 Isomorphisms

We have seen that the set $\mathrm{T} = \{0, 1, 2, 3\}$ forms a group with respect to addition modulo (4). We can form a table for this binary operation like that for the set $\mathrm{S} = \{1, \mathrm{i}, -1, -\mathrm{i}\}$ with respect to multiplication given in Sect. 16.5. You should verify that for the group based on T the table relevant to addition modulo 4 is

	0	**1**	**2**	**3**
0	0	1	2	3
1	1	2	3	0
2	2	3	0	1
3	3	0	1	2

For convenience we repeat the table for the set $\mathrm{S} = \{1, \mathrm{i}, -1, -\mathrm{i}\}$ with respect to multiplication:

	1	i	**−1**	−i
1	1	i	−1	−i
i	i	−1	−i	1
−1	−1	−i	1	i
−i	−i	1	i	−1

In the latter table, let us make what might seem only a change in notation and put 1 = 0, i = 1, −1 = 2, −i = 3. Doing this gives a table identical to that for the binary operation addition modulo (4) applied to T. This change in notation is a one-to-one mapping of the set S onto T, and what the correspondence of the tables under that mapping tells us is that the operation of multiplication in S corresponds to the operation of addition modulo (4) on the transformed data. This mapping is an example of a *group isomorphism.*

More generally, if $(\mathrm{G}_1, \circ)$ is a group with respect to the binary operator $\circ$ on G_1, and $(\mathrm{G}_2, *)$ is a group with respect to the binary operator $*$ on G_2, then an isomorphism between these groups is a one-to-one mapping, f, say, from G_1 to G_2, such that for all e_i, e_j in G_1, $f(e_i \circ e_j) = f(e_i) * f(e_j)$.

In the above example G_1 was $\mathrm{S} = \{1, \mathrm{i}, -1, -\mathrm{i}\}$, G_2 was $\mathrm{T} = \{0, 1, 2, 3\}$, and $\circ$ was the operator *multiply* and $*$ the operator *addition modulo* (4).

Your reaction at this stage may be that isomorphism is little more than a mathematical curiosity, but you will have met an important realization of a group isomorphism at school, although that term may not have been used to describe it. The situation is that where G_1 is the group of positive real numbers, f is the mapping onto the group G_2 of all real numbers that corresponds to the operation *taking logarithms.* That is, any element y of G_2 takes the form $y = \log x$, where x is a member of G_1. The operation $\circ$ is multiplication, and the operation $*$ is addition. What isomorphism expresses here is the property that if x_i, x_j are any two positive real numbers, then

$$\log(x_1 x_j) = \log x_i + \log x_j.$$

Group algebra has applications not only in mathematics itself, but in many areas of science ranging from crystallography to quantum theory. It is often relevant to pattern in an algebraic or geometric sense. Thus, in crystals where patterns are often repeated, certain characteristics of these patterns may exhibit a group structure.

You may well feel that a group such as $\mathrm{T} = \{0, 1, 2, 3\}$ with respect to addition modulo (4) is little more than a curiosity, but it is a particular case of a group based on a set $\mathrm{T} = \{0, 1, 2, \ldots, n-1\}$ with respect to congruence modulo (n) under addition. This group plays a key role in the statistical topic of experimental design, in planning what are called *confounded factorial experiments.* By its use an experimenter can decide what treatment combinations to include in an experiment so as to make it as efficient as possible in helping to determine, for example, how much of each of several fertilisers should be applied to maximize a crop yield, or at what temperature, pressure and humidity, an industrial chemical reaction should be carried out to maximize yield at minimum cost.

16.7 Try your Skills

If you have not met the notion of permutations of a set of objects you should study Sect. A.4 in the Appendix before continuing to read this section.

If we consider the set of the first n natural numbers, each ordered arrangement of those numbers is called a permutation. In Sect. A.4 we show that there are $n!$ distinct permutations or orderings of that set. If $n = 2$ the permutations are $(1, 2)$ and $(2, 1)$. If $n = 3$ they are $a_1 = (1, 2, 3)$, $a_2 = (1, 3, 2)$, $a_3 = (2, 1, 3)$, $a_4 = (2, 3, 1)$, $a_5 = (3, 1, 2)$, $a_6 = (3, 2, 1)$. If we call the operation of forming a permutation *permuting*, then the set of six permutations of the set $\{1, 2, 3\}$ form a group with respect to this operation. The set is closed with respect to the operation. We leave it as an exercise, and it is not a trivial one, to confirm that the set $\{1, 2, 3\}$ does form a group under permutation. To get you started here is a suggested approach. In the form given above if $\circ$ is the operation permute, then, for example, we interpret $a_2 \circ a_3$ to mean that $a_2 = (1, 3, 2)$ is to be permuted so that the first element 1 permutes to the value of the first element in a_3 which 2. The second element 3 permutes to the value of the third element in a_3 which is 3. The third element 2, permutes to the second element in a_3 which is 1. Thus $a_2 \circ a_3 = (2, 3, 1) = a_4$. You should not find it too hard to establish that a_1 is the identity element. With a little practice at the permuting operation, one may establish all the other requirements. For example, can you see that $a_4 \circ a_5 = a_1$?

We have looked at the special case $n = 3$. The set of the first n natural numbers also form a group under permutation. This group property of the permutation of the first n natural numbers is the tip of an iceberg. It extends beyond numbers to a wide class of groups known as *permutation groups*.

16.8 Loose Ends

A paradox. The definitions and axioms for set theory given in Sect. 16.2 generally work well in practice, but there are cracks in the logical structure. One was highlighted about 100 years ago by the logician Bertrand Russell (1872–1970) and is usually described as *Russell's paradox.*

The paradox arises in this way. It is possible to have sets where the set itself may or may not be a member of that set. For example, if S is the set of all animals which are elephants, then S is not a member of that set because it is not an elephant. It is a collection of elephants, but this is not the same thing as an elephant. It is also possible to have a set that belongs to itself. For, example, if T is the set of all things that are not elephants, then T belongs to itself, because it is not an elephant. There is no paradox so far.

Russell considered a set R, the elements, x, of which are those sets that are not members of themselves, often abbreviated to $x \notin x$, i.e.,

$$R = \{x : x \notin x\}.$$

For this set R we have a contradiction, for if R is an element of R, (i.e., R $\in$ R). this implies R $\notin$ R. On the other hand if R is not a member of itself, i.e. R $\notin$ R this implies $R \in R$. That is the contradiction.

Fortunately, this logical flaw can be repaired by introducing additional, but by no means 'obvious', axioms. One such set of additional axioms that we shall not describe here was proposed by the German mathematician Ernst Zermelo (1871–1953) and these axioms, with some further modifications, are now often referred to as the Zermelo–Frankel axioms of set theory.

An impossible goal. Another loose end it is appropriate to mention here is that at the beginning of the twentieth century pure mathematicians harboured the dream that it would some day be possible to formulate an axiomatic system for which all possible arithmetic operations could either be shown to be consistent or not consistent. Kurt Gödel (1906–78) established by logical argument that this goal could never be achieved, since any system must give rise to undecidable propositions about itself — i.e., to propositions that by their very nature can never be proved or disproved. While this finding may have disappointed some pure mathematicians, it in fact removed a potential limiting factor in that it implied mathematics could not become a potentially closed subject simply by finding a set of all-embracing axioms.

Other algebras. Dealing only with sets and groups in this chapter leaves out a lot of modern algebra, including in particular the concept of *rings*, and also the algebra of *symbolic logic.*

There is a brief note on rings in Sect. 25.2.

In the middle of the nineteenth century mathematicians, including the eminent and innovative English mathematician George Boole (1815-64), became interested in formulating systems of axioms, and for any set of axioms exploring what rules of operation could be set up to be consistent with those axioms. Boole set up an algebra — *Boolean algebra* — that has proved valuable in the logical design of modern computer systems.

Prior to Boole's time, logic had developed independently from mathematics. Boole, and one or two others, explored the possibility of sets of axioms on which algebras could be built that led to logical systems of deduction about truth, falsity etc. There is a brief description under the heading *truth functions and truth tables* in Sect. 25.2 of a few of the notions of symbolic logic.

17

Probability

17.1 Helping a Gambler

Gambling — a stimulant for mathematicians

Probability theory is a branch of pure mathematics. It provides a framework for both the theory and practice of the science of *statistical inference*, often abbreviated to *statistics*. In this context *statistics* should not be confused with the same word applied to what are just collections of data.

Like so many other mainstream ideas in mathematics, probability had its origin in attempts to solve practical problems — in this case to do with gambling.

Gamblers often sought help from mathematicians after bitter experience taught them that a game that they thought should be favourable to them was beneficial only to their adversary.

Fair game

A classic and oft quoted example is that of the gambler Chevalier de Méré who approached Blaise Pascal (1623–62) for help. From practical experience de Méré found it was good for his finances (i.e., to his advantage) to place an even money bet on at least one six in 4 casts of a die (die is the singular of dice). However, it was not a good idea to bet on at least one double six in 24 casts of a pair of dice.

De Méré thought that both should be equally favourable, arguing that there are six possible outcomes for a single die, only one of which is favourable, and 36 possible outcomes with the cast of a pair of dice, again only one on which is favourable. He reasoned that the odds of success should be the same, because the ratio of the number of attempts to the number of possible outcomes was the same, i.e., 4 when there are 6 possible outcomes, and 24 when there were 36 possible outcomes, i.e., $4/6 = 24/36 = 2/3$.

You should write down all possible outcomes when a pair of dice are cast to convince yourself that there are indeed 36 possibilities, and that only one of these would favour de Méré. We return to this in Sect. 17.6.

Pascal was able to show that there was a less than even money chance of winning with the second bet, and a greater than even money chance of winning with the first one. This was in line with de Méré's experience, but not with his reasoning. De Méré's dilemma warns us that the laws of probability are not obvious.

The simplest notion of probability stems from situations where there are sets of 'equally likely' outcomes. This is sometimes intuitively reasonable. If we toss a coin it is equally likely to land heads or tails. For practical purposes these are the only possible outcomes. The coin just might land on its rim and stay there, but that possibility is so remote that it can be ignored in practice. We define the probability of heads to be 1/2, and the probability of tails to be also 1/2, arriving at these probabilities as the ratio of the number of outcomes giving rise to the named event to the number of equally likely possible outcomes.

Likewise, if we cast a die, then, if it is a 'fair die' each of the outcomes 1, 2, 3, 4, 5, 6 is equally likely. Thus there are six equally likely outcomes and the probability of each is 1/6. For the problem de Méré took to Pascal the only favourable outcome at a single cast was a six, with probability 1/6. Using modern notation we write Pr(a six) = 1/6, where 'Pr' is an abbreviation for probability. We call each specific outcome an *event*. If A is any event we write Pr(A) for the probability that that event occurs.

Nature is not always so kind

Life is not always so simple. It may be impossible to define sets of equally likely outcomes or events. If we toss a drawing pin there is no reason to believe that it is equally likely that it will land *point up* as in Fig. 17.1(**a**), or point down as in Fig. 17.1(**b**).

To get some idea of the long term performance of a particular drawing pin we might toss the pin many times to see how often it lands point up. This will depend on the physical characteristics of the pin. How long is it? How heavy is the head? For a particular pin one hopes for reasonable consistency in the proportion of outcomes 'point up' and 'point down' after a great many tosses. With one drawing pin I got the outcomes shown in the table on the next page.

Based on this experiment, an intuitively reasonable estimate (to two decimal pleaces) of the probability for 'point up'— called a *limiting relative fre-*

Fig. 17.1. A drawing pin when tossed may land point up (**a**) or point down (**b**)

quency probability — would be 0.66 (or perhaps 0.67). The 'limiting' notion is here attached to n, the number of tosses, becoming increasingly large.

Number of tosses(n)	*Number of point up landings* (a)	*Frequency* (a/n)
10	6	0.60
50	31	0.62
100	64	0.64
500	327	0.654
1000	668	0.668
2000	1321	0.6605
5000	3329	0.6658

17.2 More Complicated Events

How do we calculate probabilities for more complicated events? One way is to break possible outcomes into sets of equally likely events, but care is needed. The three possible outcomes when we toss two coins are that we get *two heads, two tails* or *one head and one tail.* If it is not immediately obvious that these outcomes are not equally likely, try about a hundred tosses of a pair of coins and note the number of times each outcome occurs. You should be surprised if you don't find that you get either two heads or two tails about equally often, but that one head and one tail occurs about twice as often as either two heads or two tails.

It is easy to explain why. Denote the event that the first coin falls heads by H_1, and that it falls tails by T_1. Label the corresponding events for the second coin H_2 and T_2. Since each of the events H_1 or T_1 have the same probability, as do each of the events H_2 and T_2, it is evident that each of the four possible outcomes (H_1 and H_2), (H_1 and T_2), (T_1 and H_2) and (T_1 and T_2) are equally likely. Thus, each has an associated probability of 1/4. Two of these, (H_1 and T_2), (T_1 and H_2), are favourable to the event one head and one tail, so the probability of that event is $2/4 = 1/2$.

Again beware of intuition

Implicit in these intuitive arguments are the concepts of multiplication and addition of probabilities. For example, the probability of H_1 is 1/2; so also is the probability of T_2. That of both (H_1 *and* T_2) is $1/4 = (1/2) \times (1/2)$. Also, as noted above, the probability of one head and one tail is 1/2, which we obtained as the sum of the probabilities of the events (H_1 and T_2) and (T_1 and H_2). The rules of addition and multiplication might appear to be that the probability that one of two events occur is the sum of their probabilities, and that the probability that both events occur is the product. These rules work for the above example, but the general rules for multiplying and adding probabilities are more complicated.

Two events are *mutually exclusive* if both cannot occur at the same performance of an experiment. When we toss a single coin we cannot get both a head and a tail, so these events are mutually exclusive. If we cast a die once we cannot get both a six and an odd number (i.e. 1, 3 or 5), so again these events are mutually exclusive. However, the events *six* and *an even number* (i.e. 2, 4 or 6) are not mutually exclusive. If the die falls with the face *six* uppermost both events have occurred.

17.3 Probability Axioms

We said a little about axioms in general in Sect. 8.2 and again in Sect. 16.1. Three axioms form the basis of probability theory. They are:

- *Axiom* 1 The probability of any event A is a non-negative number between 0 and 1 inclusive. (i.e., $0 \leq \Pr(A) \leq 1$).
- *Axiom* 2 If an event A is certain $\Pr(A) = 1$.
- *Axiom* 3 If A and B are mutually exclusive then

$$\Pr(A \cup B) = \Pr(A) + \Pr(B).$$

The set notation $A \cup B$ introduced in Sect. 16.2 is used in *Axiom* 3 because events may be looked upon as sets. In this context the union covers all elements that are in either or both sets. The restriction that A, B are *mutually exclusive* means they have no common elements. We shall later use the set notation $A \cap B$ to mean that *both* A *and* B *occur.*

Axiom 3 is described as the *addition rule for mutually exclusive events.* In Sect. 17.5 we give a modification needed if events are not mutually exclusive. This modification is not a new axiom. It may be deduced from the three given above.

Under normal circumstances if we toss a coin and cast a die any outcome with the coin is said to be *independent* of any outcome with the die. This is because the physical setup is such that whether the coin happens to land heads or tails does not make any of the possible outcomes with the die more or less likely.

Here is a situation where events are not independent. Suppose we have four identical boxes. Their contents are respectively (i) a prize of £1000, (ii) a prize of £500, (iii) a box of chocolates and (iv) a mouse-trap. There is no visible external means of knowing which box contains a particular item. If we choose one box there is a probability of 1/4 of selecting the one with the £1000, since there is only one favourable outcome among the four possible equally likely outcomes. There is the same probability of obtaining each of the other prizes.

Suppose, after we have made our first choice, we are then allowed to select a second box from those remaining. If the box already selected had contained the mouse trap, there are three boxes left that contain £1000, £500 or a box of chocolates. So we have one chance in three, or a probability of 1/3, of selecting the £1000 box at our second choice. The probability of getting the £1000 prize at the second selection is always 1/3 if we have selected some other box at the first attempt. However, if we selected the £1000 box the first time, the probability would then be zero of our getting £1000 at the second selection, because the prize is no longer there.

The probabilities of getting the £1000 prize at the second attempt therefore depend on the result at the first selection. The outcomes are not independent, and further, we refer to the different probabilities, 1/3 and 0, associated with the second draw as *conditional probabilities*. This is because their value is conditional upon the first selection outcome. In this example it is 1/3 if the first outcome is anything other than £1000. It is zero if the first outcome is £1000.

Here is another example of conditional probability. If we draw cards without replacing them from a well-shuffled deck of playing cards, the concept of probability as a ratio of the number of favourable cases to the total number of equally likely outcomes, implies there is no reason why any particular card should appear at the first draw. Each of the 52 cards is equally likely to be drawn. This, in essence, is what is implied by the phrase 'well-shuffled'. Since 13 of these possibilities involve the drawing of a heart, the probability of drawing a heart is $13/52 = 1/4$.

Now draw a second card. What is the probability that it is a heart? This will depend on whether or not the first card was a heart. If the first card was a heart there are 12 hearts among the 51 remaining cards. Therefore the probability that the second card is a heart is 12/51. If the first card were not a heart, there will still be 13 hearts among the 51 cards that remain, making the probability of a heart at this second draw 13/51. Probabilities are *condtional* if they depend on what happened on one or more previous occasions.

If we call the first event A and the second event B, we call the probability of the second event happening when we know that the first event has occurred *the conditional probability of* B *given* A. A convenient shorthand for this is Pr(B|A). In speech we call this the *probability of* B *conditional upon* A.

A common mistake made by people new to the conditional concept is to assume that Pr(B|A) is the same as Pr(A|B). This is not generally true. If the event A is a person drinks a bottle of whisky on Monday, and the event B is that they wake up with a headache on Tuesday, then Pr(*that a person wakes up with a headache given they have drunk a bottle of whisky*) is Pr(B|A). This is an almost certain event. The exact probability may depend slightly upon the person's constitution, but it will be close to 1. However, Pr(*that a person has consumed a bottle of whisky given they wake up with a headache*) is Pr(A|B). This will be somewhat less than Pr(B|A), since there are many causes of headaches other than excessive consumption of whisky.

We return to the example of drawing two cards. We have Pr(*second card a heart* | *first card a heart*) $= 12/51 =$ Pr(B|A), say. This ratio is equivalent to the ratio of the number of cases where both A and B occur among those cases where A has already occurred. Using set theory notation the event that both A and B occur is written $\mathrm{A} \cap \mathrm{B}$. Thus the probability of observing B conditional upon A is also equivalent to $\Pr(\mathrm{A} \cap \mathrm{B}) / \Pr(\mathrm{A})$, so

$$\Pr(\mathrm{B}|\mathrm{A}) = \Pr(\mathrm{A} \cap \mathrm{B}) / \Pr(\mathrm{A}) .$$

This is easily seen to be equivalent, by the rules of algebra, to

$$\Pr(\mathrm{A} \cap \mathrm{B}) = \Pr(\mathrm{A}) . \Pr(\mathrm{B}|\mathrm{A}) . \tag{17.1}$$

This is the basic *multiplication rule of probabilities.*

In our card example the probability that both the first and second cards drawn are hearts is

$$\Pr(\mathrm{A} \cap \mathrm{B}) = \frac{13}{52} \times \frac{12}{51} = \frac{1}{17} \approx 0.05882.$$

17.4 Independent Events

Suppose that in a card-drawing experiment we replace the first card drawn, reshuffle the deck, and then draw a second card. Because we are again drawing from a full deck, the probability of the event, B, that the second card is a heart, would then be the same as the probability of the event, A, that the first card is a heart. In this case, Pr(B|A) = Pr(B). Now (17.1) becomes

$$\Pr(\mathrm{A} \cap \mathrm{B}) = \Pr(\mathrm{A}) . \Pr(\mathrm{B}).$$

When Pr(B|A) = Pr(B) we say the events A, B are *independent.* In the case considered above for drawing two cards, if we replace the card after the first draw we also have Pr(A) = Pr(B). This equality is not generally true for independent events. For example, if we toss a coin and cast a die, the

Table 17.1. Numbers of boys with different eye colour/shirt colour combinations

	Eye colour	*Blue*	*Brown*	Total
Shirt	*Coloured*	50	30	80
type	*White*	15	5	20
	Total	65	35	100

event that the coin lands heads (event A), and the die cast results in a 5 or a 6 (event B), are independent. Here $\Pr(A) = 1/2$ and $\Pr(B) = 1/3$, so $\Pr(A \cap B) = (1/2) \times (1/3) = 1/6$.

17.5 Events that are not Mutually Exclusive

When two events are not mutually exclusive the addition rule is modified. Suppose that in a group of 100 boys 65 have blue eyes and the remaining 35 have brown eyes. In the group 80 wear coloured shirts and the remaining 20 wear white shirts. One of the group is selected at random to present a farewell gift to a retiring teacher. Given this and the additional information in Table 17.1 what is the probability the boy selected is either brown eyed (event A) or wears a white shirt (event B)? The phase 'selected at random' is here a shorthand way of saying any member of the group is equally likely to be selected.

From Table 17.1 the probability of A is easily seen to be $\Pr(A) = 35/100$, since that is the proportion of brown-eyed boys. Similarly , $\Pr(B) = 20/100$. Again, by similar reasoning, the probability that the boy chosen has both brown eyes and wears a white shirt is 5/100. This is the event $A \cap B$.

We exclude only the 50 boys who wear coloured shirts and have blue eyes, because these are the only ones who meet neither of the required criteria. It follows that the probability that at least one of the specified criteria is met (i.e., the event $A \cup B$ occurs) is

$$\Pr(A \cup B) = (15 + 5 + 30)/100 = 50/100 = 0.5\,.$$

On the other hand $\Pr(A) + \Pr(B) = 35/100 + 20/100 = 55/100$. The addition rule for mutually exclusive events, i.e., $\Pr(A \cup B) = \Pr(A) + \Pr(B)$, no longer holds. This is because we have counted the 5 boys that satisfy both A and B (i.e., $A \cap B$) twice — once in the count of the numbers with brown eyes and once in the count of the numbers with white shirts. We need to deduct one of these counts and this leads to the rule

$$\Pr(A \cup B) = \Pr(A) + \Pr(B) - \Pr(A \cap B).$$

This is the *addition rule* for any pair of events (i.e. the probability that at least one occurs). For mutually exclusive events $\Pr(A \cap B) = 0$, since 'mutually exclusive' implies that *both happening* is an event with zero probability.

The attraction of opposites

A useful spin-off from the addition rule concerns *opposite events*. The events A and B are opposite if they are mutually exclusive and one or other must happen at any trial. Thus, when casting a die the outcomes 'score 5 or more' and 'score less than 5' are opposite events, and one of these must happen. The probability of a certain event is 1 (probability axiom 2). Thus, if A and B are mutually exclusive and opposite events, we have

$$\Pr(\mathrm{A} \cup \mathrm{B}) = \Pr(\mathrm{A}) + \Pr(\mathrm{B}) = 1 ,$$

whence

$$\Pr(\mathrm{B}) = 1 - \Pr(\mathrm{A}) .$$

A notation often used is to write $\mathrm{B} = \overline{\mathrm{A}}$ where $\overline{\mathrm{A}}$ is called the opposite event to A, i.e., it is the event that A does not occur. The set of events A and $\overline{\mathrm{A}}$ are often described as a *mutually exclusive and exhaustive* set of outcomes. This result is useful if it is easier to calculate from first principles the probability associated with one of A or $\overline{\mathrm{A}}$, than it is to calculate the other. Sometimes A^* is used instead of $\overline{\mathrm{A}}$ for the opposite event to A.

The addition and multiplication rules may be extended to any number of events. Their form becomes more involved in general situations, but if we confine attention to mutually exclusive events. where only one of them may occur at a given trial, then in a fairly obvious notation the addition rule for, say, 5 events A, B, C, D, E takes the form

$$\Pr(\mathrm{A} \cup \mathrm{B} \cup \mathrm{C} \cup \mathrm{D} \cup \mathrm{E}) = \Pr(\mathrm{A}) + \Pr(\mathrm{B}) + \Pr(\mathrm{C}) + \Pr(\mathrm{D}) + \Pr(\mathrm{E}) .$$

Since the events are mutually exclusive this is the probability that exactly one of them occurs. The general form of the addition rule when events are not mutually exclusive is more complicated and we do not consider it here.

Similarly, if events A, B, C, D, E are mutually independent, i.e., the probability of each of them occurring is not influenced by whether or not any of the others occur, then

$$\Pr(\mathrm{A} \cap \mathrm{B} \cap \mathrm{C} \cap \mathrm{D} \cap \mathrm{E}) = \Pr(\mathrm{A}).\Pr(\mathrm{B}).\Pr(\mathrm{C}).\Pr(\mathrm{D}).\Pr(\mathrm{E}).$$

That is, the probability all occur is the product of the probabilities that each occurs. The form of the extension to numbers of events other than 5 is obvious. If events are not independent the extension to more than two events 2 events is more complicated. We shall not consider this here.

The probability axioms, and the addition and multiplication rules, provide useful tools for determining probabilities of complicated events. That is, in situations where the equally likely outcomes concept, or that of a limiting relative frequency, may be difficult, or even impossible, to apply directly.

Here is an example. A die is cast 10 times, what is the probability of the score being *five* on exactly 6 out of the 10 casts? The usual physical set-up implies all casts are independent. Let A_i be the event *five* scored at the ith cast where i may be any integer between 1 and 10 inclusive. It is easily seen that $\Pr(\mathrm{A}_i) = 1/6$ for all i. Also, for the event $\overline{\mathrm{A}}_i$ that the outcome is *not a five*, $Pr(\overline{\mathrm{A}}_i) = 5/6$. Suppose that in 10 casts we get the following pattern of outcomes:

$$\mathrm{A}_1, \mathrm{A}_2, \overline{\mathrm{A}}_3, \mathrm{A}_4, \overline{\mathrm{A}}_5, \mathrm{A}_6, \overline{\mathrm{A}}_7, \overline{\mathrm{A}}_8, \mathrm{A}_9, \mathrm{A}_{10} \, .$$

This tells us that we score a *five* at six particular casts, i.e., the 1st, 2nd, 4th, 6th, 9th and 10th cast, and some other score at the remaining 4 casts. Since each of the events A_i or $\overline{\mathrm{A}}_i$ are independent for different value of i, it follows from the multiplication rule that

$$\begin{aligned} &Pr(\mathrm{A}_1 \cap \mathrm{A}_2 \cap \overline{\mathrm{A}}_3 \cap \mathrm{A}_4 \cap \overline{\mathrm{A}}_5 \cap \mathrm{A}_6 \cap \overline{\mathrm{A}}_7 \cap \overline{\mathrm{A}}_8 \cap \mathrm{A}_9 \cap \mathrm{A}_{10}) \\ &= \frac{1}{6} \times \frac{1}{6} \times \frac{5}{6} \times \frac{1}{6} \times \frac{5}{6} \times \frac{1}{6} \times \frac{5}{6} \times \frac{5}{6} \times \frac{1}{6} \times \frac{1}{6} = \left(\frac{1}{6}\right)^6 \left(\frac{5}{6}\right)^4 . \end{aligned}$$

It is equally likely that 6 scores of *five* might have arisen with any other selection of 6 from the 10 casts. For example, at the 1st, 4th, 5th, 7th, 8th and 10th casts, with some score other than *five* at the remaining 4 casts. For each such arrangement the associated probability, by the multiplication rule for independent events, is again $(1/6)^6(5/6)^4$. Further, two different orderings are mutually exclusive, so the probability that one of them occurs is obtained by adding, i.e., it is $(1/6)^6 \times (5/6)^4 + (1/6)^6 \times (5/6)^4 = 2[(1/6)^6 \times (5/6)^4]$.

There are many other possible orderings corresponding to 6 scores of *five* and 4 of some other outcome in 10 casts.

Working out how many possible orderings there are is a counting problem. It is solved by using the concepts of *permutations* and *combinations*. If you are not already familiar with these topics, a brief outline is given in the Appendix, Sect. A.4. The general problem is one where we are given n items and want to know in how many ways we can select sets of r of these with a specified property, while the remaining $n - r$ have some other property.

In our example $n = 10$, and $r = 6$, the property associated with r being a score of *five*. We show in the Appendix that the number of different ways we can select r from n recognizably distinct objects of two different kinds is given by the binomial coefficient

$$\binom{n}{r} = \frac{n(n-1)(n-2)\ldots(n-r+1)}{1 \times 2 \times 3 \times \ldots \times r} .$$

The denominator is our friend $r!$ (factorial r).

For the case $n = 10$, $r = 6$ substitution shows the number of mutually exclusive orderings to be

$$\frac{10 \times 9 \times 8 \times 7 \times 6 \times 5}{6 \times 5 \times 4 \times 3 \times 2 \times 1} = 210 \, .$$

The probability we calculated above for each of two specific orderings is the same for each of these 210 possible and mutually exclusive orderings. So, by the addition rule, the required probability of exactly 6 scores of *five* in 10 casts is

$$210 \times \left(\frac{1}{6}\right)^6 \left(\frac{5}{6}\right)^4 .$$

Attacking this with a pocket calculator you should find that

$$210 \times \left(\frac{1}{6}\right)^6 \left(\frac{5}{6}\right)^4 \approx 0.002171.$$

Similar arguments can be applied to calculating each of the probabilities of 0, 1, 2, 3, 4, 5, 7, 8, 9 or 10 scores of five in 10 casts of a die. The only difficulty is the tedious arithmetic. By generalizing the argument above, the probability of r casts from 10 resulting in a score of *five* is easily found to be

$$\binom{10}{r} \left(\frac{1}{6}\right)^r \left(\frac{5}{6}\right)^{10-r} . \qquad (17.2)$$

Substituting each value of r in turn in this expression, and using a pocket calculator to complete the otherwise tedious arithmetic, gives the probabilities in Table 17.2. Each probability is rounded to 6 decimal places. You must either take my word for it, or slog through the arithmetic. With ingenuity the slog can be reduced, because increasing r in unit steps only results in simple computational changes to successive probabilities. Since we have a mutually exclusive and exhaustive set of outcomes the probabilities should sum to unity. The slight discrepancy in Table 17.2 is due to rounding all probabilities to 6 decimal places.

17.6 The Binomial Distribution

We may generalize the type of probability given in (17.2) by replacing 10 by any integer n, and the proviso that r does not exceed 10 by the proviso that r does not exceed n. We may also replace the probability, 1/6, that an outcome of interest occurs at each trial by a general probability p, and the probability that that the event does not occur (the opposite event), which in the above example was 5/6, by a probability $q = 1 - p$, since $p + q = 1$.

If we call the event with probability p a *success* and this probability, p, remains constant in n independent trials, then the probability of r successes in n trials is given by

Table 17.2. Probabilities of scoring r fives in 10 casts of a true die

r	*Probability*
0	0.161506
1	0.323011
2	0.290710
3	0.155045
4	0.054266
5	0.013024
6	0.002171
7	0.000248
8	0.000019
9	0.000001
10	0.000000
Total	0.999999

$$\Pr(r \text{ successes in } n \text{ trials}) = \binom{n}{r} p^r q^{n-r}. \tag{17.3}$$

This is a term in the binomial expansion for $(p+q)^n$ in the form given in Sect. 12.2. The number of successes in n independent trials, at each of which there are two possible outcomes with constant probabilities p, q respectively, is an example of a *random variable.* Although it is common to call the event with probability p a *success*, and that with probability q a *failure*, in a particular context these terms may not be literally appropriate.

If we repeat the experiment of casting a die 10 times, and record for each repetition the number of times a *five* appears, this number will vary from experiment to experiment. In the long run each possible number of successes between 0 and 10 should occur with relative frequencies that approach the probabilities in Table 17.2. A random variable with the properties given here is said to have a *binomial distribution.*

When casting a fair die the probability of each possible outcome is $p = 1/6$. If we are interested in repeated casts of a pair of dice, what is the probability of casting a double six? It is reasonable to assume that all possible score pairs, (x, y), where x is the score on the first die and y the score on the second die, are equally likely. As we mentioned at the beginning of this chapter, there are 36 different, but equally likely, pairings. This follows because each of the 6 possible scores on the first die may be associated with each of the 6 possible scores in the second die, giving $6 \times 6 = 36$ possible pairings. If you have any doubt, here are the 36 pairings:

$$
\begin{array}{cccccc}
(1,1) & (1,2) & (1,3) & (1,4) & (1,5) & (1,6) \\
(2,1) & (2,2) & (2,3) & (2,4) & (2,5) & (2,6) \\
(3,1) & (3,2) & (3,3) & (3,4) & (3,5) & (3,6) \\
(4,1) & (4,2) & (4,3) & (4,4) & (4,5) & (4,6) \\
(5,1) & (5,2) & (5,3) & (5,4) & (5,5) & (5,6) \\
(6,1) & (6,2) & (6,3) & (6,4) & (6,5) & (6,6)
\end{array}
$$

The probability of scoring a double six at each cast of a pair of dice is $p = 1/36$, since only one of the 36 possible equally likely outcomes is favourable to this event.

We pointed out in Sect. 17.5 that it is sometimes easier to calculate the probability of the opposite event to that of the one we are interested in. We then obtain the probability we want by subtracting the calculated probability from 1. This dodge is useful with a binomial distribution when we want the probability of at least one occurrence of a favourable event. For example, if we want the probability of *at least one double six in 4 casts of a pair of dice* we could find this by calculating the probabilities of each of the mutually exclusive events 1 pair, 2 pairs, 3 pairs, 4 pairs of sixes. It is easier to recognise that the opposite event to *at least one* is *none*, and to simply calculate this one probability. For 4 casts the probability of *no double six* is easily seen to be $(35/36)^4 = 0.8934$. This implies that the probability of the opposite event, *at least one double six*, is $1 - 0.8934 = 0.1066$.

Back to the Start

We now have the tools to answer the poser in Sect. 17.1 put to Pascal by de Méré. The two gambles to be compared are even money bets on at least one six in 4 casts of an ordinary die, and at least one double six on 24 casts of a pair of dice. From the arguments above, it is clear that in the first case the probability of the opposite event (no six) is given by a binomial distribution term with $n = 4, p = 1/6$, $q = 5/6$, and $r = 0$, i.e.,

$$\Pr(\text{no six in 4 casts}) = (5/6)^4 \approx 0.4823\ .$$

Thus, the probability of the opposite event (at least one six) is $1-0.4823 = 0.5177$. Since this probability exceeds 0.5, an even money bet is favourable. By similar reasoning, for 24 casts of a pair of dice the probability of no double six is $(35/36)^{24} \approx 0.5086$. Thus the probability of at least one double six is $1 - 0.5086 = 0.4914$. This means an even money bet is unfavourable. Since de Méré's intuitive reasoning led to the wrong result, this is a good example of intuition being a bad master.

17.7 Random Variables

The binomial distribution illustrates the use of a *mathematical model* to represent the distribution of a random variable. In statistics it is conventional to denote a random variable by X. Any specific value of that variable that is, or might be, observed is denoted by the corresponding lower case x. When, as is usual in experiments, we have more than one observation, we denote these by $x_i, i = 1, 2, \ldots, n$, where n is the number of observations. In this notation, for the binomial distribution in Sect. 17.6 involving 10 casts of a die, X is the number of *fives* in 10 casts and an observed x_i may take any integer value between 0 and 10. A random variable that takes only integer values is an example of a *discrete random variable.* The corresponding distribution is called a *discrete distribution.*

The binomial distribution is one of many discrete distributions. For any given n, p, with a slight alteration to the notation used in Sect. 17.6, the binomial distribution is completely determined if we know for all possible values of x

$$\Pr(X = x) = \binom{n}{x} p^x q^{n-x}.$$

The right hand side of this expression is a function of x. It is called the *probability density function*, or sometimes the *frequency function* or *probability mass function*, of X. This function determines the probability distribution of X. The quantities n and p, which we must know to determine the precise function of x, are called *parameters.*

Data that simply take discrete values such as integer values are not the only data type showing random variability. In manufacturing processes there is often a target measurement for some characteristic of a product such as weight, area or length. Because of variability due to imperfections in the manufacturing process, that target is not often achieved exactly. For example, in packing sugar or flour or tea, the aim is usually to achieve a certain weight in each container. The containers are filled mechanically.

A common practice is to indicate a 'nominal weight' e.g., 1 kg for each container if, for instance, these are bags of sugar. Because the weights of packages delivered by the packing machine fluctuate, the machine may be set to deliver a target weight of say 1.005 kg. This is done because the manufacturer wants to be reasonably sure that few, if any, packets will be less than the stated nominal weight. To see whether this is the case samples may be taken and weighed. The weights, in kg, of 10 bags in a random sample may then turn out to be

$$1.001, 1.007, 1.005, 1.003, 1.008, 1.006, 1.006, 1.002, 1.004, 1.006.$$

The concept of a *random sample* is an important statistical notion. To explain it even sketchily we need another notion — that of a *population.* To a statistician, an appropriate population in this situation might consist of a set of all items in which the statistician, or his or her client, is interested. In the sugar bag example the population may be the weights of all bags in one day's production. Here the population would consist of a finite set of weights. Often one envisages somewhat hypothetical unlimited or infinite populations, such as weights of all possible bags of sugar that may be packed by a particular machine, or group of machines. A true random sample from a finite population is one obtained by the method of random selection. This is a process where every possible sample of a given size is equally likely to be selected. This process cannot be defined precisely for a hypothetical infinite population, but there are ways of overcoming this difficulty in practice.

In our sugar bag-filling example, a manufacturer usually wants to reduce by as much as possible the variability about the mean weight delivered. The manufacturer also wants to give away as little 'free' sugar as is consistent with some guarantee that the percentage of items weighing less than the nominal weight will be small. The aim is to set the packaging machine to deliver accordingly. Guidance in this situation is often given by the *normal law of errors*, first formulated by Carl Friedrich Gauss (1777–1855). This distribution is only one of many associated with continuous random variables.

Basically, a *continuous random variable* is one that may take any real value either within a finite interval, or sometimes may take any real-number value. For continuous random variables we cannot associate a probability with each of the infinite number of possible values that may be taken by X. We can, however, associate a probability with any arbitrary small interval $(x, x + \delta x)$ that lies entirely in the range of values that X may take. This latter interval may be written (a, b) if the interval is finite. If X may take any real number as its value, it is convenient to write this interval formally as $(-\infty, \infty)$. The probability an observed value of X lies in a specific interval $(x, x + \delta x)$ will depend on the particular value of x, i.e., it will be a function of x which we write $f(x)$. This function is so chosen that

$$\Pr(x \leq X \leq x + \delta x) = f(x)\delta x \text{ when } \delta x \to 0.$$

We often want to know the probability that X takes some value less than a particular value x. For a continuous random variable this is a problem of integration and we have

$$\Pr(X \leq x) = \int_{-\infty}^{x} f(x)\mathrm{d}x. \tag{17.4}$$

This formula incorporates the case where X only takes values in a finite interval (a, b) if we define$f(x)$ to be zero if $x < a$ or $x > b$. Since a random variable must take some real value it follows that $\Pr(X \leq \infty)$ implies an exhaustive set of events, so that

$$\int_{-\infty}^{\infty} f(x)\mathrm{d}x = 1. \tag{17.5}$$

The integral (17.4), often denoted by $F(x)$, is called the *cumulative distribution function* for X. The function $f(x)$ is called the *probability density function*, or *frequency function*, of X.

Mathematically, one of the simplest continuous distributions is the uniform distribution over a finite interval (a, b). From the statistical viewpoint the assumption is that a value of X is equally likely to fall anywhere between a and b. More formally, this means that for a given δx, the value of X is equally likely to fall in any subinterval $(x, x+\delta x)$ lying entirely in the interval (a, b). This means that $f(x)$ is a constant over the interval (a, b), and zero everywhere outside that interval. Further, it is easily verified by integration that if (17.5) is to hold then $f(x) = 1/(b-a)$ for all x in the interval (a, b).

Substituting in (17.4) we get

$$F(x) = \int_a^x \frac{1}{b-a}\,\mathrm{d}x = \frac{x-a}{b-a},$$

Please take this on trust if you are not happy with integration, although all we are doing is integrating a constant. If $x = b$ we have $F(b) = 1$, as it should since the random variable can only take values in the interval (a, b). It cannot take values less than a or greater than b.

Statistically, a more interesting and practically useful distribution is the already mentioned normal law of errors, referred to as the *normal distribution*, or sometimes as the *Gaussian distribution*.

The normal distribution is often relevant in situations where a large number of factors contribute to variation about some target value. Some of these tend to push observations above, and others below, the target. In these circumstances a theorem called the *central limit theorem*, which we do not cover, comes into play. It tells us that the probabilities that observations will be off-target by various amounts closely follows the normal law. This reflects the fact that, due to a compensating effect of positive and negative errors, most observations have relatively small deviations from the target. Only a few exhibit gross deviations.

Suppose the target value is μ. In the sugar bags example we suggested a target weight of 1.005 kg., so there $\mu = 1.005$. How much variation there is about μ influences the spread of observations about the target value. The greater the magnitude of the various factors contributing to deviations from the mean, the greater the spread of possible observed deviations under the normal law of errors will be.

For the normal distribution Gauss showed that $f(x)$ is given by the somewhat formidable function

$$f(x) = \frac{1}{\sigma\sqrt{2\pi}}\,\mathrm{e}^{-[(x-\mu)/\sigma]^2/2} \qquad (17.6)$$

where σ is a constant that reflects the spread of observations.

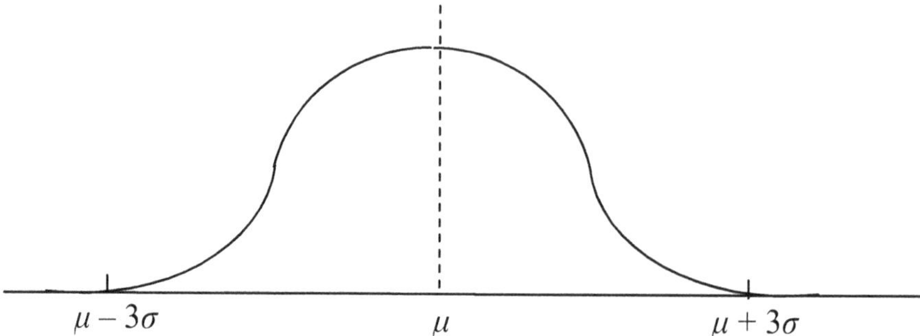

Fig. 17.2. The normal distribution function with parameters μ, σ

There is no simple explicit function $F(x)$ that has (17.6) as its derivative. However, for any given x, numerical methods (Sect. 15.8) can be used to determine the value of

$$\int_{-\infty}^{x} f(x)\mathrm{d}x.$$

Even with a modern computer this is tedious for a direct approach, since it is highly dependent on the value of μ and σ. We do not give details as to why this is so, leaving a major loose end. In practice all we need to obtain numerically is the value of the somewhat simpler integral

$$F(x) = \frac{1}{\sqrt{2\pi}} \int_{-\infty}^{x} \mathrm{e}^{-x^2/2}\mathrm{d}x \,.$$

Not only are there many computer programs to calculate this integral for any given x, but it is widely tabulated. We have already met something rather like this integral in Sect. 15.8. We indicated there how to calculate the integral over finite intervals such as (0, 1).

For any given μ and σ, the function $f(x)$ in (17.6) has a curve that is often described as bell-shaped. The maximum occurs when $x = \mu$. It is symmetric about this value of x. For all practical purposes $f(x)$ takes values approximately zero for x outside the interval $(\mu - 3\sigma, \mu + 3\sigma)$. A sketch of the curve is given in Fig. 17.2.

The quantities μ, σ are the *parameters* of the normal distribution. They are known respectively as the *mean* and *standard deviation*. Basically, the mean is the theoretical (random variable) analogue of the arithmetic mean of a set of observations, while the standard deviation is a measure of spread. Both are important concepts in applied statistics.

An increasingly important statistical problem is a study of survival times after some medical treatment, or in more general biological contexts.

For example, in the comparison of treatments for a particular type of cancer, one of the more important measures of relative effectiveness of several treatments is information on how long recipients of each treatment survive. Unless a treatment is highly efficient, typically many patients survive only a short to moderate time. A few survive much longer, giving rise to what are described as *long tail* distributions of survival times. The normal distribution is no longer appropriate.

Survival data are often modelled by the *exponential distribution.* A survival time, t, must be non-negative, so for its distribution the probability density function is only defined for $t \geq 0$, being zero if $t < 0$. We use t rather than x to remind us that we are thinking about *time* as an observed value of a random variable T.

For the exponential distribution the probability density function is

$$f(t) = \lambda e^{-\lambda t}, t \geq 0.$$

The precise shape of this function is determined by the parameter λ. The physical interpretation of this is that $1/\lambda$ is the mean survival time if this distribution is appropriate. In practice, more complicated models with extra parameters are often needed in survival time studies. The exponential and other related distributions are also used by engineers interested in the working lifetime (i.e., time to breakdown) of machine components.

In the next chapter we explain how statisticians use information gleaned from a sample to make inferences about population characteristics. For most inferences to be valid they should be based on a *random sample.* One reason for this is to guard against samples being unrepresentative of the population. For example, suppose a bag-filling machine has a periodic fault that results in every tenth bag having a weight above the target weight. If we choose every tenth bag that was produced we may get a sample that consisted only of these over-weight bags, or more likely one that contained no overweight bags. Neither of these samples would be typical of a large daily production run.

When different treatments are applied to different units — these could be different fertilisers to each of a number of fields all growing the same crop, or alternative treatments given to different patients suffering from the same ailment — the device of *random allocation* is widely used to decide which treatment is given to each unit. This means that, usually subject to certain restrictions, any unit has the same probability of being allocated to each specific treatment.

17.8 Bayes's Theorem

My grocer informs me he gets half his eggs from a supplier P and half from a supplier Q. He also assures me that one in 10 eggs from supplier P has a double yolk, whereas only 1 in 500 from supplier Q has a double yolk. If the egg I have for breakfast has a double yolk, intuition tells me the source of the egg is more likely to be supplier P. Given this 'experimental' observation that my egg has a double yolk, how do I use the 'prior' information given me by the grocer to work out the probability that my egg comes from supplier P? We may calculate it using a result we now explain.

If A_1 and A_2 are opposite events, i.e, they are mutually exclusive and $\Pr(A_2) = 1 - \Pr(A_1)$, and B is an event that may be associated with either A_1 or A_2, Thomas Bayes (1702–1761) showed that

$$\Pr(A_1|B) = \frac{\Pr(A_1)\Pr(B|A_1)}{\Pr(A_1)\Pr(B|A_1) + \Pr(A_2)\Pr(B|A_2)}.$$

This result, a simple form of Bayes's theorem, is not difficult to establish although we omit details. To apply it to the egg problem, let the event A_1 be that the egg came from supplier P, the event A_2 that it came from supplier Q, and B the event an egg is double yolked. Since half the grocer's eggs come from each supplier, we have $\Pr(A_1) = \Pr(A_2) = 0.5$. Also it is easily seen that $\Pr(B|A_1) = 0.1$ and $\Pr(B|A_2) = 0.002$. Substituting these values in Bayes's result we find

$$\Pr(A_1|B) = \frac{0.5 \times 0.1}{0.5 \times 0.1 + 0.5 \times 0.002} = 0.9804.$$

.

Without the information that the egg has a double yolk, since the grocer gets half his eggs from each source, all we can say is $\Pr(A_1) = 0.5$. This is often referred to as the *prior probability* because it is based on information prior to our discovery that the egg has a double yolk.The probability $\Pr(A_1|B)$ calculated by applying Bayes's theorem to make use of the additional information about the probabilities of eggs from each supplier having a double yolk is called the *posterior probability* . This name reflects the fact that it is computed after we have performed the 'experiment' that showed our egg had a double yolk. The probabilities $\Pr(B|A_1)$ and $\Pr(B|A_2)$ reflect the relative probabilities of eggs being double-yolked for each source, and each is referred to as a *likelihood* .

Bayes's theorem has applications in medical diagnostics. Here is a simple example. It is proposed that a mass screening test be used for 30 000 individuals to try to detect presence of a rare disease with a known incidence rate of 3 cases per 10 000 individuals. The probability is 0.99 that a person with the disease will give a positive response to the test, but there is also a probability of 0.001 of a positive response (called a *false positive*) for a healthy person. Two questions of interest are (i) if a person tests positive what is the probability that they have the disease, and (ii) how many people will be expected to give a false positive?

Adopting the notation used in our statement of Bayes's theorem, we let A_1 be the event a person has the disease, and A_2 the opposite event that they are healthy, and B the event a positive reponse. The given information implies $\Pr(A_1) = 0.0003$, $\Pr(A_2) = 0.9997$, $\Pr(B|A_1) = 0.99$, $\Pr(B|A_2) = 0.001$. Bayes Theorem then gives

$$\Pr(\mathrm{A}_1|\mathrm{B}) = \frac{0.99 \times 0.0003}{0.99 \times 0.0003 + 0.001 \times 0.9997} = 0.2290.$$

. Since 3 in 10 000 have the disease, the expected number having the disease in a population of 30 000 is 9. Effectively, since the probability of a positive response from an infected person is 0.99, it is likely that all 9 will give a positive response. More precsiely, the expected number is $9 \times 0.99 = 8.91$. Thus, we would expect that there are approximately 29 991 healthy people given the test, and since the probability of a positive response from a healthy person is 0.001 approximately 29.991 (i.e. about 30) people might be expected to give a false positive.

Screening programs are often criticised for giving an unacceptable number of false positives. However, if the screening test can be followed up by costly but more definitive tests, whereas the preliminary screening test is relatively cheap to perform, it may be a largely economic and social decision, to decide upon the best diagnostic procedures.

Bayes's theorem extends to a mutually exclusive and exhaustive set of n events A_i, $i = 1, 2, \ldots, n$, when it takes the form

$$\Pr(\mathrm{A}_i|\mathrm{B}) = \frac{\Pr(\mathrm{A}_i)\Pr(\mathrm{B}|\mathrm{A}_i)}{\sum_{k=1}^{n}\Pr(\mathrm{A}_k)\Pr(\mathrm{B}|\mathrm{A}_k)}.$$

Further extensions are possible to situations involving prior probabilities that have a continuous distribution, but these are well beyond the scope of this simple introduction. These last extensions are commonly used in procedures known as *Bayesian inference*. We consider this briefly in Sect. 18.6.

17.9 Loose Ends

Oddities of odds

Odds. Bookmakers quote *odds* like '7 to 1 against' or '5 to 4 on' as a basis for accepting bets on a horse-race or other event. Odds are closely related to probabilities. If the probability of an event is known to be one half, or 0.5, then an even money bet — odds of *one to one on*, or what is the same thing *one to one against*, is usually referred to as *evens*. More generally, if the probability of an event such as a particular horse winning a race is p, then a fair bet, in the sense that a punter who places a bet, and a bookmaker who accepts it, should in the long run each expect to break even, is one in which the *odds on the event* are $p/(1-p)$. The reciprocal, $(1-p)/p$ are the *odds against occurrence of an event* that has probability p. Thus, if the probability of a horse winning a race is $p = 0.2$ fair odds on a win are $0.2/0.8 = \frac{1}{4}$, or 1 to 4 on. The odds against are $0.8/0.2 = 4/1$, or 4 to 1 against. It is common, though not universal, practice to quote odds in the form *odds on* if $p > 1/2$ and in the form *odds against* if $p < 1/2$.

Any statement of a probability may be expressed in terms of odds and vice-versa, but the odds offered by a bookmaker for or against a horse winning a particular race are in general not based on any relative frequency or similar concept of probability. When a bookmaker taking bets on a horse race offers certain odds, these are based on factors such as the amount already wagered on that horse, and on other horses in the same race, and are modified in many cases by a personal assessment of how likely the bookmaker thinks each horse is to win. The odds offered are calculated to minimize the risk of the bookmaker being forced to make a large payout. Yet, they must be sufficiently attractive to encourage punters to place a bet, while at the same time allowing the bookmker to make a profit in the long term. In that sense bookmaking might be regarded as a mixture of art and science.

Because of his need to make a profit, a bookmaker's odds are not 'fair' odds in the sense that both bookmakers and punters might expect to break even in the long run. One consequence of this is that probabilities corresponding to bookmaker's odds don't obey rules based on the probability axioms! In particular, they don't comply with the rule that the probabilities associated with a set of mutually exclusive and exhaustive events should add to 1. Typically, the sum will exceed 1.

Here is an example. For a game between two teams, A and B, in which the outcome had to be a win for one or other of the teams, a British bookmaker recently offered odds of 8 to 15 against team A winning, and odds of 11 to 8 against team B winning. This is a mutually exclusive and exhaustive set of outcomes. Remember that 'odds against' take the form $(1 - p)/p$, where p is the probability of a win. The probability team A will win, as calculated from the bookmaker's odds, is found by solving the equation $(1 - p)/p = 8/15$, giving $p = 15/23$. You should check that the analogous probability that team B will win is $8/19$, and that the sum of these probabilities is approximately 1.073.

Subjective probabilities. Probabilities arrived at on the basis of expected frequencies, such as those considered in this chapter, are *objective probabilities.* In practice, most of us base some of our decisions on a different concept of probability, using what are called *subjective*, or sometimes *personal*, probabilities. These probabilities are based on degree of belief.

Most of us use this notion even though we do not go as far as associating a particular number with the probabilities we conceive. If we say that there is a 'high probability that it will rain today' we express a stronger belief that it will rain than if we claim there is a 'a moderate probability that it will rain today'. That in turn represents a stronger belief than the statement 'it is unlikely to rain today'.

Many everyday actions are largely controlled by such assessments of probabilities. However, on any particular day you and I might not agree that there is 'a high probability of rain'. I might think so, but you might think there is 'an extremely high probability that it will rain'. Rational people would agree that your statement represents a higher subjective probability than that implied by mine. Thus, subjective probabilities can at least be ordered. If one agrees that a statement of the form 'certain to' as in 'certain to rain' or 'certain to be sunny' implies a probability equal to 1, and a statement of the form 'it won't' implies a probability zero, it is possible to manipulate subjective probabilities using the axioms of probability, and the concepts of mutual exclusive and conditional probabilities, in much the way one operates with objective or frequency-based probabilities.

The concepts of objective and subjective probabilities are logically distinct. While everyone reaches the same conclusions when operating with frequency based probabilities, these may be open to criticism in practice on the grounds that we are using the wrong frequency base. For example, if we apply the binomial distribution, we are assuming that the probability, p, of 'success' remains constant throughout the experimental procedure, giving observations which we assume to follow that distribution. We also assume that the outcome for one observation does not depend on what has happened at previous observations. If this is not the case any inferences based on such assumptions may well be, probably will be, erroneous.

In the next chapter we deal with techniques for drawing inferences from experimental data where assumptions about distributions play a key role. In Sect. 18.6 we give a brief indication of how many statisticians apply subjective probability to the interpretation of observed data.

Utility. Probability is the basic mathematical tool for handling uncertainty, but it is not the only measurable concept used to make decisions when there are uncertainties. The UK national lottery, and similar lotteries in other countries, offer a prize of more than 1 million times the stake paid to play. Usually the probability of winning that prize is less than 1 in 10 million. In the current UK lottery, where the prime aim is to raise money for 'worthy' causes, only about one half the stake money is returned as prizes. These include not only the jackpot prize, but smaller prizes, most of these being only ten times the stake money. Thus, if we invest £1 each week in the lottery we can expect a return of less than half our stake money in the long run. Yet something like half the adults in the UK buy a lottery ticket each week. Probability considerations suggest they are foolish to do so, for there are virtually overwhelming odds against winning the jackpot prize. That, and several other large prizes, are the only ones most ticket holders are really interested in. If we invest in only one ticket each week we would have to live a few million years to make the probability of just one jackpot win anything near a certainty. Even if we discovered the elixir of life, and survived that long, it would be likely that by the time we won the jackpot we had invested more than the amount we won.

If people based their decision to invest in lotteries on probability alone they would be foolish, for it is an almost certain way to lose money. But are investors being foolish? Maybe not, for there are other measurable concepts concerned with uncertainty. One such is *utility*. When we invest £1 in a lottery we are going to lose that money with a probability that is greater than 0.975. There is a small probability of less than 0.025 of a minor win, and a miniscule probability of less than 1 in 10 million of a jackpot win. The idea of utility comes in when we consider (a) what else we might do with that £1 stake if we did not invest it in the lottery, and (b) what we could do with the one million pounds plus jackpot if we won the lottery. After all, there is a possible jackpot win every time the lottery is drawn, and our only chance of being a jackpot winner is to buy at least one ticket. The utility value of our £1 stake depends on what alternative use we could put that £1. We might buy an ice-cream. Our utility decision depends upon whether we prefer to buy our weekly ice-cream, and thus forego any chance of winning the lottery jackpot, or whether we are content to forego the luxury of a weekly ice-cream to give ourselves the small chance of being a lottery jackpot winner.

We have only touched on the germ of the utility concept, but it is obviously something most of us have in mind in deciding certain actions. It is particularly relevant in the field of insurance. We are prepared to pay a moderate premium to insure against a large loss, such as that of our house being burnt down, even though that event has a relatively small probability. If people did not pay more in premiums than the average payout per policy, insurance companies would not remain solvent.

Utility is not devoid of probabilistic ideas. Most people who consider that the utility aspect favours a stake of £1 per week in the lottery would put a different utility interpretation on investing £1000 per week (especially if the latter were twice their weekly income). They would increase the probability of winning the jackpot by a multiple of 1000 (although that probability would still be miniscule), but it is likely they have better things to do with £1000 than buying ice-creams.

18

Statistical Inference

18.1 Decision Making Under Uncertainty

Statistical inference is an application of probability theory developed in the last 100 years or so. It has revolutionised the way data are interpreted in fields as diverse as agriculture, medical research, engineering, psychology, commerce and archaeology.

Table 17.2 gave the probabilities of observing each possible number of *fives* in ten casts of a fair, or true, die. A 'fair' die is one where any of the six faces is equally likely to land uppermost. A table like this may be used as a starting point for making statistical inferences. Table 18.1 repeats Table 17.2 with probabilities rounded to 3 decimal places, sufficient for present purposes.

In this chapter we shall often refer to the number of times a *score of five occurs in 10 casts of a die*. As explained in Sects. 17.6 and 17.7, this number is a *random variable* that takes integer value between 0 and 10. We use X

Table 18.1. Probabilities of scoring each number of *fives* in 10 casts of a die

r	*Probability*
0	0.162
1	0.323
2	0.291
3	0.155
4	0.054
5	0.013
6	0.002
7	0.000
8	0.000
9	0.000
10	0.000
Total	1.000

to denote this variable. If the die is true, the probabilities associated with each possible value of X are those in Table 18.1, subject to the restriction of rounding to three decimal places. Thus $\Pr(X = 4) = 0.054$, and by the addition rule for probabilities

$$\Pr(X \leq 2) = 0.162 + 0.323 + 0.291 = 0.776 ,$$

since the outcomes $X = 0$, $X = 1$ and $X = 2$ are mutually exclusive, and their probabilities are those in Table 18.1. From that table we see that $\Pr(X = 1) = 0.323$, or approximately $1/3$. This is greater than the probability associated with any other possible value of X. Taking the limiting relative frequency view of probability, this implies that in a large number of experiments, in each of which we cast a true die 10 times, we should observe exactly one occurrence of a score *five* in roughly one third of the experiments.

It is nearly, but not quite, as likely (probability 0.291) that we would observe the event $X = 2$, but only about half as likely that we would observe either $X = 0$ or $X = 3$. However, the probability of observing one or other of the mutually exclusive events $X = 0$ or $X = 3$ is $0.162 + 0.155 = 0.317$, only slightly less than that of observing $X = 1$, and slightly more than that of observing $X = 2$.

For other possible outcomes the probabilities fall off rapidly, being 0.054 for $X = 4$. This implies that we would expect this value in the long run in only a little more than 1 in 20 experiments in which we cast a die 10 times. The addition rule for probabilities tells us that

$$\Pr(X \leq 4) = 0.162 + 0.323 + 0.291 + 0.155 + 0.054 = 0.985 .$$

This means, as is easily verified from the table, that there is only a probability of $1 - 0.985 = 0.015$ of observing the opposite event, $X \geq 5$. In particular, going back to the more accurate Tab. 17.1, we see that $\Pr(X = 10)$ is less than 0.000 000 5 since, after rounding, it is zero to six decimal places. A more precise calculation establishes that$\Pr(X = 10) \approx 0.000\ 000\ 017$, implying that in the long run we should expect 10 *fives* in 10 casts in less than 2 in 100 million such sequences of 10 tosses.

Implausible outcomes

If, therefore, we do this experiment just once and observe a score of *five* at each of the 10 casts, we might reasonably:

- (i) write to our local newspaper and say something very unusual has happened;
- ii) take a close look at the die to see whether it might be unfairly weighted, or misshapen in a way that would make the face *five* more likely than any other to appear;

- (iii) repeat the experiment a number of times and see if there is a tendency to get more *fives* than might reasonably be expected with a fair die.

Option (i) may win short-term notoriety among friends, but options (ii) and (iii) are practical ways of deciding whether there is something unusual about the die. If we cannot repeat the experiment (a souvenir hunter may have stolen the die), then all we can say is that if the die is fair, our one experiment has produced an outcome to be expected only about once in 50 million such experiments. The alternative is that the die is not a fair or true die.

Even if we observe $X = 5$ we might be mildly surprised because, as already pointed out, $\Pr(X \geq 5) = 0.015$. This implies that in the long run we expect 5 or more occurrences of a score *five* in only about 15 experiments in every 1000. One would, on the basis of this one experiment, perhaps have some doubts about the die being true. If we could repeat the experiment we might be able to decide whether this was a 'one off' extreme outcome, or whether the die had a tendency to give too many *fives*.

Can we say more about what such an outcome means if the experiment cannot be repeated? Obviously, if the die is prone to giving too many *fives*, the probability that X takes a value 5 or greater will be higher than it is for a true die. For example, suppose that, due to a fault in manufacture, the number of spots on each face were not respectively 1, 2, 3, 4, 5 and 6; but by accident the factory making it had put 5 spots on the face that should have had 3. Then the spots on the faces would be 1, 2, 4, 5, 5 and 6. The probability of scoring a *five* at any one cast would increase from 1/6 to $2/6 = 1/3$. If we set $n = 10$ and $p = 1/3$ in the binomial formula (17.3), relevant computer software, or tedious calculations, give the probabilities associated with each number of *fives* in 10 casts. These are given in Table 18.2, from which we see that

$$\Pr(X \geq 5) = 0.137 + 0.057 + 0.016 + 0.003 + 0.000 = 0.213.$$

Thus, in the long run we expect values for X equal to 5 or more to occur in about 21 in every 100 experiments, or approximately 1 experiment in 5. If we get one of these outcomes in one trial with the faulty die we may be mildly surprised, but not unduly so.

18.2 Testing Hypotheses

Statistical inference is the science of making broad-based inferences from experimental data. The above discussion of the outcome of an experiment involving casting a die introduces ideas that are fundamental to one of the earliest statistical ways of assessing results of scientific experiments. It is called *hypothesis testing*.

Table 18.2. Probabilities of r *fives* in 10 casts if $p = 1/3$.

r	*Probability*
0	0.017
1	0.087
2	0.195
3	0.260
4	0.228
5	0.137
6	0.057
7	0.016
8	0.003
9	0.000
10	0.000
Total	1.000

Modern developments have made hypothesis testing of decreasing practical importance. Some, but not all, statisticians actively discourage its use. Nevertheless, the ideas behind it are basic to more sophisticated methods of inference. For that reason we explore it in some detail, using the die casting example, before illustrating its relevance in a more realistic scientific context.

When forming Tables 17.2 and 18.1 we assumed that the die was a true, or 'fair' die, calculating the probabilities on that basis. In formal statistical hypothesis testing we start with some such basic hypothesis called the *null hypothesis*. This is a hypothesis we are prepared to accept until we have evidence to raise serious doubts about it. In this example the null hypothesis is that the probability of a score of *five* at any cast is $p = 1/6$. The adjective *null* has historic associations. It reflects the fact that in many experiments where hypothesis testing was originally used, the data consisted of measurements relevant to two or more treatments. The basic hypothesis was that some characteristic of these data did not differ from treatment to treatment, i.e., the hypothesis was one of *null* difference.

To perform a hypothesis test we need to specify an *alternative hypothesis*. In the die-casting example we suggested a probability $p = 1/3$ as the alternative, this being appropriate if the die had two faces with 5 spots and none with 3, but was otherwise 'true'. In many practical situations a broader class of alternative hypotheses might be $p > 1/6$, or even the more general $p \neq 1/6$.

To save writing we denote a null hypothesis by $\mathrm{H_0}$, and an alternative hypothesis, by $\mathrm{H_A}$. With this notation for the die casting example, where $p = 1/6$ is the null hypothesis, we express this as

$$\mathrm{H_0} : p = 1/6.$$

The alternative hypothesis that $p = 1/3$ is written

$$H_A : p = 1/3.$$

Belief and truth are not always the same

The first step in hypothesis testing is to decide for what outcomes of an experiment we agree to accept H_0. That decision implies we reject H_0 for other outcomes. Accepting H_0 does not mean we have proved it to be true. It means no more than that there is not strong enough statistical evidence to claim that it is *not* true. What this implies will become clearer as we develop more tools for hypothesis testing.

We noted above that if H_0 holds, then $\Pr(X \geq 5) = 0.015$. Further, we showed that the corresponding probability was greater when H_A held, i.e. that $\Pr(X \geq 5) = 0.213$. So $\Pr(X \geq 5)$ is more than fourteen times greater under H_A than it is under H_0.

We could make such comparisons for $\Pr(X \geq r)$ for values of r other than $r = 5$, but to make useful progress we need new concepts. We introduce these in the context of classical hypothesis testing, although as we show below, this has some limitations. Many of these can be overcome in more modern procedures, though we won't explore these in detail.

A key step is to set up a rule for deciding whether to accept or reject H_0. To do this we define a *critical region.* This consists of a set of outcomes which, in toto, have an associated probability not exceeding some predetermined value when H_0 holds. Conventionally, we denote that value by the Greek letter α. These outcomes associated with the critical region are normally extreme values, each with small probability under H_0, but some or all of which, have greater probabilities under H_A, and the total probability associated with the critical region when H_A holds is greaster than that when H_0 is true. If we set $\alpha = 0.05$, then if H_0 is true we expect in the long run to get values in the critical region in not more than one experiment in twenty, since $1/20 = 0.05$. Thus, if we perform an experiment many times, and for each of them H_0 is true, we will get a result in the critical region in only about 1 experiment in 20.

What does it mean if we get a result in the critical region in just one experiment, and decide on that basis to reject H_0? One possibility is that we may be rejecting H_0 when it is true. If we do reject H_0 when it is true we make an *error of the first kind.* The probability of making such an error, in our one experiment does not exceed α. Naturally, we want this probability to be small. However, there is a price to pay if we make α too small.

Turning again to the die casting example, recall that if we choose as the critical region the set $X \geq 5$, it has a size (associated probability) 0.015 which is less than $\alpha = 0.05$. Had we chosen $X \geq 4$ the critical region would have size 0.069, which is larger that $\alpha = 0.05$. On the other hand, if we chose $X \geq 6$ the critical region would (Table 18.1) have size 0.002.

What happens if we have a faulty die with 5 spots on two faces for which H_A holds? From Table 18.2 it is easy to see that $\Pr(X \geq 4) = 0.441$, $\Pr(X \geq 5) = 0.213$ and $Pr(X \geq 6) = 0.076$.

Now, if we choose $X \geq 5$ for the critical region, we have seen that the probability that we make an error of the first kind is 0.015. Also, if the result lies in this critical region, we reject H_0 and accept H_A. If H_A is indeed true, we have $\Pr(X \geq 5) = 0.213$. This means that if we carry out the test when H_A is true, the probability of getting a result in the critical region is still no more than 0.213. If we reduce the critical region to $X \geq 6$, we reduce the probability of an error of the first kind to 0.002, but we also reduce the probability of getting a result in this smaller critical region when H_A is true to $Pr(X \geq 6) = 0.076$.

Thus, if we proceed in this way when H_A is true and we get a result that is not in our chosen critical region we continue (wrongly) to accept H_0. To do that is described as making an *error of the second kind.* It is not very hard to see (but pause a moment, to be sure you see why) that when the critical region is $X \geq 5$ the probability of an error of the second kind is $1 - 0.213 = 0.787$. And further, if the critical region is $X \geq 6$, the probability of an error of the second kind increases to $1 - 0.076 = 0.924$.

This, then, is our dilemma – if we make the critical region smaller, we reduce the probability of making an error of the first kind, but we also decrease the probability of accepting the alternative hypothesis when it is indeed the correct one. That is, we increase the probability of making an error of the second kind.

18.3 Some Limitations

You can't have your cake and eat it too — trite but true

A weakness of hypothesis testing is that we cannot control both the probability of an error of the first and the probability of an error of the second kind. In practice, the best we can do is to decide what is the maximum acceptable probability for an error of the first kind. For a particular experiment, the probability of an error of the second kind then depends upon what the alternative hypothesis is.

It is almost universally accepted, partly for pragmatic and partly for historic reasons, that we should set α, the probability of an error of the first kind, at or near to $\alpha = 0.05$. As the die-casting example has shown, this will not give a very high probability of rejecting H_0 in favour of H_A, even when the latter may well be true.

All is not lost in such circumstances, for we can often increase the probability of correctly accepting H_A by doing a larger experiment. The probability of accepting H_A when it is true is called the *power* of the test. If we denote the probability of making an error of the second kind by β it is not hard to see that the power is equal to $1-\beta$. We omit the arithmetic detail of power calculations in general. That is best left to standard statistical computer software.

The power of a test is strongly influenced both by the size of an experiment, and what the alternative hypothesis is. We might increase the size of our die casting experiment from 10 casts to 20 casts. Then if Y is the number of *fives*, it can be shown that if we choose our critical region to be $Y \geq 7$, the probability of an error of the first kind is approximately 0.037. The probability of an error of the second kind is approximately 0.480, i.e., the power is now $1 - 0.480 = 0.520$.

The above discussion of hypothesis testing in the die casting example is somewhat artificial. It is unusual to have so specific an alternative hypothesis as $p = 1/3$. If a die is not true, it may still have 1, 2, 3, 4, 5, 6 spots each occurring on only one face, but it may have an uneven density, or even a small piece of lead embedded in it, so that some scores have a probability greater than 1/6, and others a lesser probability. If p is the probability that a *five* is scored we may, as already suggested, then wish to test the hypothesis $H_0 : p = 1/6$ against an alternative $H_A : p > 1/6$, or the even more general $H_A : p \neq 1/6$. The last implies that we concede the possibility that p may be either greater than or less than the value specified in H_0.

Details of how these tests are formulated and the choice of appropriate critical regions are beyond the scope of this brief introduction. It is intuitively clear that the power of any test is likely to be small if p differs little from 1/6, but hopefully the power will increase as the actual departure from this null hypothesis value becomes larger.

If we get a result in the chosen critical region, and so reject H_0, this is said to indicate *significance*, or to imply a *significant difference* from the null hypothesis.

You cannot prove anything with statistics

Two aspects of hypothesis testing cause a lot of confusion, especially among nonstatisticians. The first is a failure to recognise that accepting a null hypothesis because a result does not fall into the critical region, does not mean the hypothesis is true. Acceptance is only an indication that, under the rules of hypothesis testing, there is insufficient evidence to persuade us to abandon a prior-specified H_0. The second is that a significant difference may not be of practical importance. Conversely, a difference of practical importance may not be statistically significant. We elaborate on the latter point in the next two sections.

18.4 Significance and Practical Importance

We look at the distinction between statistical significance and practical importance in a real data situation.

Significance and importance — uncomfortable bedfellows

In fields as diverse as anthropology, archaeology, biology, chemistry, psychology, engineering and marketing, a great deal of experimental and observational data consist of measurements on groups of individual units. Within any one group, the measured characteristic differs from individual to individual. What is of interest is whether there are differences between some of the groups, over and above those that simply reflect differences between individuals. Each group may consist of individual units subjected to a particular treatment, that treatment being different for each group. In another situation each sex might represent a group.

Higham et al [30] gave these data for mandible lengths in millimetres for some male and for some female remains of jackals from Thailand.

Males	120	107	110	116	114	111	113	117	114	112
Females	110	111	107	108	110	105	107	106	111	111

The variation in length within each sex represents typical biological variation of the kind we mentioned in Sect. 14.1. What might be of interest in broad terms is whether there is strong enough evidence to reject the null hypothesis that the 'average' mandible lengths for males and females are the same. To explain what an experimenter has in mind in asking that question, we must be more specific about what is meant by 'average' in this context.

The data above may be the only measurements available from what was a large population of jackals in the area where the remains were found. Then, the average of interest is the 'unknown' mean for that larger population. We may want to test the hypothesis

H_0 : the population mean mandible length for males and females are equal

against the alternative

H_A : the population mean mandible length differs between the sexes.

If the population mean length for males were greater than that for females, we would expect this to be reflected in samples by the sample mean for males being greater than that for females. With a pocket calculator, we may quickly verify for the above data, that the respective arithmetic means are 113.4 for males and 108.6 for females. These means suggest a higher average for males in the population. This is reinforced by the fact that only for one male is the length (107) less than the female average, while all females, lengths are less than the male average. Intuitively, it seems unlikely this state of affairs would exist in the sample if H_0 held.

I am stepping on dangerous ground in making that last comment. It carries an implication that the samples are fairly representative of the populations. This would be a reasonable assumption for proper random samples, as described briefly in Sect. 17.7. It would not be valid if the jackals came from an archaeological dig where, for some reason, all females were young immature animals, while all male jackals were fully developed adults. The difference then may represent a difference between immature and mature animals, rather than a difference between sexes.

If you have already studied statistics, you may know that a common hypothesis test for differences between population means for data like those for the jackals is called the *t-test* (or more fully *Student's t-test*). There is an alternative called the *Wilcoxon rank sum test* that you may also have encountered. We are not going to describe either in detail, but both give strong evidence of significance, i.e., that we should reject H_0. For either test if we reject H_0, the probability of an error of the first kind is only about 0.003.

18.5 Confidence Intervals

A technique we have not yet discussed allows us to specify an interval known as a *confidence interval.* This is a subtle concept, and we consider only one aspect of its interpretation in terms of the jackal data. A confidence interval tells us what hypothetical differences between treatment means would be acceptable in a test at significance level α. If $\alpha = 0.05$ the corresponding interval is called a 95 per cent confidence interval. One interpretation of such an interval is that there is, in a certain sense a 95 per cent probability, that this interval contains the true (but unknown) population mean difference. What this implies is that if we were to perform a lot of experiments with different samples each of the same size and from the same populations, and formed such intervals for each data set, then 95 per cent of these intervals would include the unknown population mean difference.

Given the jackal data in Sect. 18.4 an appropriate 95 per cent interval turns out to be the interval from 1.91 to 7.69. If you are familiar with the t-test you may know how this is calculated, but do not worry if you have not met this. We are only using the result for illustrative purposes.

The sample mean difference, $113.4 - 108.6 = 4.8$, is at the midpoint of this interval. Most programs for a t-test will give the probability of an error of the first kind, and this confidence interval, in less than 1 second computing time for data sets like these

In Sect. 18.4 we said that the null hypothesis of no difference in length might be of interest. Many anthropologists looking at such data would be interested in evidence of any difference (no matter how small). Here there is a pretty clear indication of some difference between males and females, since the confidence interval does not include the value zero. Had it included zero we would not have rejected a hypothesis of no difference.

There are situations where the hypothesis of no difference is of little interest. Suppose, for another area, experimental evidence had suggested a mandible sex-related length difference between 1 and 3 mm. Then it might be of interest to know if the population from which our samples came were also likely to show a similar difference. A confidence interval is useful here. The confidence interval of 1.91 to 7.69 obtained above implies that a difference between 1.91 and 3 mm is possible for the population, but the difference may well exceed 3 mm. If it were important to reach a firmer conclusion, one would need more data. In practice, this might or might not be available.

Whenever a hypothesis of no difference between means is accepted this implies the confidence interval for the population difference includes zero. If, in data like that for the jackals, the confidence interval had been $(-0.3, 4.7)$, then there would be a definite possibility that the present population showed a similar difference to the population where the difference was fairly firmly established to lie between 1 and 3. However, the true difference for the new population may lie outside these limits (and may even be zero!).

Larger experiments, or larger samples, generally result in shorter confidence intervals. We may well have found that if the investigators had a much larger number of specimens of both males and females in the above hypothetical example, that the confidence interval reduced from $(-0.3, 4.7)$ to, say, $(0.8, 1.9)$, an indication of a difference rather similar to that for any data that indicated a difference between 1 and 3, though probably towards the lower part of that range.

Situations where only differences above a certain size are important arise frequently in a medical context. For example, drugs for reducing blood pressure are only of clinical value if they reduce systolic pressure by at least 15 mm Hg. A new drug is only of interest if the average reduction is 15 mm or more. Here, confidence intervals for the difference in mean values before and after treatment for a number of patients may help to decide if this is likely. If the confidence interval is (3, 11) the difference is statistically significant, since zero is not included in the interval. It is highly likely to be less than 15, so it is not of practical importance.

If the difference has confidence interval (18, 43) it is not very precisely determined, but, being almost certainly greater than 15, it is of practical importance. If the interval had been (7, 21) the difference might or might not be important. In this situation if we really must know whether there is a practically important difference, we need more data.

18.6 Loose Ends

Bayesian inference. Many statisticians think it reasonable to modify interpretation of experimental results by taking account also of prior beliefs of a probabilistic nature, that hold without reference to the experimental data. In our discussion of Bayes's Theorem in Sect. 17.8 we did something rather more concrete in our example with the eggs. There we took into account prior information to the effect that the grocer got half his eggs from each of two sources. The associated prior probabilities, $\Pr(A_1) = \Pr(A_2) = 0.5$, were objective probabilities. Prior beliefs replace such objective probabilties by subjective, or personal, probabilities.

Given prior beliefs expressed as subjective probabilities, and associated likelihoods, calculation of a posterior probability using Bayes's theorem is mathematically sound. This is the basis of *Bayesian inference.* Its use is controversial mainly because of the subjective element. Different statisticians may not agree on a choice of priors, and therefore may reach different conclusions from the same experimental evidence. In practice this is not a serious problem with a reasonable amount of experimental evidence, because as evidence accumulates, the influence of the prior decreases.

The approach used in Bayesian inference is logically quite distinct from the frequentist approach adopted in this chapter. In practice, when both have some claim to relevance, inferences based on the Bayesian and on the frequentist approach usually, but not invariably, lead to similar conclusions.

Decision theory. Inferences — whether based on the frequentist concepts of probability outlined in the bulk of this chapter, or using Bayesian methods — are often the basis for real world decisions. The basis on which decisions are made is a mathematical subject in its own right. It is called *decision theory.*

There are many logical bases for making decisions. One aim in commercial situations is to make decisions to maximize possible profit or minimize possible loss. When there is uncertainty (e.g., economic uncertainty) achieving these goals may be virtually impossible. A goal that is often attainable, given appropriate information on the distribution of the random variables involved, is maximization of the minimum possible gain, or minimization of the maximum possible loss. Such procedures are called *maximin* or *minimax* procedures respectively. To determine the appropriate minimax or maximin procedure in industrial or commercial situations often requires an appropriate choice among several possible actions, often at each of a number of stages.

Here is a simple example showing how, when there are several options, one may find an optimum strategy. In this example the strategy is one of maximizing expected profit.

> A builder is allowed to tender for only one of two contracts, A or B. If he tenders for contract A there is a probability of 0.8 that he will win the contract. If he does he will make a profit of £12000. If he tenders

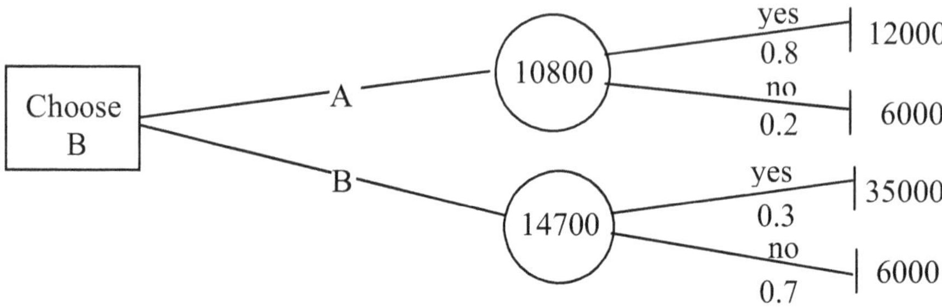

Fig. 18.1. A decision tree for choosing between two strategies so as to maximize expected profit.

for contract B there is a probability of 0.3 that he wins. If he does he will make a profit of £35000. If he fails to win the contract for which he tenders, he will do other work that will give him a profit of £6000.

If a strategy gives a probability p of making a profit of M and a probability $q(= 1-p)$ of making a profit N, the expected profit if that strategy is followed is defined as $(Mp + Nq)$.

The information given above may be represented diagrammatically using what is called a *decision tree.* Fig. 18.1 is a decision tree for this problem. The word 'tree' is used because the diagram has the form of a branching tree (in this particular illustration a tree lying on its side). The square box (or node) on the left is called a decision node. The two branches emanating from it, labelled A and B, represent the possible strategies the builder may adopt, i.e., to tender for A, or to tender for B. The circular nodes at the other end of these branches are sometimes called random nodes. The numbers we will put in them are the expected profits for the associated strategy. The two lines emanating from each random node show beside them the probabilities of acceptance (labelled 'yes'), or of rejection (labelled 'no'), for each strategy. Finally at the right of the tree is a number representing the profit for each possible outcome of either of the strategies. We work backward from these profit figures to determine the appropriate entries for 'expected' profits for each strategy. These, as mentioned above, are placed in the random nodes. The optimum strategy is that associated with the higher entry in the random nodes.

We explained above how to obtain expected profit in this situation. From the data we calculate this in the random node for strategy A to be

$$12000 \times 0.8 + 6000 \times 0.2 = 10800\,,$$

indicating an expected profit of £10800 if strategy A is followed. Similarly, for strategy B the expected profit is

$$35000 \times 0.3 + 6000 \times 0.7 = 14700.$$

The expected profit is higher for strategy B, so this choice is entered in the decision node.

The above problem led to a very simple decision tree. In real life problems one often has to consider more than two strategies. There may also be more than two alternative outcomes associated with any strategy.

19

Operations Research

19.1 Mathematics in Business

Mathematics faces reality

Mathematics has had a role in commerce since the days when bartering was the main method of trade. It was to meet the needs of traders that the natural number system was extended to include first negative integers and then rational fractions.

During the twentieth century sophisticated accounting, and drives for business efficiency, stimulated the use of a cocktail of mathematical ideas. Some of the ingredients had been around for 2000 years or more. Others were contemporary. Many strands come together in a discipline called *operations research* in America, and *operational research* in Britain, often abbreviated to OR. OR consists of diverse techniques applicable to business, construction and economic problems.

Mathematically, many of these are *constrained optimization problems*, which may be described less formally as making the best use of (optimizing) limited resources (the constraints).

Constrained optimization problems range from finding how to maximize profits or minimize costs, to determining a policy that makes best use of storage space, or minimizes delivery costs between warehouses and retail outlets.

In a production context the optimum policy is subject to constraints. The latter might include restrictions on the amount of raw materials available, the capacity of a plant, or limited labour resources. The price that can be charged may depend both on customer demand, and what competitors charge for similar goods. Warehouses have only a finite storage capacity, and perishable goods have limited storage or shelf life.

Linear programming is a well-known technique for solving many problems in these categories. A simple example for a manufacturing process illustrates some basic ideas.

19.2 A Production Schedule

A firm produces two types of microwave oven — standard and deluxe. For each standard model components cost £23 and it takes 45 person-minutes to assemble. For each deluxe model components cost £28, and it takes one person-hour to assemble. With the current workforce, there are a maximum of 160 person-hours of labour available each day. The labour cost per person-hour is £12. Overheads amount to £1200 per day. There is a further constraint that only 180 oven doors are available daily. Each may be fitted to either a standard or deluxe model, but every oven must have a door! The wholesale price for each standard oven is £42, and for each de luxe oven £52. It is possible to sell no more than 140 standard and 120 de luxe models each day. How many of each model should be produced daily to maximize profits?

It would be tedious, and time consuming, to look at all possible combinations of production levels and to work out the profit for each. To do so would not be impossible with modern computing power for a relatively simple problem like this. For more complicated real-world problems, with many production constraints to be taken into account, we need a more sophisticated mathematical approach. A linear programming model may be used.

The first step is to translate the information given into mathematical statements. To do this we need to specify two key variables — the numbers of each type of oven to be made in a day. Let these be x standard and y deluxe ovens. The information given enables us to work out the profit on each type of oven. Apart from overheads, which remain fixed for each day, irrespective of how many ovens of each type are produced, the profit for a unit of each type is

profit = *wholesale selling price* − *labour cost* − *material cost.*

Using the information given, it is easily verified that for each standard oven the profit, ignoring overheads, is

$$\mathrm{P_s} = 42 - 9 - 23 = 10\ ,$$

while for each de luxe oven it is

$$\mathrm{P_d} = 52 - 12 - 28 = 12\ .$$

The total profit, ignoring overheads, from a production of x standard and y de luxe ovens is therefore

$$V = 10x + 12y\ ,$$

while the profit after deduction of overheads drops to

$$U = 10x + 12y - 1200 .$$

The function, U, that we want to maximize is called the *objective function.* The constraints — available labour, and a limit on the number of oven doors and on daily sales — restrict the possible feasible values of x and y. Because we only have 180 doors, the total of x and y cannot exceed 180. Maximum daily sales for each type may further restrict the possible output. A manufacturer does not want to produce items he cannot sell.

Mathematically, constraints are expressed as inequalities. You should verify that the information given in the statement of the problem implies the following constraints:

Sales constraints

$$x \leq 140, \qquad y \leq 120$$

Labour constraint

$$0.75x + y \leq 160 .$$

In the labour constraint the coefficients of x and y are the time in person-hours required to make a unit of each type. Since an inequality is unaltered by multiplying both sides by the same positive constant, we avoid fractions by rewriting this inequality as

$$3x + 4y \leq 640 .$$

The constraint for doors is

Door constraint

$$x + y \leq 180 .$$

Almost trivially, since we never make a negative numbers of ovens, we add the constraints $x \geq 0$, $y \geq 0$. A terse summation of the problem is:

Maximize

$$U = 10x + 12y - 1200 \tag{19.1}$$

subject to

$$x \leq 140, y \leq 120 , \tag{19.2}$$
$$3x + 4y \leq 640 , \tag{19.3}$$
$$x + y \leq 180 , \tag{19.4}$$
$$x \geq 0, y \geq 0 . \tag{19.5}$$

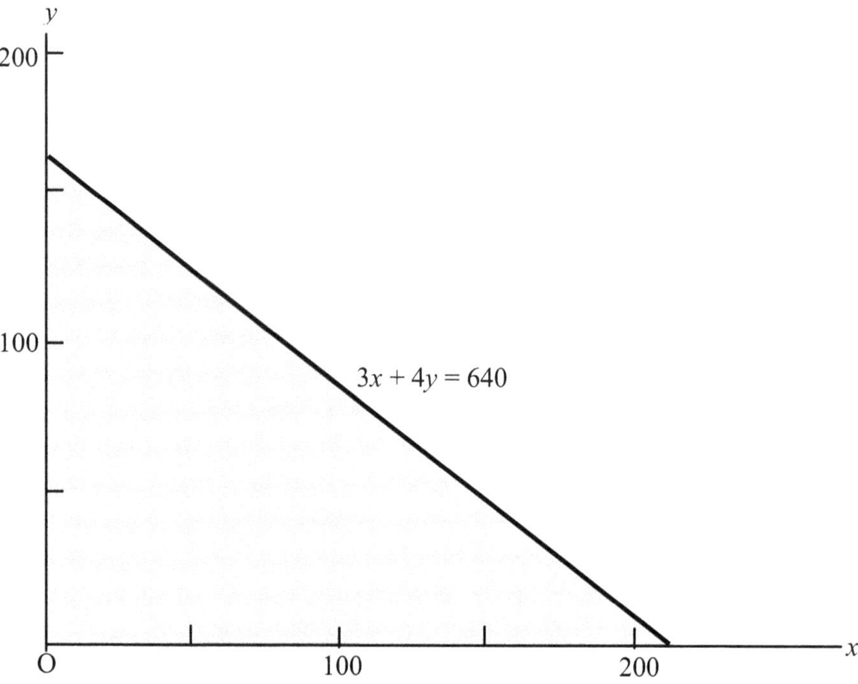

Fig. 19.1. The constraints (19.3) and (19.5) hold in the shaded region

19.3 A Graphical Solution

With only two variables it is easy to solve this problem graphically. We need more sophisticated computational methods if there are many constraints, or more than two items in production.

> **Graphing inequalities — a shady business**

The 'equality' elements in (19.2)–(19.5) are represented by straight lines on a graph. What about the inequalities? The inequalities (19.5) imply we are only interested in points with nonnegative coordinates (x, y). These points all lie in the first, sometimes called the north-east, quadrant of the graph.

The remaining inequalities further restrict the area of interest. For example, (19.3) means that we are restricted to points lying either on the line $3x + 4y = 640$ or to one side of it, for a moment's reflection shows that points on one side of this line satisfy the inequality $3x+4y < 640$. Points on the other side satisfy the inequality $3x + 4y > 640$. Since the inequality $3x + 4y < 640$ is satisfied for the point $(0, 0)$, it is points in the first quadrant to the left of

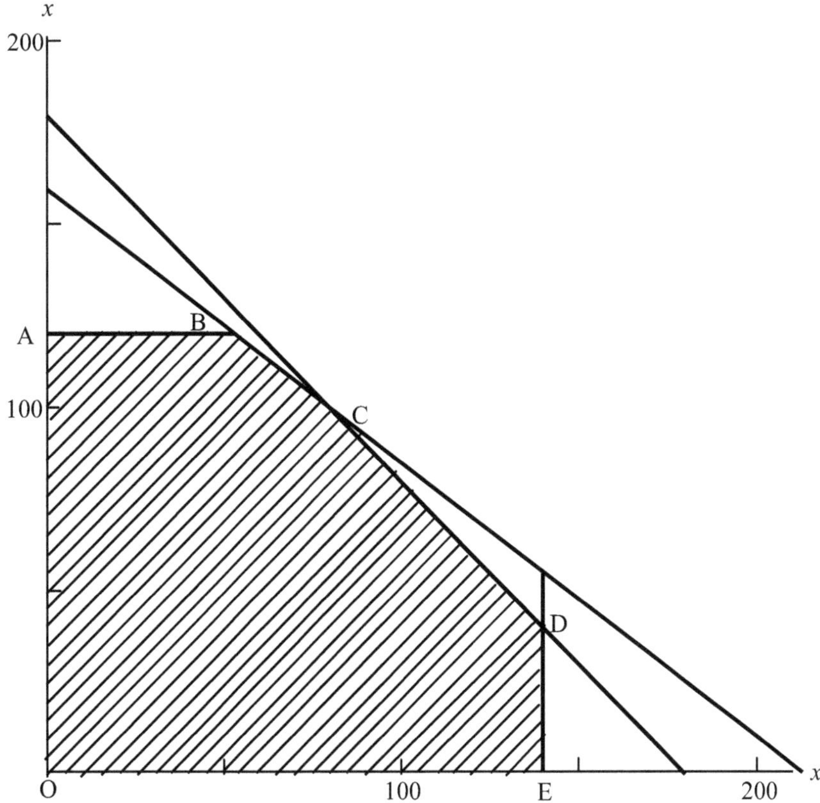

Fig. 19.2. The hatched polygon OABCDE is the feasible region consisting of x, y values meeting all constraints

and below the line $3x + 4y = 640$ that satisfy both (19.3) and (19.5). This is the shaded area in Fig. 19.1.

The line $3x + 4y = 640$ is easily positioned on the graph by noting that when $y = 0$, the intercept on the x-axis is at the point $(213.33, 0)$. Similarly, the intercept on the y-axis is at $(0, 160)$. In like manner, we can insert on a graph lines corresponding to the equality part of the constraints in (19.2) and (19.4). We then establish on which side of each relevant line the inequality part holds for non-negative x and y. These lines are drawn in Fig. 19.2. The cross-hatched area is the region in which all the constraints (19.2) to (19.5) hold. This region is called the *feasible region* because it includes all possible values of x and y consistent with the constraints. How do we decide which point or points in this region give the maximum of U, the profit?

Recalling that $U = 10x + 12y - 1200$, it is evident from Sect. 7.3 that if we substitute various values of U in this equality we get a series of parallel lines. For example, if $U = 0$ the equation of the relevant line is

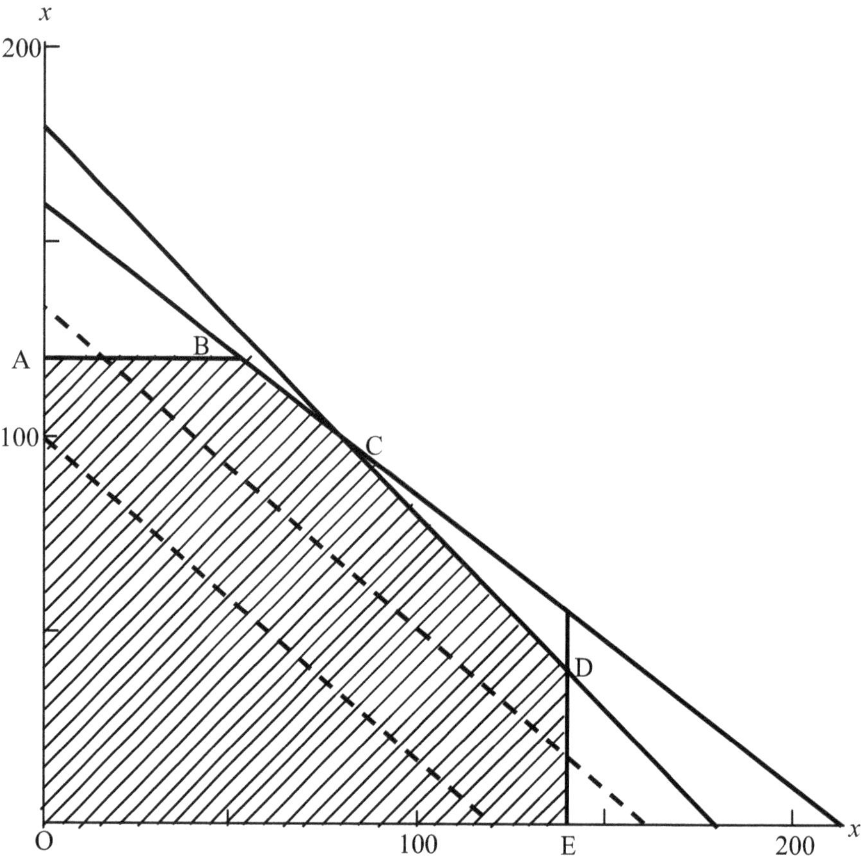

Fig. 19.3. The dashed lines represent profits of zero and 400. Clearly any (x, y) on these lines lying in the critical region will produce these profits

$$10x + 12y = 1200\ , \tag{19.6}$$

and if $U = 400$ the equation is

$$10x + 12y = 1600\ . \tag{19.7}$$

Equation (19.6) is represented by a line through the points (120, 0) and (0, 100). Equation (19.7) is represented by a line parallel to (19.6) through (160, 0) and (0, 133.33). In Fig. 19.3 both these lines are superimposed as dashed lines on the diagram in Fig. 19.2.

As the line representing U shifts upward and to the right it is evident that the profit will increase. Also, after that shifted line passes through C it will move out of the feasible region if we continue to increase U. Thus, the greatest

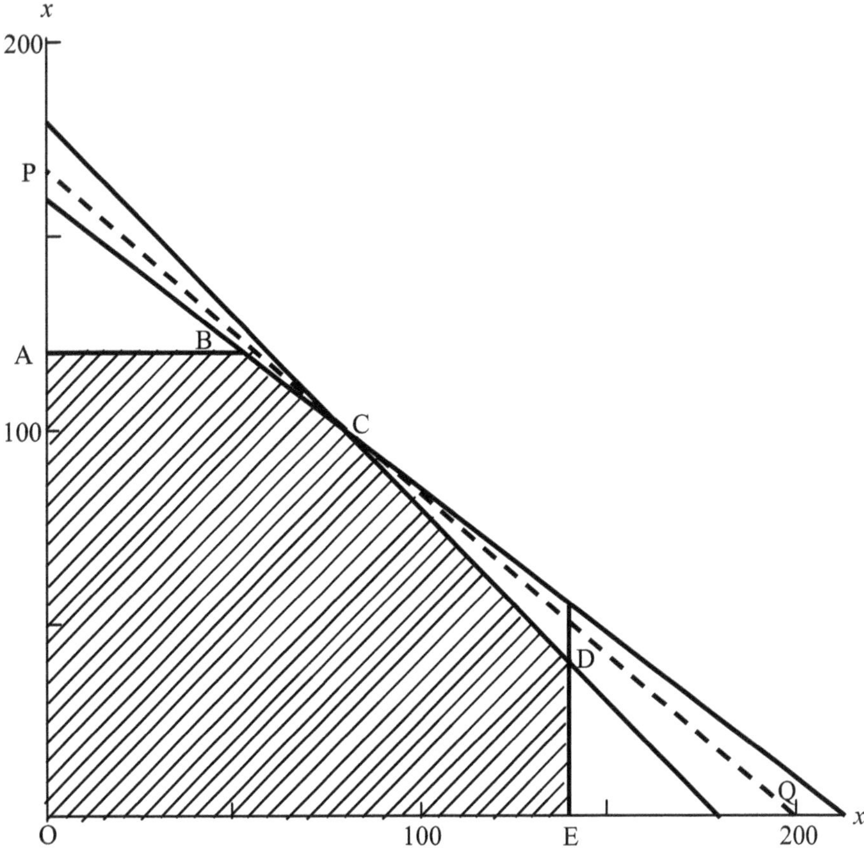

Fig. 19.4. The function $U = 10x + 12y - 1200$ has a maximum value in the feasible region at C where $x = 80$ and $y = 100$

profit is obtained when the objective function line passes through C. Now, C lies at the intersection of the lines with equations

$$3x + 4y = 640 \tag{19.8}$$

$$x + y = 180\ . \tag{19.9}$$

These represent the equality elements of the constraints (19.3) and (19.4). Solving (19.8) and (19.9) we find $x = 80$ and $y = 100$, whence

$$U = 10 \times 80 + 12 \times 100 - 1200 = 800.$$

Thus, the maximum daily profit is £800. The line $10x + 12y - 1200 = 800$ (i.e., $10x + 12y = 2000$) passing though C is shown by the dashed line PQ in Fig. 19.4. All points on this line, other than C, lie outside the feasible region. They are unattainable without breaching the constraints.

Only the constraints (19.3) and (19.4) reach equality at C. For that reason these are often referred to as the *active constraints.*

In general, there will be at least one vertex of the polygon forming the feasible region at which the linear function U representing profit is a maximum. In this example the relevant vertex is a unique point representing maximum profit. However, had the set of lines representing profit been parallel to BC, then all points (x, y) lying on the segment BC between B and C would yield the same profit. An analogous argument applies if the profit lines had been parallel to CD.

A possible complication is that the maximum profit may not correspond to integer values for x and y. This is unlikely to be a serious practical problem for the manufacturing process here. If the optimum solution had left an icomplete oven at the end of the day, this could probably be completed as part of the following day's production, or by working a small amount of overtime. However, we show in Sect. 19.6 that if we apply a further constraint that an integer number of units are an essential feature of the solution, then the situation may, though it does not always, become complicated.

19.4 Sensitivity Analysis

Constraints are not always hard and fast. Manufacturers often want to know what happens if one or more are modified. Some modifications will affect the solution, others may leave it unchanged. For example, the constraint (19.2) that limits the maximum numbers of standard and deluxe ovens to 140 and 120 respectively do not come into play in our solution because, for optimality, only 80 standard and 100 de luxe ovens are made. These constraints could be reduced to $x \leq 80$ and $y \leq 100$ without affecting the solution. In terms of the graphs in Figs. 19.2, 19.3 and 19.4, the effect of these tighter constraints would be to make shifts in AB and DE that kept them parallel to the axes, but now B and D would coincide with C. Any further restrictions imposed by these constraints would alter the solution. For example, if we reduced the constraint on x to $x \leq 70$, keeping all other constraints as in (19.2) to (19.5), the graphical solution is that indicated in Fig. 19.5.

Study this figure carefully, and verify that the optimum output is now given by point D, with $x = 70$ and $y = 107.5$. This leads to a maximum profit of $U = 790$.

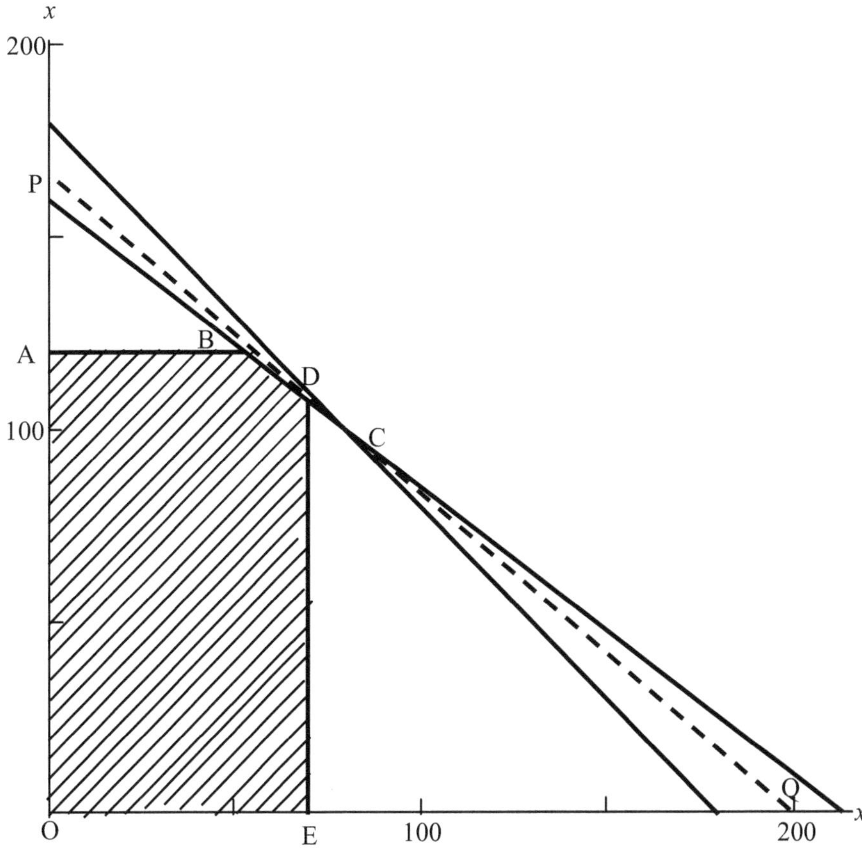

Fig. 19.5. The effect of reducing the constraint $x \leq 140$ to $x \leq 70$

19.5 Limitations of Graphical Methods

In the microwave oven example the constraints and the functions to be optimized involved only two variables x and y. In many real-life problems 100 or more variables may be involved, and graphical solutions are impossible.

A step towards reality

In a realistic extension of the example on the microwave ovens, the manufacturer may well make more than two products. As well as deluxe and standard ovens, the company might also turn out a simpler popular model. In addition to ovens, they might make two different kinds of ceramic hobs, and also a washing machine, all at this same factory. They then want to allocate resources subject to appreciably more complicated constraints than those discussed in our example.

We shall not cover extension to more than two variables, but we indicate briefly the basic idea behind an algebraic approach that may be used with any number of variables. For large numbers of constraints where many different items are produced, modern computing facilities are virtually essential for its application.

The method, called the *simplex method*, is tedious with only a pencil and paper, even for just two variables. All we shall do here is outline, in very simple terms, the way it works. For the reader who wishes to take up the challenge of following the process through to a complete solution of the microwave oven example, the detailed algebra is given in the Appendix, Sect. A.5.

We summarized the data for the oven problem in Sect. 19.2 in equations (19.2) to (19.6). For convenience we give these again here. We want to maximize

$$U = 10x + 12y - 1200$$

subject to

$$\begin{aligned} x \leq 140, y &\leq 120 \,, \\ 3x + 4y &\leq 640 \,, \\ x + y &\leq 180 \,, \\ x \geq 0, y &\geq 0 \,. \end{aligned}$$

The key to the simplex method is to introduce new variables called *slack variables*. This is a neat dodge to convert constraints from inequalities to equalities. In the context of maximizing profits, we may think of slack variables as representing zero profit items that are never produced. In this example we need one slack variable for each non-zero inequality constraint. There are four such constraints after excluding the zero inequality constraints $x \geq 0$, $y \geq 0$. We denote the slack variables by s_1, s_2, s_3 and s_4, and use one in each constraint to give

$$x + s_1 = 140 \tag{19.10}$$

$$y + s_2 = 120 \tag{19.11}$$

$$3x + 4y + s_3 = 640 \tag{19.12}$$

$$x + y + s_4 = 180. \tag{19.13}$$

In the sense that they are making up a 'deficiency', it is apparent that all slack variables must be non-negative when included in a '$\leq$' inequality. Thus, x, y, s_1, s_2, s_3 and s_4 are all zero or positive.

The equations (19.10) to (19.13) are 4 linear equations in 6 unknowns. We saw in Sect. 13.4 that when the number of equations is less than the number of unknowns, we can usually find more than one set of solutions that satisfy all equations. For example, you should verify by direct substitution that $x = y = 0$ and $s_1 = 140, s_2 = 120, s_3 = 640$ and $s_4 = 180$ satisfy (19.10) to (19.13). Again, $x = 80, y = 100, s_1 = 60, s_2 = 20, s_3 = 0$ and $s_4 = 0$, and also $x = 140, y = 40, s_1 = 0, s_2 = 80, s_3 = 60$ and $s_4 = 0$ satisfy these equations. Two features are common to these three solutions:

- In each the x, y values correspond to a vertex of the feasible region in Fig. 19.4.
- Two from the six variables take the value zero in each solution.

You should verify that for any vertices of the feasible region in Fig. 19.4, the second property holds. Such solutions form what are known as *basic feasible solutions*.

A procedure known as the *simplex algorithm*, explores basic feasible solutions systematically until we find a solution for which U is a maximum. The algorithm works like this:

- Step 1. Determine any basic feasible solution.
- Step 2. Examine U to see if it can be increased by moving along any boundary away from the vertex chosen in step 1. If not, U is the maximum. If U can be increased proceed to step 3.
- Step 3. If U can be increased by moving along one or more boundaries, choose one of these and determine the new basic feasible solution corresponding to the next vertex reached. Take this as a new basic feasible solution. Return to step 2 and continue the cycle until a maximum is reached.

Since each slack variable appears in one constraint equation only, it is easy to see that if we set $x = y = 0$ equations (19.10)–(19.13) are satisfied if we set the slack variables equal to the numerical values on the right-hand side of the corresponding equation. This gives a basic feasible solution for step 1. We proceed to step 2. We express the non-zero variables, usually called the *basic variables*, and also U, in terms of the variables (here x and y) that were zero in the basic solution.

Thus

$$s_1 = 140 - x \tag{19.14}$$

$$s_2 = 120 - y \tag{19.15}$$

$$s_3 = 640 - 3x - 4y \tag{19.16}$$

$$s_4 = 180 - x - y \tag{19.17}$$

$$U = 10x + 12y - 1200. \tag{19.18}$$

For the basic feasible solution $x = y = 0$, we have $U = -1200$. Physically, since the slack variables may be looked upon as representing phantom ovens produced at no profit, this solution says that if we make nothing we make no profit, but are left with a bill of £1200 for overheads. Because x and y are never negative we can increase U by increasing either x or y. This is evident since each has a positive coefficient in (19.18), We move to step 3.

At this step we have to choose to move to another vertex by increasing either x or y. Since we increase profit by £12 for each increase of one unit in y, but only by £10 for each increase of one unit in x, a reasonable action is to increase y. From (19.15), however, we see that we can only increase y to 120. With any greater increase s_2 would become negative, and this is not permitted. From (19.14)–(19.17), we find this gives a basic feasible solution where $y = 120, x = 0, s_1 = 140, s_2 = 0, s_3 = 180, s_4 = 60$.

We now return to step 2 to see whether we can increase U by increasing x. To check, we express the constraints and U in terms of the non-basic variables which are now x and s_2. This, and the later tedious steps are set out in the Appendix, Sect. A.5. We leave it to the reader to decide whether to work through that detail with pencil and paper, but the equation (A.25) given there has the form

$$U = 800 - 4s_4 - 2s_3. \tag{19.19}$$

Since slack variables must always be nonnegative, (19.19) shows that we cannot further increase the profit. Also from (A.22) and (A.23) we see that the maximum profit is obtained by making 80 standard and 100 deluxe ovens, in accord with the solution we got by the graphical approach.

Using the simplex method in this example looks like using a sledge-hammer to crack a nut. The important feature is that (i) it generalizes to cases with m unknowns and n constraints, and (ii) there are clear and unambiguous instructions on how to proceed, and a defined stopping rule when we attain a maximum.

In this brief introduction to the *simplex method* we have glossed over some additional points that have to be considered in developing an all embracing software algorithm. For example, in some problems constraints may take the inequality form *greater than or equal to* rather than *less than or equal to.* There is also a possibility that some basic variables, as well as nonbasic ones, may take the value zero at some stage. There may also be problems relating to a stopping rule if there is no optimum, or more than one optimum.

Even if we use only pencil and paper methods, we do not need to write down all the algebra given here and in the appendix. Most books on operational research that deal with linear programming show that the arithmetic of the simplex method can be reduced to almost mechanical applications of a few simple rules using what is called a *simplex tableau.* Such a tableau is a basis for computer software for linear programming.

The method can be extended to cover sensitivity analyses by considering what are known as the solution of *dual problems*, a development not covered here.

19.6 Integer Programming

We indicated in Sect. 19.3 that linear programming models sometimes lead to inappropriate non-integer solutions. This is not a serious problem when making microwave ovens. As already pointed out, it is usually feasible to finish a part-manufactured oven either by working overtime or by completion the next day. If these are not practical alternatives, 'rounding down' to an integer solution that loses one unit of production will often give a near optimum solution.

Sometimes only an integer solution is acceptable, and rounding down may not be optimal. That situation may arise if an airline has ample aircraft of two types, A and B, to operate a route, but there are limitations on available crew and on fuel supplies. Suppose aircraft of type A each carry 295 passengers and require 7 cabin crew and 8400 litres of fuel per flight; aircraft of type B each carry 94 passengers and require 4 cabin crew and 2600 litres of fuel per flight. There are 65 cabin crew and 60 800 litres of fuel available. Each aircraft/crew can only make one flight per day. On a day when passenger demand exceeds seating capacity (ensuring each aircraft takes off fully loaded), how many aircraft of each type should be used to maximize the number of passengers carried?

This has the makings of a simple linear programming problem, but an integer solution is needed in practice. Formulating it as a linear programming problem, if x is the number of type A aircraft, and y is the number of type B aircraft used, the problem is to maximize

$$U = 295x + 94y$$

subject to the constraints

$$\begin{gathered} x \geq 0, \quad y \geq 0\,, \\ 7x + 4y \leq 65\,, \\ 8400x + 2600y \leq 60800\,. \end{gathered}$$

A graphical approach like that in Sect. 19.3 gives the feasible region as the rectangle OABC in Fig. 19.6. As an exercise, check that the maximum feasible solution occurs at the vertex B, which is the intersection of the lines

$$7x + 4y = 65\,,$$

and

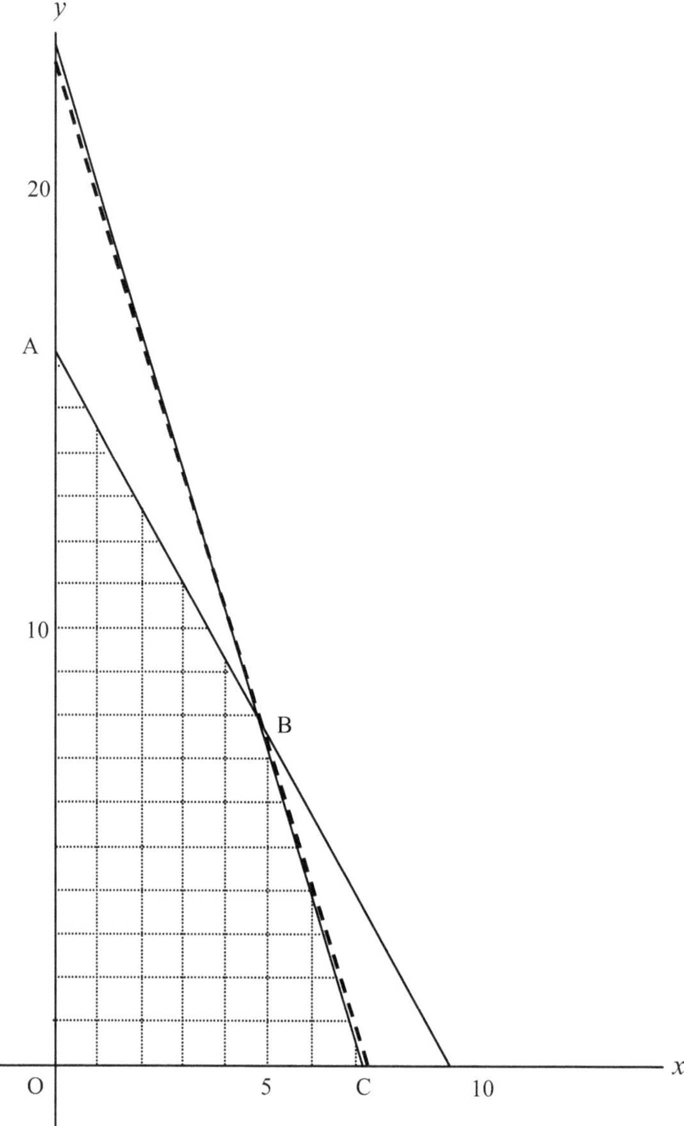

Fig. 19.6. The linear programming optimum is the vertex B (4.82, 7.82) but the integer programming solution is the point (6,4)

$$8400x + 2600y = 60800 .$$

This gives a solution (rounded to 2 dec. pl) $x = 4.82, y = 7.82$, whence $U = 2156.98$ passengers.

This is no practical solution. We can only use integer numbers of aircraft. It is tempting to hope that the optimum solution, when x and y are confined to integer values, will be obtained by rounding to the nearest integer values, subject to the condition that the result is within the feasible region. To help examine these, and other possible integer solutions, we have added a grid of dotted lines on an integer scale in Fig, 19.6. Possible integer solutions are points where grid lines intersect.

The nearest integer points to B that lie within the feasible region have co-ordinates (5, 7), (4, 8) and (4, 7). We may ignore the last of these, as U necessarily has a greater value at (5, 7) or (4, 8) than it has at (4, 7). The values of U at (5, 7) and (4, 8) are respectively 2133 and 1932. So it is better to use 5 aircraft of type A, and 7 of type B. However, inspection of Fig. 19.6 suggests (6, 4) is also a feasible integer solution. It is easy to verify that all constraints are satisfied by this point, that for fuel being satisfied as an equality. For this point $U = 2146$. The point is not particularly close to the vertex B, and we have not eliminated the possibility of a better solution. We have included in Fig. 19.6 a broken line representing the equation $295x+94y = 2156.98$. This is the objective function line corresponding to the linear programming solution.

Flying bits of aeroplanes is not a good idea

In fact, there is no other integer solution within the feasible region that improves on the one above. This is because the perpendicular distance from the point $(6, 4)$ to the line with U set equal to the linear programming optimum, is less than that for any other integer feasible solution. This implies it is as close as possible to what the optimum solution would be without the additional constraint that the solution must be integers. Because of scale limitations, it may not be quite clear from Fig. 19.6 that the point (6, 4) is indeed the closest such point. The point (7,1) may look to be a close competitor, but it is easily verified that it does not satisfy the fuel constraint.

You may feel that we have chosen a freak case, because the line corresponding to optimum U lies close to the constraint line for fuel. This is not unrealistic in practice. It happens because each type of aircraft is almost equally fuel efficient per passenger carried. Type A requires $8400/295 \approx 28.47$ litres per passenger, while type B requires $2600/94 \approx 27.66$ litres per passenger. By contrast, type B aircraft are less cabin crew efficient, requiring $4/94 \approx 0.043$ crew per passenger, compared to type A which require only $7/295 = 0.024$ crew per passenger.

A more formal approach to integer programming is to use the simplex algorithm as a first stage to get a linear programming solution. If this is not an integer solution, then to use a further algorithm to apply the additional integer requirement. One algorithm for this latter phase was given by Ralph Gomory [26].

The example is simplified but moderately realistic. However, the solution only nibbles at the kind of operational problems facing airlines. There will almost certainly be constraints on available flight crews as well as cabin crews. There is unlikely to be an unlimited supply of each type of aircraft as we assumed. Resources may have to be shared with other routes. Optimal allocations to each may be incompatible because these, when totalled, would exceed available resources. Some aircraft types may only be suitable for specific routes due to fuel capacity, restricted runway lengths, etc.

More subtle factors may also come into play. When all demands on competing routes cannot be met by optimal allocations, different profit in cash terms for opposing routes may influence decisions. There may also be questions of goodwill, or claims for compensation, if appreciable numbers of potential passengers cannot be accommodated on a specific flight or route.

What we have shown is that the basic spade-work for optimum allocations can be handled mathematically. This leads to better solutions than those based on guess work or trial and error, the latter approach being virtually impossible for larger problems.

19.7 Loose Ends

Assignment problems. A group of operations research problems involving optimal allocation of resources are called *transportation problems.* These are linear programming problems, but their structure leads to special methods of solution. We omit details, but a few simple examples hint at the sort of problem involved.

The key to solving more realistic problems is usually to devise a procedure — typically an algorithm — that leads to the optimum solution as rapidly as possible.

Transportation problems, as well as being linear programming problems, are also a subset of a more general class of problems called *assignment problems.* The simplest assignment problem is that of one-to-one assignment. Here is an example.

In an award scheme for physical prowess each competing club enters a team of four. Each member of that team is allowed to tackle one and only one of four challenges. Each team aims to complete all four challenges in the minimum possible total time. Challenge (a) is to collect from one spot, and closely stack at another, 512 concrete blocks each of which is a cube of side length 10 cm. The stacked blocks are to form a larger cube. You might like to amuse yourself working out what the length of each side of this larger cube will be! Challenge (b) is to run 1500 metres. Challenge (c) is to scale a number of high fences and clear 3 other obstacles. Challenge (d) is to saw through a tree-trunk 30 cm in diameter. The four members selected for the team by one club are each given practice attempts at all four tasks. The times taken by each to complete each task are recorded. It is assumed they will take the same

Table 19.1. Times (minutes) taken to compete challenges (a)-(d) by each of four competitors

	Task			
Competitor	a	b	c	d
I	18.32	4.96	5.83	7.12
II	15.45	5.12	4.97	7.23
III	15.42	4.82	6.94	6.99
IV	19.24	5.87	7.02	7.25

time if entered for that task in the competition. If the object is to minimize the total time taken for all four challenges, how should the club allocate these competitors to challenges? Remember each competitor is allocated to one and only one challenge.

The recorded practice times in minutes are given in Table 19.1. That table shows that competitor IV is the least useful, being the slowest at all challenges. On the other hand, competitor III is the fastest in all challenges except (c). In allocating competitors to minimize the total time we must take into account the time saved by putting a generally good competitor in a given event, as opposed to time lost by not using him or her in some other event. Similar considerations arise over the time lost in the various events by entering a really bad competitor in that event. If you are a glutton for punishment, you may care to use a trial and error process to decide the best strategy to minimize total time to complete all challenges. Algorithms not given here exist to find this. If you know any of these, or have used a trial and error method and want to check if you have got the optimum allocation, it is given at the end of this section.

A transportation problem. A transportation problem, at its simplest, involves the situation where there are m depots and also n sites such as retail outlets. At each depot known quantities of a specific commodity are stored. At each outlet specified quantities of that commodity are required. The cost per unit of transportation from each depot to each site is known. The problem is to find the strategy to meet the requirements in a way that minimizes transport costs. Here is an example:

Whisky galore

A whisky distiller has to supply three bottling plants from two distilleries. At distillery I there are 30 barrels of whisky, and at distillery II there are 65 barrels of whisky. Bottling plants A, B and C respectively require 20, 35 and 40 barrels. Table 19.2 gives the cost in £UK of transporting one barrel of whisky from the distillery in that row to the bottling plant in that column.

Table 19.2. Costs per barrel of transporting whisky between distilleries and bottling plants

	Plant			
	A	*B*	*C*	*Available*
Distillery I	6	4	2	30
Distillery II	4	5	3	65
Required	20	35	40	

For example, the cost of supplying a single barrel from distillery I to plant B is £4. The column on the right of the table gives the number of barrels available at each distillery, and the row at the bottom the requirement at each bottling plant. How many barrels should be moved from each distillery to each plant to minimize transport costs?

This example is particularly simple in that the total number of barrels available equals the total demand. This is basically a linear programming problem in which we have to minimize costs, subject to constraints both on maximum availability at each distillery, and on requirements at each plant. The nature of these constraints are such that they enable special algorithms to be used. The basic idea behind special algorithms for solution of this, and more general, transportation problems is first to determine any feasible solution. The second step is to see if changes to some of the numbers of barrels transported in that solution will reduce costs. Algorithms in common use systematize this search.

Optimal assignment. How did you assign competitors to the challenges using the data in Table 19.1? Total time is minimized by this assignment: For (a) choose competitor III, for (b) choose I, for (c) choose II and for (d) choose IV. In this four-competitor situation it is feasible, but tedious, to solve the problem by a trial and error approach. There are only 24 possible allocations of 4 competitors to 4 events. If there are 10 competitors to be allocated to 10 events, trial and error is not on without a fast and appropriately programmed computer, for there are $10! = 3\,628\,800$ possible allocations.

20

Networks

20.1 Putting Graph Theory to Practical Use

In Chap.9 we touched on the subject of graph theory, using the word graph not in the Cartesian sense, but as it is associated with networks and with the Euler formula for polyhedra. In this chapter we look at some roles that graphs and networks play in operations research (OR).

How to get your house built — fast

Critical path analysis is a classic OR method to solve a problem by using a network of lines joining points usually referred to as *nodes*. These are essentially the *vertices* we met in Chap.9. For many projects in engineering, construction, and in other branches of technology, and also in commercial operations, we want to know the minimum time needed to complete a number — often hundreds or even thousands — of interlocking activities.

For solving such scheduling problems a computer is essential when, as is usual, many operations are constrained by the need to complete some before others can be started. There will also be restrictions on the rate at which each can proceed.

The logic for dealing with problems of this type can be illustrated graphically by simple examples like building a boat or a house, writing and publishing a book, landscaping a garden, or even baking a cake.

But a nuclear power station will take longer

Designing and building an airport, a nuclear power station or a space satellite is a major scheduling problem if the project is to be completed on time, and at reasonable cost.

Features common to situations where critical path analysis is appropriate are:

- there is an order of precedence for certain activities;
- some activities may be carried out simultaneously;
- the duration of each activity is either known precisely, or can be estimated within certain limits, or can be adjusted within some specified range (often at additional cost if the job is to be speeded up).

Building a house illustrates these points. Foundations must be laid before the walls are erected, the roof put on before the interior walls are plastered. Depending upon the mode of construction, it may or may not be necessary for certain plumbing operations and electric wiring to be completed before walls are plastered. However, both electrical and plumbing work may often proceed simultaneously. A specific time may be laid down for some operations like digging and laying foundations. It may in practice be virtually impossible to reduce this time if it embodies a substantial 'settling period' before further work can proceed. On the other hand, the time to build walls might be reduced by employing extra bricklayers, but usually only at additional cost.

If certain key operations fall behind schedule, completion of the whole project will be delayed — a familiar experience for those waiting to move into a new home. Flexibility may be possible with some operations because their delay, at least within certain limits, will not affect the final completion date.

Critical path analysis (often abbreviated to CPA) uses a network to represent activities and their inter-relationships. It tells us which key operations must not fall behind schedule if a job is to be completed on time.

As an example of a small CPA problem, we look at how author and publisher must plan to ensure publication of a book entitled *The Making of a Plutonian President* on the date a new president of the republic of Plutonia is to be inaugurated.

Table 20.1 is a list of activities that must be completed before the book appears. It shows the minimum times needed for each task, and gives information about which must be finished before certain other activities start. Such preceding tasks are indicated in the column headed *Precedents*. It is convenient to designate tasks — or activities as we call then in the table — by a capital letter.

The information in the table enables us to find the minimum possible elapsed time between the author beginning to write the book, and its eventual availability in bookshops. If it is to be on sale the day the new President is installed, this tells us the latest possible date for the author to start writing.

The times given in Table 20.1 for the various steps reflect to some extent the nature of the book and the techniques used to produce it. For example, the times taken for steps B and D reflect in part the fact that a book on a

Table 20.1. Schedule for publishing the book *The Making of a Plutonian President.*

Activity	*Description*	*Time(days)*	*Precedents*
A	Researching and preparing the MS	150	
B	Clear MS with publisher's solicitors for libel, copyright queries, etc.	10	A
C	Copy edit MS and agree any proposed alterations with author	30	A
D	Negotiate with author to alter any potentially libellous material	7	B
E	Final electronic type-setting	8	C, D
F	Prepare page proofs	4	E
G	Author checks and corrects proofs	10	F
H	Production editor checks proofs, layout	12	F
I	Author prepares index	6	G
J	Production editor collates author's proof corrections with his own and checks index	5	H, I
K	Design and prepare cover	25	C, D
L	Prepare publicity material	15	J, K
M	Printing, binding and delivery to warehouse	20	J, K
N	Prepublication delivery to shops, distribution of review copies	20	L, M

political topic may contain libellous material. For a technical book with a lot of formulae and detailed proofs, some or all of the steps E to J might take longer.

Publisher's schedules are based on estimated times. All too often there is a slippage at some vital stage. In addition to determining the minimum possible time for publication, CPA indicates whether or not slippage in specific

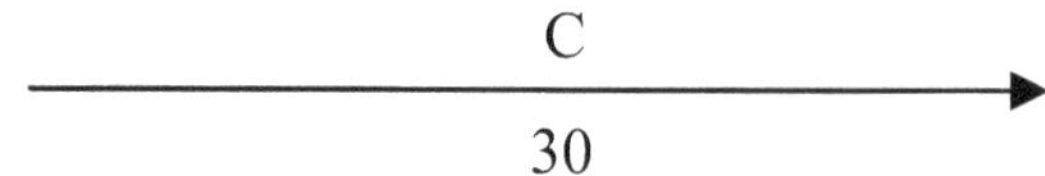

Fig. 20.1. A labelled arrow to represent an activity and associated resource requirement

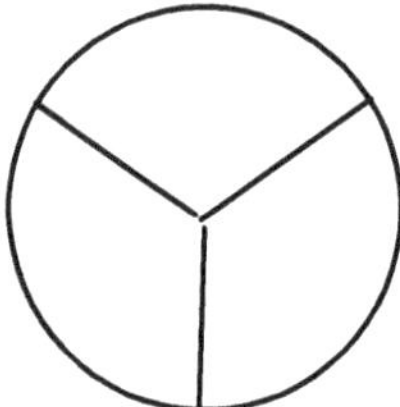

Fig. 20.2. A circle or node representing an event. Data are entered into the 3 sectors

activities, and if so how much slippage, can be tolerated without increasing total production time. There are many elaborations, such as determining the possible time-saving by speeding up certain steps. In essence, all these problems can be solved diagrammatically by a logical representation of the process. Computer programs exploit this logic by using an algebraic and arithmetic translation of features of these diagrams.

20.2 Basic Concepts of CPA

We need to define and explain some conventions before using the information in Table 20.1 to set up a CPA diagram. Not all these conventions are essential to the logic, but it is wise to adhere to them because they are widely used. For example, it is usual to represent, as far as possible, logical sequences progressing diagrammatically from left to right across a page.

An *activity* is a task requiring certain resources. In Table 20.1 the resource is time to perform. In the network an arrow is used to represent an activity. The tail indicates where it begins, and the head where it ends. The arrows are not drawn to scale. An activity is designated by a name, letter or number. We use a letter. We write this designator and associated resource (time in this example) alongside each arrow in the manner indicated in Fig. 20.1 for activity C.

A *dummy* activity is an activity that requires no resources. In setting up a network we use these either to meet the logic of the problem, or to help us comply with established conventions. A dummy activity is usually denoted by a broken or dotted arrow and needs no label.

An *event* corresponds to the start, or finish, of one or more activities. It is conventionally denoted by a circle, usually referred to as a node. Each node

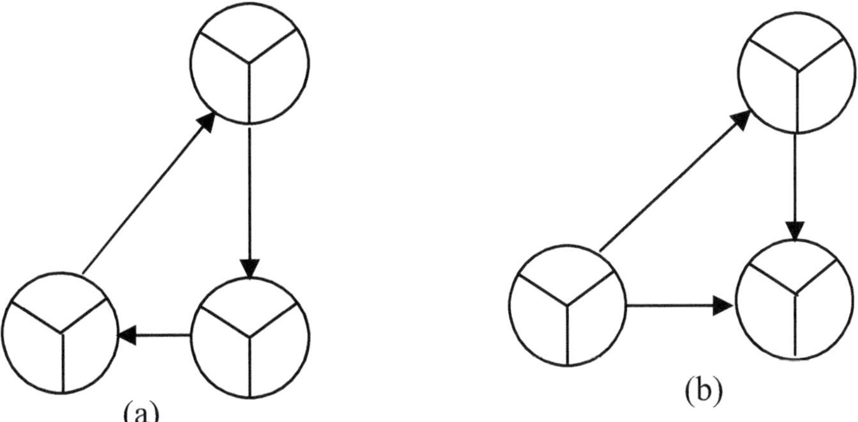

Fig. 20.3. In CPA (**a**) is not but (**b**) is permissible

is divided into three sectors as shown in Fig. 20.2. During a CPA data are entered into these sectors.

A CPA network is a logical combination of activities, dummy activities and nodes which conforms to certain rules. There is only one point of entry to a network (the starting event), and one end point (the finishing event).

Every activity, including a dummy activity, joins two nodes, one at the tail or start of the activity and the other at the head or finish of that activity. More than one activity may use any tail or head node, but no two activities may jointly use the same tail and head nodes. The latter possibility can be avoided by introducing a dummy activity if necessary. The restriction is not essential to the logic, but it originated in an era when its neglect led to difficulties with computer programs. The starting event is a tail node only, and the finishing event is a head node only. All other nodes are head nodes for some activity or activities and tail nodes for others, perhaps including some dummy activities. An activity, other than the start activity or activities, uses a tail node that is a head node for some previous activity. An appropriate tail node must be chosen to ensure completion of all activities which must be finished before the current activity can start. If there are constraints on the order of activities, these are reflected in the choice of tail nodes. For example, in Table 20.1 activity C cannot start until A is completed, but there is no further constraint on C. It follows that the head node for A is an appropriate tail node for C.

Loops which lead back, following the direction of the arrows, to a node previously visited are not allowed. Thus Fig. 20.3(**a**) would not be allowed as part of a CPA network, but 20.3(**b**) would be permitted, because the arrow directions are not all clockwise or all anti-clockwise.

All activities must contribute to the progression through the network. An activity that does not contribute is sometimes called a *dangler* and should be

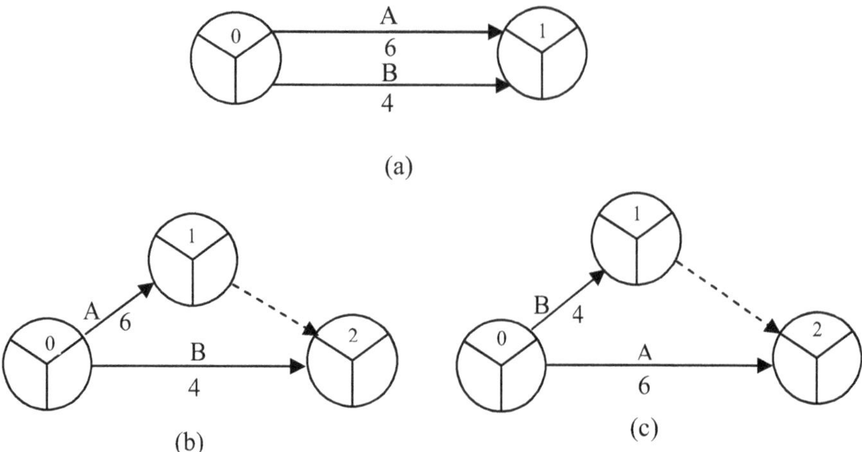

Fig. 20.4. Figure (**b**) and (**c**) represent two ways of avoiding the nonpermissible (**a**).

omitted. In a well formulated problem there should not be danglers, because in essence these correspond to irrelevant information.

Events are usually numbered continuously in the top sector of corresponding nodes, labelling the start node 0, or sometimes S (for start). The first head node for which the start node was a tail is labelled 1, and so on. The final node is either labelled F (for finish), or has the highest node number in the network.

Playing with dummies

Dummy activities are useful to help conform to the convention that two events should not share common head and tail nodes. Thus, if I were to go out and make a cup of tea (activity A taking 6 minutes) while the printer prints out the draft of this chapter (activity B taking 4 minutes), this should not be indicated on a two-node diagram like Fig. 20.4(**a**). We overcome the difficulty by introducing a new node 1 and relabelling the node 1 in Fig. 20.4(**a**) as node 2. We then link nodes 1 and 2 by a dummy activity in either the way shown in Fig. 20.4(**b**) or in Fig. 20.4(**c**).

Dummy activities are also used to complete logical paths in a network. Suppose we have to perform the tasks in Table 20.2. For simplicity in this illustration we have omitted times, but an order of precedence is given.

Fig. 20.5 is a network representation and is an example of a complete network (apart from times). Node 0 is the starting node for the network, and also the tail node for activities A and B, that have head nodes 1 and 2 respectively. The precedence list required C to follow B, so we may use node

Table 20.2. A simple table of events and precedents

Event	*Precedents*
A	
B	
C	B
D	A
E	A, C
F	C, D

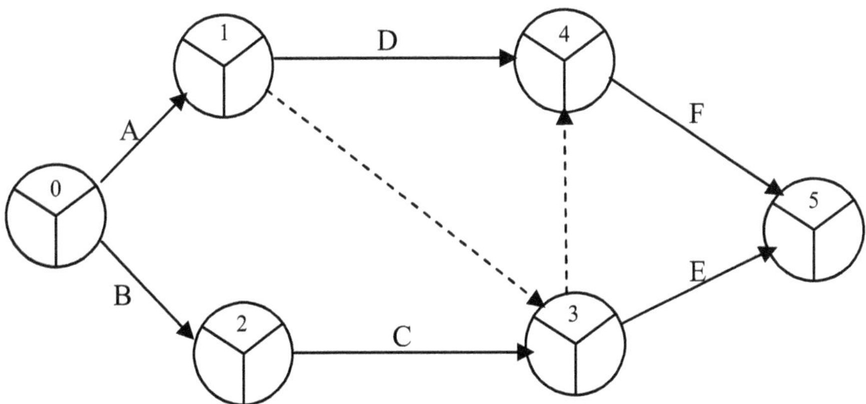

Fig. 20.5. A simple illustration of using dummy activities to preserve the logic

2 as a tail node for C and node 3 as a head node. Since D must follow A it has tail node 1, and a new head node 4. Now E must follow A and C. This is indicated as a logical path by joining node 1 to node 3 by a dummy activity. Since F must follow both C and D we need a dummy activity linking node 3 to node 4. Since E and F must both be performed to finish the task we link these to a final node 5. Logical diagrams may not be unique. Although it is more complicated, and in practice unnecessary, it would still be logical to link nodes 3 and 4 by dummy activities to a new node 5 and proceed as shown in Fig. 20.6.

If these ideas are new, you may find that they take time to digest. You may have to refer back to them as you build up a real network. Now try your hand at setting up a network based on the information in Table 20.1. If you have trouble setting this up, you should study Fig. 20.7, together with the comments below, about how we form this network.

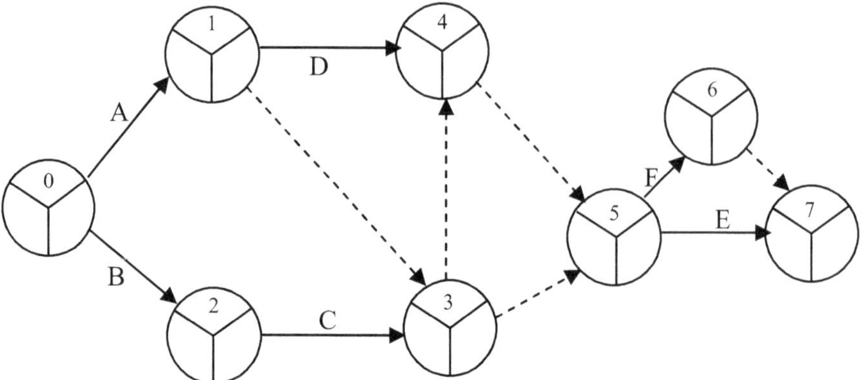

Fig. 20.6. A permissible but not essential use of dummies connecting nodes 3 and 4 to node 5

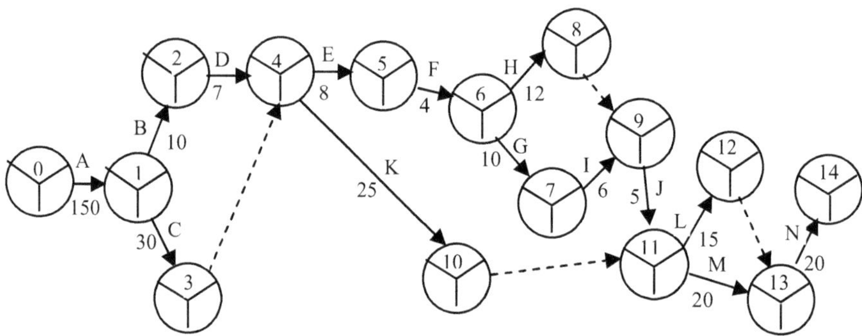

Fig. 20.7. The basic CPA diagram using the information in Table 20.1

We show how to use the network to determine the shortest possible production time, and to answer questions about which job ordering and timings are 'critical' to produce the book as quickly as possible.

In Fig. 20.7 the nodes are numbered from 0 to 14. The tail node for any activity determines the earliest time at which it can start. For example, no activity subsequent to A (researching and preparing the manuscript) can start until that activity is completed, but both B and C may start not earlier than the completion of that task. They may commence at any time thereafter. Similarly, activity K (design and prepare cover) cannot commence until both C and D are completed, implying that node 4 must be the tail node for K. The dummy activity starting at node 3, and ending at node 4, is needed to avoid node 3 appearing incorrectly as a dangler. It is important to note that the total time required to reach node 4 from node 1 is at least 30 days (the time taken for task C). This is because all of tasks B, C and D must be completed

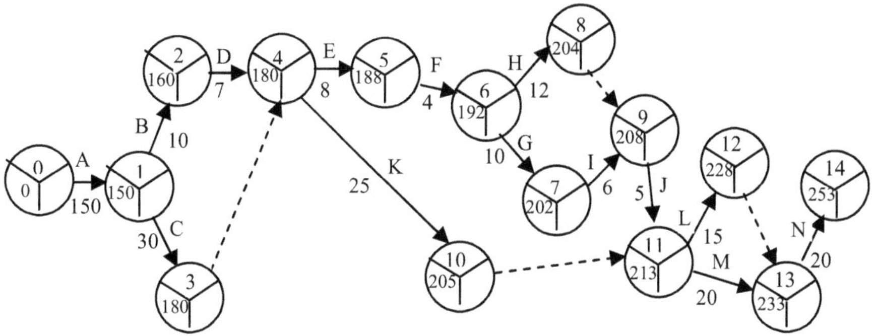

Fig. 20.8. Use of a forward pass to determine shortest possible publishing time

before either tasks E or K can commence. Be careful to see that you can justify the other paths indicated. In some cases there are possible alternative linkages that do not alter the logic.

20.3 Minimum Publication Time

The critical path is the key to determining the minimum time for publication. To find it we need to know, for each activity, not only the earliest time it can be finished, but also the latest possible finishing time if we are to keep to schedule. We determine the earliest finishing times by working forward through the network, referred to as making a *forward pass*. We write these times in the lower left-hand sector of each node. The earliest finishing time at any node becomes the earliest possible starting time for activities commencing at that node. These times are included in Fig. 20.8, which represents the same network as Fig. 20.7. The 0 in bottom left sector of node 0 is the starting time for the project.

From the data, since A takes 150 days it cannot finish until time 150. This is written in the bottom left sector of node 1. Since B takes a further 10 days, the corresponding entry in node 2 is $150 + 10 = 160$. Because C may be performed in parallel with B, it may start at the same time as B, so the minimum time entry in node 3 is $150 + 30 = 180$. Care is needed when we come to node 4, because there are two routes leading to it. The earliest starting time for any following activities is the greater of the times taken by the two routes. Proceeding from node 2 gives a total time of 167 days, but proceeding from node 3 the total time is 180 days, the same as the time taken to reach node 3. No additional time is needed for the dummy activity linking nodes 3 and 4. You should check the steps for the remainder of the network to confirm the entries given in the bottom left sectors.

From Fig. 20.8 we see that the earliest possible time after starting the project that the book can be in the shops, the so-called publication date, is 253 days after the author starts researching the project. This is given by the entry in the bottom left of the finishing node, node 14. If the date of inauguration of the incoming president of Plutonia is known in advance, this tells us when the author must start writing the book if it is to be ready for that important date. Late publication is likely to reduce potential sales.

The situation so far described is oversimplified. Authors are notoriously bad at estimating how long it will take them to write a book. There are always possible hold-ups such as the author falling sick during the writing process, or industrial action holding up printing, distribution, etc. Nevertheless, an exercise such as this is useful for indicating the sort of planning needed to get the book out by the critical inauguration date.

Late starters allowed?

Why is the word *critical* used in *critical path analysis*? It is easy to see that some operations need not be started at the earliest possible starting date without delaying the whole project. For others it is essential they be started as early as possible. Inspection of Fig. 20.8 shows that although activity D can start on day 160, it need not start until day 173. It would then still be completed on day 180, and so will not delay step E or later steps, because E cannot start until C is completed. The earliest possible day for that is day 180. It would not matter if. instead of taking only 7 days, D took anything up to 20 days. However, if it took 20 days it would have to be started at the earliest possible time. If task D were spread over more than 20 days, this would hold up the whole production process.

To decide which activities must be kept up to schedule if production in 253 days is to be achieved, we need to find the latest possible starting dates for each activity. To do this we work backward (called making a *backward pass*) from the finishing node 14. We put appropriate entries in the bottom right sector of each node. Since the book must be finished on day 253, we write 253 in the bottom right sector of node 14. We proceed to obtain appropriate entries at earlier nodes, and relevant ones are recorded in Fig. 20.9.

Since activity N takes 20 days it must be started 20 days before completion. i.e., on day $253 - 20 = 233$. This is entered on node 13. Since the step from node 13 back to 12 is a dummy activity that takes no time, it may be started on day 233 also, so that is the entry for node 12. Care is needed for node 11. We must be able to start from there 20 days before we reach node 13. although we could reach node 12 if we started only 15 days earlier. So that we can reach node 13 on time, we must therefore enter $233 - 20 = 213$ in node 11. Such care is needed when there are two paths with tails at a common node. You should check the entries for the remaining nodes at the backward pass.

For some nodes the earliest and latest times coincide, specifically this is so for nodes 0, 1, 3, 4, 5, 6, 7, 9, 11, 13, 14. These determine a *critical path* because there is no freeway, or float time, at these nodes. Any delay in the activities

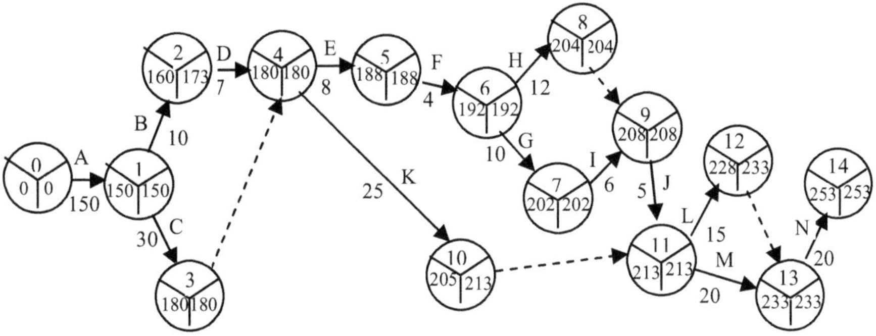

Fig. 20.9. The final CPA diagram using the information in Table 20.1

A, C, E, F, G, I, J, M, N joining these nodes will delay publication. A critical path need not be unique. For example, if activity H had taken 16 days instead of 12, then nodes 6, 8, 9 and 6, 7, 9 would both be part of critical paths. We have already indicated possibilities of delaying the start of, or extending the time for, some activities without delaying completion. The differences between the earliest and latest possible start times at each node indicate the degree of flexibility. These differences are referred to as *floats*. There is no float at nodes on a critical path. Further analysis of float is possible; for instance, sometimes a permissible late start on one activity may not place any constraints on later activities, whereas in other cases it will. An example of the latter is provided by activities B and D. If the start of B is delayed by 13 days, then D must be started as soon as B is finished if node 4 is to be reached in 180 days. Whereas if B is started 10 days late, there is still a possible float of 3 days in the start time of D.

This is only a basic introduction to CPA. Many variants are grouped under the general title of *program evaluation and review techniques*, or PERT for short. PERT covers problems where activity times may be varied (often only at additional cost), and assessment of cost changes if we depart from, or modify, a critical path.

20.4 Other Networks in Operations Research

Motoring organizations often provide best routes between any two among many hundreds of towns. What is a 'best route'? Is it the one involving the least travel distance in miles, kilometres or in whatever unit distance is measured? Is it the fastest route? Often the two are different. The shortest distance route might be heavily congested, while a fast freeway may involve a greater distance but ensure a quicker journey.

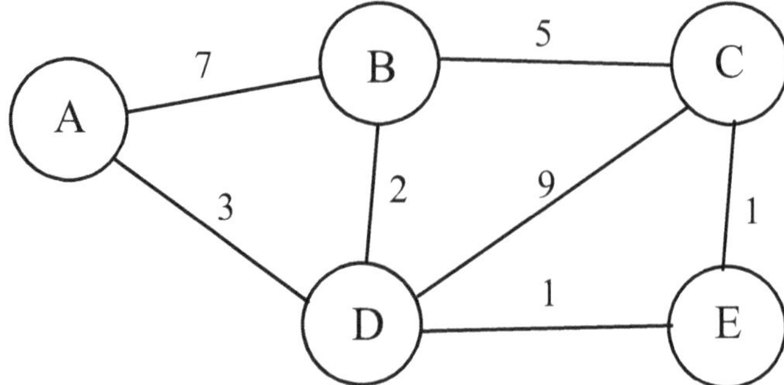

Fig. 20.10. Possible routes between 5 towns A, B, C, D, E and associated penalties

High road or low road?

Figure 20.10 refers to a situation where 5 towns A, B, C, D, E are connected by roads shown in the diagram by straight lines. This is not meant to imply that the connecting roads are straight, but simply that they exist. The lengths of the joining lines need not correspond to physical distance between the towns. If we want a fastest route, what is relevant is the time taken on any sector consistent with our driving at permissible and safe speeds, and allowing for any known or likely hold-ups.

If we wanted a scenic route some 'score' assessing the scenic merit of each sector would be appropriate. A sector passing through an industrial wasteland, or a highly industrialized area, might receive a very low score. One passing through pleasant, but not spectacular, country may be given a moderate score. A route of great natural beauty would get a high score, one passing through really spectacular mountain scenery, an even higher score. With this scoring system, if our prime aim was to admire the scenery, we would seek a route with a highest possible total score. For minimum distance or fastest routes we seek the lowest total score. When scores are to be minimized, rather than maximized, it is common to refer to them as *penalties*.

Once the appropriate scoring system is decided it is convenient to record the score for each sector on the line joining the towns. In Fig. 20.10 we have denoted towns by nodes labelled A, B, C, D, E.

In Fig. 20.10 it is easy to spot by inspection the minimum penalty route between any two towns. It may not be the direct link. The direct route from A to B carries a penalty of 7, while if we travel from A to D then D to B the penalty is only $3 + 2 = 5$. Similarly, the optimum route from B to C is not the direct route with penalty 5, but from B to D to E to C with penalty only 4. The optimum route from A to C is from A to D to E to C.

If there are a large number of towns a trial and error method of finding routes is tedious, if not impossible. Computer algorithms exist to do the job.

A slightly modified problem that may be solved by network analysis is that facing a freight company that has to make a choice between several potential routes between two cities, when these involve roads that each traverse several mountain passes. The higher any pass is, the more likely it is to be blocked by snow in winter, so it is desirable to choose a route where the highest pass that lies on that route has an altitude less than that of the highest pass occurring on any other route. A modification of an algorithm known as Dijkstra's algorithm is one network approach to solving this problem, but we do not describe this here.

21

Order or Chaos?

21.1 Stable or Unstable?

Before about 1800 mathematics was concerned primarily with orderly or well behaved phenomena.

From its foundations in the natural number system, through to the development of sophisticated notions such as calculus and its applications in Newtonian dynamics, mathematics dealt mainly with *deterministic systems*. These are systems where, once appropriate equations and formulae to describe their behaviour have been obtained, these are adequate to describe each system.

Theory and practice do not always co-habit easily

Deterministic systems are often idealized. An implication of Newtonian dynamics is that a one tonne and a one kilogram stone dropped from the same height will take the same time to reach the ground. This would prove untrue if time were measured with great accuracy, because simple Newtonian equations ignore air resistance. Mass differences are only one factor affecting the time of fall if we take air resistance into account. Shape and density must also be considered. For stones of mass of 50, 100, 200 or 1000 kg, and similar shape, the effect of air resistance is small enough to be ignored in most practical situations. If we replace one such stone by a feather then air resistance, convection currents, wind direction, all affect markedly the path and speed of descent for the feather. This path is called in dynamics the *trajectory*. One way to remove the effect of the extraneous factors like air resistance, or wind, is to conduct the experiment in a vacuum.

Phenomena like air resistance may be small irritants when formulating simple dynamic laws about falling objects in a gravitational field. In other circumstances they are of practical importance. Were it not for air resistance and wind currents, parachuting or hang gliding would be ways to commit suicide, rather than popular sports.

In the physical sciences the gap between the mathematical idealization to deterministic behaviour, and what happens in the real world, is often narrower than it is in the life sciences. Deterministic models may be inappropriate, or not entirely adequate, for biological phenomena. This is because random elements may be of as great, or even greater importance, than deterministic elements.

Population dynamics — a study of the growth and decline of populations — cannot be adequately described by deterministic models, even though there are often clear signs of regular patterns. In many parts of Europe, the wasp population drops to near zero in winter and rises to a peak in July and August. Just when the population will peak, and how large that peak will be in a given year, or at a particular place, is virtually unpredictable. It will depend on a host of factors ranging from the weather to local attractors, such as available food supplies. A plum tree in one's garden may attract wasps; a thick hedge may well tempt a colony to build their nest there. Removing a hedge, or cutting down the plum tree, may diminish a local wasp population in future years.

In the last couple of centuries, particularly the twentieth, probability theory has provided a basis for *stochastic*, as distinct from *deterministic*, models for predicting — at least in the short term — things like trends in insect populations, the spread of infectious diseases and other processes where erratic fluctuations are as, or even more, important than some deterministic elements.

Stochastic, or random, models are widely applicable. Traffic flow on major roads varies. Traffic densities are functions of the time of day and the time of year. They may also change suddenly, in either the short or long term, due to what are often, in part at least, unpredictable factors. These could be an increase in fuel costs, outbreak of a war, a change in economic conditions that reduces the amount of commercial traffic, the building of a new road that diverts traffic from an existing one. On a particular day traffic may fall due to a change in people's domestic habits, such as deciding to stay at home to watch an important football match, or a state occasion, on TV. In addition, there may be traffic build-ups due to irregular and unpredictable events such as an accident.

Patterns of queues in banks, post offices or at airline check-ins are other examples of stochastic processes. Stochastic models are used to predict the likely effect on reduction in length of queues by putting on extra staff, or speeding up the average time it takes to serve customers. This might be achieved by cutting the amount of paper work needed to provide a service.

In simple queues, if the average time between the arrival of successive customers is greater than the average time taken to serve a customer, long queues will be rare. If the average time taken to serve is longer than the average interval between each arrival, the queue will keep getting longer. What happens if the average time between arrivals, and the average service time are the same is not obvious. For one simple model the answer surprises many people. Intuition may suggest that the queue length will tend to fluctuate only slightly about some small average length. In fact, it will become longer and longer as time progresses!

Long range forecasting – a dream is shattered

Until the latter half of the twentieth century it was widely believed that either deterministic or stochastic models broke down only because they were not refined enough to reflect all real-world conditions in sufficient detail. It was believed these failures could, in theory at any rate, be eliminated by using more sophisticated mathematical models and collecting more data.

Increasingly sophisticated mathematical models used in weather forecasting were proving more reliable for forecasting one or two days ahead, even in regions were weather was notoriously changeable over a matter of hours. It was confidently expected that further refinement of the mathematical models to take account of more atmospheric factors that affect weather, coupled with more detailed records of actual weather conditions at more weather stations, would achieve the ultimate goal of long-range forecasting months and even years ahead. Given the relevant information, it would then just be a matter of solving quickly enough appropriate differential equations with the right initial conditions, and long range forecasting would become possible. Ever faster computers would provide the necessary solutions.

In the last 50 years the self-same computers that made rapid solution of large systems of differential equations feasible, exposed a problem hitherto unrecognised with the solutions to some of these. As we hinted in Sect. 1.2, there is an inherent instability described by mathematicians as *chaos*.

This phenomenon does not arise when applying Newton's laws of motion to dynamic problems with projectiles like falling stones, or shells fired from a gun. Once the initial conditions are known, the behaviour of the projectile — its position and velocity at future times — can be predicted accurately. Tne system is stable once the requisite initial conditions are given.

This concept of stability is, in general, harder to pin down than would appear from familiar examples in Newtonian dynamics. In many systems stability may not exist, or may be easily upset.

The way the mathematics behaves is mirrored by many physical phenomena, such as turbulence in fluid dynamics. If a tap or faucet is opened carefully and slowly there is at first a regular series of drips. These gradually become faster and closer. At some stage, they merge almost imperceptibly into a single smooth flow something like a liquid columnar tube. This gradually becomes wider. Increase the flow a little further, and small indications of turbulence appear. Slight wisps of water break away from the main column. As the tap is opened further the column virtually breaks up, and the flow becomes chaotic. Even in such flows, if one looks closely there are usually regular, or more or less regular, features that recur. These may be little spurts in a particular direction, or apparent surges at regular intervals.

Similar phenomena may be observed in the atmosphere. A cloud builds up slowly, gradually growing as a sphere or perhaps some more complicated, but fairly regular, shape. After it reaches a certain size the shape distorts, wisps of cloud breaking away from the main body. If the shape were originally a sphere it may start to elongate in certain directions, and not at the same rate in all directions. The cloud gradually takes on a complicated form which can no longer be described, even approximately, by simple terms like spherical or elliptical.

21.2 Long Term Behaviour of Recursive Relations

Sect. 15.9 gave a brief introduction to difference equations. We wrote a few simple equations in a form that was essentially

$$y_{t+1} = f(y_t) \tag{21.1}$$

We shall be interested here in functions, or mappings, that have the property that there is a unique value y_{t+1} for any given y_t. We may repeat this mapping by replacing y_t by y_{t+1} on the right, to calculate a new value y_{t+2} on the left, and so on. Equation (21.1) forms the basis of a *recursive* or *recursion* formula. The name highlights the feature that repeated application, or iterations, using the formula, produce successive values of the function when $t = 1, 2, 3, \ldots$, once we are given the initial value when $t = 0$.

The behaviour of recursive relations needed to solve some difference or differential equations is often complicated. One of the surprising discoveries of the twentieth century was that even comparatively simple recursive formulae do not always behave in what was once regarded as an intuitively reasonable manner. The reason it took so long to establish this was that the hitherto unexpected behaviour was not easy to detect. It was even harder to study in detail without the power and speed of modern computers.

An obvious 'nice' kind of behaviour for recursive formulae is that, in the long run, the y_{t+1} should settle down to some fixed, or limiting value, or else exhibit periodic or other patterned behaviour.

Some possible behaviour patterns for simple mappings are easily illustrated with a pocket calculator. We look at the relations

$$y_{t+1} = y_t^2; \qquad y_{t+1} = \sqrt{y_t}; \qquad y_{t+1} = 1/y_t;$$
$$y_{t+1} = \sin y_t; \qquad y_{t+1} = \cos y_t; \qquad y_{t+1} = \tan y_t.$$

All are simple mappings that can be used to study some of the things that happen after many recursions from a given initial condition y_0.

We look first at $y_{t+1} = y_t^2$, starting with the initial value $y_0 = 0.517$. Your pocket calculator probably has a key labelled x^2, or something similar, to indicate that hitting it will square an entered number. If there is no such key, use the multiplication key to multiply any number, x, by itself. The recursion formula effectively tells us to keep pressing the x^2 key to give $y_1, y_2, \ldots y_t, \ldots$. The successive values (to 3 dec. pl.) that I got, starting with $y_0 = 0.517$, were 0.267, 0.071, 0.005, 0.000, 0.000 .... After a few more iterations I got a value exactly zero, which was repeated in all further iterations.

The same happens if we start with any value of y_0 between 0 and 1, but excluding 1 itself. If we set $y_0 = 1$ then, not unexpectedly, repeated pressing of the x^2 key retains the value 1. If y_0 is set at any real value greater than 1, again not surprisingly, the recursions lead to larger and larger values of y_t, eventually running over the limit the calculator can handle. This limit will depend on the particular machine. We say the iterations diverge, i.e., they tend in the limit to infinity. Try it yourself with any y_0 greater than 1 that takes your fancy.

The recursions lead to a limiting value $y_t = 0$ for any positive initial value less than 1. In deterministic mathematics we call zero the *limit.* Another term that has appeal in *chaos theory* came into use with the development of that topic some 30 or 40 years ago, namely *an attractor.* We shall soon see that while an attractor is sometimes a point limit, this is not always so.

In a recursive procedure, an attractor is a descriptor of long term behaviour of the iterative process. In the mapping $y_{t+1} = y_t^2$, if $y_0 < 1$ and positive the attractor is 0, if $y_0 = 1$ the attractor is 1. If $y_0 > 1$ the attractor is infinity, or in this last case, mathematicians prefer to say the recursion diverges. The value $y_0 = 1$ forms a boundary between a set of initial values that recurse to zero and a set that diverge. Such boundaries are especially relevant to situations we meet later in studying chaos.

It is easy to guess what happens for the recursion $y_{t+1} = \sqrt{y_t}$. The attractor is 1 if y_0 is positive and less than 1. It is also 1 if $y_0 \geq 1$. Check with your calculator to confirm this.

What about the mapping $y_{t+1} = 1/y_t$? If you have a reciprocal i.e., $1/x$ button on your calculator you will soon confirm that for all positive y_0 the recursions alternates between y_0 and its reciprocal, $1/y_0$. For example, if $y_0 = 0.4$, successive values are $0.4,\ 2.5,\ 0.4,\ 2.5,\ 0.4, \ldots$. If $y_0 = 25$ successive values are $25,\ 0.04,\ 25,\ 0.04, \ldots$. Here we have periodicity, or alternations between two values, for any starting value except 1, where the value remains at 1. Thus, for any positive starting value except 1, there are two attractors. The values of those depends upon the initial value, which is itself always one attractor, the other being its reciprocal. As y_0 approaches 1 the difference between the periodic values approaches zero. For example when $y_0 = 0.99$ the successive values are 0.99 and 1.01010101. The difference is 0.020101..., while if $y_0 = 0.999$ the difference between the attractors is easily verified to be 0.002001001....

Beware of vague terms

For $y_{t+1} = \sin y_t$, where the angle is measured in radians, we can experiment to see what happens when we repeatedly hit the sine key on a calculator starting with $y = 0.5$, say. Rounded to 3 decimal places, I got 0.479, 0.461, 0.445, 0.431, 0.417, 0.405, 0.394, 0.384, 0.375, 0.366, 0.358, 0.350, 0.343, 0.337, 0.330, 0.324, 0.319, 0.313, 0.308, 0.303, 0.299, 0.294, 0.290, 0.286, 0.282, ... Things looked to be settling down, though rather slowly. The number is decreasing at each iteration, so there is some pattern. Try for yourself with the same starting value, but persevere a bit longer. If you go on a long time and your calculator battery doesn't get flat, the recursive process will eventually take you near to zero. Can you see why? It may help if you think about the limit as $x \to 0$ of $\sin x/x$ and also how we got that result.

Try for yourself the recursive relation $y_{t+1} = \cos y_t$ starting with some y_0 between 0 and 1. Things again may proceed slowly at first, but after 50 or so recursions I got the number 0.739085133. I am not telling you what starter I used, and I don't know what starter you used, but you should end up with essentially this number. There may be minor differences in the number of decimal places, depending on the calculator you use. Why is this number the attractor for this recursion? Here is a hint; find what angle in radians has this value for its cosine.

Next, try for yourself recursions using the formula $y_{t+1} = \tan y_t$ starting with $y_0 = 0.312$. To 3 decimal places the first few iterations I got were 0.323, 0.334, 0.347, 0.362, 0.379, 0.398, 0.420, 0.447, 0.479, 0.519, 0.572, 0.643, 0.750, 0.931, 1.345, 4.353, 2.659, −0.523, −0.577, −0.651, −0.762, −0.954, −1.409, −6.135, 0.150, 0.151, 0.152, 0.153, ... At this stage, in the light of what we saw with earlier examples, intuition may suggest that we are approaching an attractor. However, after a further 50 iterations I got a value 0.315. This was almost, but not quite, the same as the starting value 0.312. Carrying on further gave a pattern not altogether unlike, but not exactly the same as the previous one. There was still some pattern, but no definite limiting value or periodicity.

If we start with $y_0 = 0$, however, we soon discover that this value itself is an attractor, and in repeated iterations we continue to obtain the value 0. Think about this. It is easy to see what is happening here if you know the value of $\tan x$ when $x = 0$. Before leaving this recursive relation here is a warning. We have asserted that except for some starting values such as $y_0 = 0$, there is in general no point or periodic attractor when mapping with the tangent function. This is true, but if we persist long enough with a pocket calculator the cycle may repeat itself after many thousand or even millions of iterations. This is because calculators round off to a limited number (often 10 or so) significant figures.

A simple illustration of an effect of rounding is given if we start the iterations with $y_0 = \pi$. Direct substitution in the iterative formula then gives $y_1 = \tan\pi = 0$, whence, as we saw above, all subsequent iterations will be zero. If your calculator has a button for entering π insert this as your starting value and see what happens. It will depend on how accurate is the approximation to π. Try again with 3.14 and again with 3.1416 as starters. The latter is the value of π correct to the 4th decimal place. Now try 3.14159265, the value correct to the 8th decimal place.

Iterations with the basic trigonometric functions have shown behaviour a little less obvious than that for simple algebraic functions like the square, square root or reciprocal. Apart perhaps from the tangent, even these functions do not do anything unexpected if one reflects carefully on the properties of the functions.

21.3 The Logistic Difference Equation

During the 1970s the physicist and biologist Robert May, who later became president of the Royal Society, studied recursive solutions to the so-called logistic difference equation

$$y_{t+1} = ky_t(1 - y_t)\ . \tag{21.2}$$

We met (21.2) as an example of a simple difference equation in (15.9), where we indicated its relevance to discrete changes in population size, y_t, measured as a proportion of the maximum possible value. This maximum corresponds to the proportion $y_t = 1$, while the population becomes extinct if $y_t = 0$. The equation also implies that if $y_t = 1$, the population will become extinct at time $t + 1$.

Its behaviour for a range of values of the constant k in the interval (0,4) and for many initial values,y_0, was explored by May, who published his results in 1976 in the journal *Nature* [36].

Restricting k to lie between 0 and 4 ensures that y_t will not fall outside the physically meaningful interval (0, 1). It is easily seen that if y_t is near zero or 1, then y_{t+1} will be rather smaller than it will be when y_t is nearly one half. Indeed, it is not hard to see that y_{t+1} will have the greatest possible value when $y_t = 1/2$.

As indicated in Sect. 15.9, the equation is relevant to modelling the size of simple biological systems where, when the population size is small, the next generation is likely to be small because there is only a limited breeding stock. Whereas, when it is large, the next generation may be small because a lack of food, or physical overcrowding, leads to many organisms dying before they reproduce. Moderate population sizes at time t look intuitively desirable. These might be expected to give reasonably stable population sizes at the next generation.

The constant k acts as a damper. If $k = 4$ the maximum value of y_{t+1} occurs when $y_t = 1/2$ and it is easy to confirm then that $y_{t+1} = 1$. In this situation $y_{t+2} = 0$, so the population is extinguished. If $k = 3$ it is easy to verify that the population proportion for appropriate initial values can only settle down, if at all, to values between 0 and 0.75.

We explore the behaviour of (21.2) for various values of k, and a few initial values y_0 between 0 and 1. We exclude the extremes $y_0 = 0$ and $y_0 = 1$, because it is obvious that the population never exists if $y_0 = 0$, and that it dies out without regeneration if $y_0 = 1$. It is easy to do a few iterations on a pocket calculator. If appropriate computer facilities are available you should check the computations outlined below.

First, set $k = 1$ and $y_0 = 0.62$. We might set the program to perform 200 recursions. With these values of k and y_0 we find that y_t decreases at first fairly rapidly and steadily, and then more slowly. After 200 recursions I got $y_{200} = 0.004825$. If you have written a program, or use one already available, you should find a similar pattern. For reasons to do with software and hardware, the numerical values may not be exactly the same, because of differences in rounding.

It is the pattern that is important. It suggests that the population appears to be heading for extinction. Indeed, if we continue for a few hundred more recursions we will find that it does eventually become extinct, though only slowly. The same applies for any other starting value when $k = 1$. In terms of *attractors*, zero is the attractor.

An intuitively reasonable hunch may suggest a faster approach to the attractor zero if $k < 1$. This is indeed the case. When $k = 0.75$ and $y_0 = 0.62$, I found after 44 recursions that the population became zero (at any rate to six decimal places), and retained that value. If you have a computer program to do this, it is worth exploring the behaviour for other values of $k < 1$ and various permissible starting values y_0. These should confirm that if $k < 1$ the attractor seems always to be zero. In terms of a population this implies it always dies out due to the damping effect of low values of k.

What happens if we set $k = 1.5$? With a starting value of $y_0 = 0.62$, I found that in less that 20 recursions the population proportion attained the value $y_t = 1/3$. It maintained that value for all later iterations. It is easy to verify directly from (21.2) that when $y_t = 1/3$, then $y_{t+1} = 1.5 \times 1/3 \times 2/3 = 1/3$. Thus, once the recursions hit this value the population proportional size stays the same. The attractor here is 1/3.

If we take $k = 2$, then with any permissible starting value there is rapid convergence to the attractor 0.5. Again, substitution in the relevant form of (21.2) confirms that once $y_t = 0.5$, it retains that value in all future recursions. There is nothing strange here, providing that with any starting value it will at some stage hit the value 0.5. It will do so with any rational starting value in the appropriate range and since, computationally, we can approximate to an irrational number by a nearby rational, we always converge to this attractor.

When $k = 3$ the iterations settle down very slowly to $y_t = 2/3$. Again, direct substitution indicates that further recursions will maintain this value.

Further exploration along the lines indicated above shows that, for all choices of k greater than 0 but less than or equal to 3, the dynamic behaviour of the population is such that it settles down to a limit. This limit is zero if $k \leq 1$, but if $1 < k \leq 3$ the limit, or attractor, increases steadily from 0 to 2/3.

What does intuition suggest if we increase k beyond 3? Should we expect a settling down to point attractors between 2/3 and 1? I set $k = 3.1$, and choose some starters in the permitted range (0, 1). If you repeat my calculations I hope you will agree with my finding that the recursions settle down to an alternation between attractors 0.558 and 0.765. Again, this alternation between two possible attractors (two phase periodicity) is a kind of behaviour we have already met with reciprocals. If you have the computing facilities, play around with k values between 3 and 3.5, and any initial values you choose from the permissible range. If you have not the facility to do this, Table 21.1 indicates what you should find for the values $k = 3.1, 3.2, 3.3, 3.4$.

Table 21.1. Two-period attractors for the recursion system (21.2) for selected k

k	*lower attractor*	*upper attractor*
3.1	0.558	0.765
3.2	0.513	0.799
3.3	0.479	0.824
3.4	0.452	0.842

From Table 21.1 we see that as k increases from 3.1 to 3.4 we have, in each case, 2 attractors. The lower one decreases as k increases, while the upper one increases as k increases. This implies that the difference between upper and lower attractor increases as k increases. We can increase detail in this picture by taking other values of k between 3 and 3.4. Starting with $k = 3.01$ the attractors are 0.634 and 0.698. Convergence to the attracts becomes slow as k becomes closer and closer to 3.0, but the paired attractors converge towards a value 2/3. This, we found, was the single attractor when $k = 3$.

Hint of mystery

If we increase k beyond 3.4, intuition may lead us to expect that the attractors will follow the trend suggested by Table 21.1, the lower one decreasing and the higher one increasing as k increases. Or are you getting shrewd by now, intuition itself warning you that it may let you down. Try a few values.

Set $k = 3.45$ and you may be in for a surprise. Convergence is slow, but after about 200 iterations you may detect some sort of alternation between four values that do not coincide completely. If we extend the program to allow, say 1000 iterations, you should detect a 4 phase periodic cycle between 0.445, 0.852, 0.434 and 0.848. This implies that the 2 phase cycle has become a 4 phase cycle, with a split of the two cycles somewhere near 0.44 and 0.85. Taking a few values of k between 3.4 and 3.45. should help us rack down where the flip from a 2 to a 4 phase periodic cycle occurs. I took a few such values. For $k = 3.42$ the attractors were 0.447 and 0.845, while for $k = 3.43$ they were 0.445 and 0.847. For $k = 3.44$ the attractors were 0.442 and 0.849, while for $k = 3.445$ they were 0.849 and 0.441. Convergence to these was slow, and working to an accuracy of 6 decimal places there was a slight indication of a possible split at the sixth decimal place.

At k = 3.447 a split at the 5th decimal place is more clearly discernable. One finds a 4 phase cycle with attractors at 0.44050, 0.84955, 0.44059, 0.84958.

While we have now moved from a single attractor, first to paired or 2 phase periodic attractors, then to cycles of 4 phase attractors, there is still nothing too alarming. The pattern of attractors is becoming more complicated, but it still has a periodic structure.

What happens as k increases? Explore this for yourselves if you have the appropriate computing facilities. In case you have not, we summarize what happens in broad terms. We find a well established 4 phase cycle at $k = 3.5$. By the time we get to about 3.55, an 8 phase cycle emerges. At about 3.6 not only is no cycle evident, but the pattern after 200 or more iterations is highly sensitive to small changes in the initial condition y_0. This is the phenomenon mathematicians describe as *chaos*. The attractors are a patternless set of numbers within some interval. To distinguish them from the more familiar point, or periodic, attractors they are often described as *strange attractors*. We discuss this type of attractor further in Sect. 21.4.

This is not the end of the story. Intuition may now lead us to think that once chaos has set in it will persist as k increases. Let us try $k = 3.835$ and $y_0 = 0.5$. With this, or any other starting value, we find not chaos, but a 3 phase cycle with attractors 0.15207, 0.49451, 0.95863. We find order in the midst of chaos, as Robert May did more than 30 years ago.

Summarizing so far, we first find a single attractor, then 2, 4, 8 (and although I did not show it, then 16, 32, 64 phase periodic attractors in quick succession), then chaos, then a 3 phase periodic attractor.

The process of splitting into 2, 4, 8, 16 phase period attractors is called *bifurcation*. Successive bifurcations have an important feature. In broad terms, each successive bifurcation has a property called *self-similarity*. We saw a hint of this in our earlier discussion where we saw that the 4 phase attractors represented splits in each of the 2 phase attractors. Considered separately, each of these proceeds as a small scale model of the earlier 2 phase attractors. Fig. 21.1 illustrates this graphically. The first bifurcation takes place when k just exceeds 3, where the single attractor becomes a 2 phase attractor.

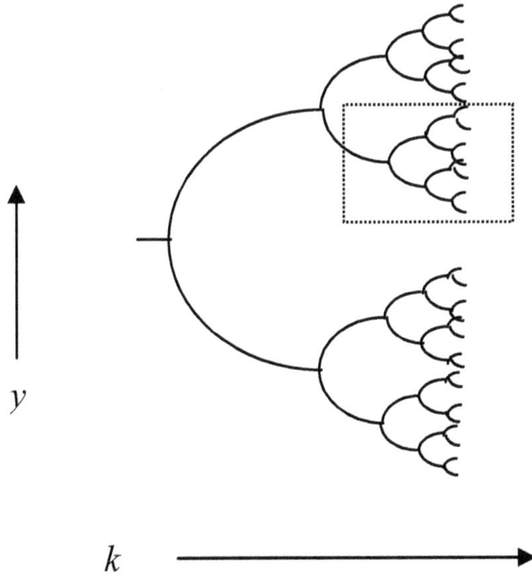

Fig. 21.1. A diagrammatic representation of attractors for the logistic recursive formula as k increases. The portion in the dotted rectangle is self-similar to the whole figure

The next bifurcation (or more precisely, pair of bifurcations) occurs, as we found above, when $k = 3.445$. The next bifurcation is to an 8 phase cycle at approximately $k = 3.55$. Bifurcation then take place rapidly as k increases.

If we denote the values of k at which successive bifurcation doubling takes place by $k_0, k_1, k_2, \ldots$, Mitchell J. Feigenbaum [21] [22] found that

$$\lim_{n\to\infty} \frac{k_n - k_{n-1}}{k_{n+1} - k_n} = 4.669201609.$$

This number has two remarkable properties. The first is that there exists such a constant associated with bifurcation. From the point of view of self-similarity it may be regarded as a magnifying factor that magnifies the later stages to look like the earlier ones. The more remarkable property is that there are many other mappings that exhibit bifurcations where the same *Feigenbaum number* occurs. Another mapping for which the same constant is relevant is

$$y_{t+1} = k \sin y_t$$

where y_t takes values between 0 and π.

A full account of the implications of this finding would require a chapter to itself. Chaps 8 and 10 in Stewart [43] [44] discuss this and related constants in some detail. In the field of chaos the Feigenbaum number has an importance akin to that of e or π in classic mathematics. We explore further the role of self-similarity in chaos, and indeed as a subject in its own right, in the next chapter.

21.4 Some Implications of Chaos

The recursive behaviour of (21.2) has important practical implications. We saw that starting with a simple mathematical expression and repeated recursions, we got patterns of behaviour that included convergence to a limit, periodicity or chaos. The strange attractors associated with chaos look not unlike a series of random numbers. By *random numbers* we mean numbers that are in essence sequences of numbers in some interval, such that any number in the interval is as likely as any other number to appear in any position in the sequence.

A dream shattered

In the real world chaos is important because it is now established that the sort of differential equations that cover, for example, fluid flow in gases or liquids, or humidity measurements, or many other factors affecting weather, often have chaotic solutions. The precise forms of these are highly dependent on initial conditions. In particular, a small change in conditions may lead to a dramatic change in the solution after 100 or so iterations. For example, in (21.2) if we set $k = 3.6132$ after some 200 iterations the chaotic pattern will bear no resemblance to that obtained if we set $k = 3.6130$. A small change in y_0 may have a similar effect.

As we have already indicated, this shattered a dream that many mathematicians had about long range weather forecasting. Many of the equations needed for this have chaotic solutions when we iterate over time (i.e., into the future).

When there is a strange attractor, the dependence of future values of y_t upon a precise knowledge of an initial condition, and its susceptibility to small changes in that condition, is encapsulated in the phrase the *butterfly effect.* In mathematical terms this means that a small change in initial conditions can alter dramatically what happens after many iterations.

The analogy is based on the possibility that a butterfly flapping its wings in the Brazilian jungle may cause a minute change in air flows that may affect the weather in continental Europe, in a more or less unpredictable way, many weeks later.

But take heart. This does not mean than macro-forecasting is hopeless. It is reasonable to expect that in temperate zones most places will have cold winters and warm summers, or that countries with wide rainfall variations throughout the year will continue to have their 'wet' seasons in more or less the same month or months each year. What it does mean is that is impossible to set up a mathematical model in the year 2009 to tell us if in London on 3 July 2012, say, it will be warm and dry or cold and wet. This may be summarized rather simply by saying that although climate is reasonably predictable seasonally, a day-by-day long-term weather pattern is not.

There is a possibility that certain physical changes e.g., global warming, may have a dramatic effect on climate. This has little to do with chaos in the mathematical sense, no matter how much social chaos such a change might create.

The example of the effect of a butterfly's wing-flap in the Amazon region on European weather is at variance with commonsense, but only because there are more important factors that influence weather, whereas small butterfly effects may often cancel one another out. Nevertheless, a few will have a profound influence, and even build upon each other, making any mathematical-model approach to weather forecasting at best short term. Chaos only becomes evident after many iterations, and the butterfly effect is unlikely to show up in the first few iterations. In meteorological terms the first few iterations might be equivalent to a few days forecasting. but no more. After a short time these changes become initial conditions for further iterations and the small change in these reflects itself chaotically.

The discovery of chaotic behaviour in equations relevant to convection in air currents, an important meteorological factor, was made by Edward Lorenz. He describes them in Lorenz [34]. His findings were made some ten years before Robert May explored chaos in the logistic difference equation, and involved a more complicated system.

In the 1960s recursive methods for solving even simple systems of differential equations took a long time even with the best available computers. Lorenz looked at three differential equations in three unknowns. A feature of a recursive process is that the solution at any time, t, provides new initial conditions for later recursions.

Because Lorenz wanted to repeat some calculations to check them, but did not want to start from the beginning each time, on one occasion he decided only to repeat the later portion of a set of iterations. He took the mid-way values on a computer printout as his starting values to repeat the last half. He found that his new iterations bore no resemblance to the original ones. The reason was that the printouts were given only to three decimal places, whereas the computer was working to six decimal places. This rounding in a printout was common in those days, and for many applications entirely appropriate. However, what Lorenz had done by using printout rather than the more accurate computer values, was to make a small change to his initial conditions — something like replacing 0.621314 by 0.621000 — and this produced a butterfly effect.

For a full description of *strange attractors* see any book on chaos. Elementary ones include Stewart [44] or Gleick [25] and Ruelle [39].

21.5 Loose Ends

At several points we have emphasized that if we use different calculators, but the same starting values, the recursions may not give exactly the same results after, say, 100 iterations. However, the general patterns should not be very different, e.g., for the logistic difference equation phase doubling should give way to chaos at a value of k that is almost identical to the one we stated. Chaos is a property of the dynamic system that is little affected by rounding processes in the numerical calculations, except in so far as the butterfly effect comes into play. Such properties are, however, often highly sensitive to small changes in the value of k or to small changes in initial values y_0, i.e., to a true butterfly effect.

It is important to distinguish between chaotic behaviour and numerical instability. The latter, often due to rounding, may lead to dramatic differences in computed stable solutions. An extreme example of numerical instability is illustrated by the effect of rounding coefficients in simple linear simultaneous equations, where small changes in coefficients may make a dramatic difference to solutions. For example, the equations

$$x + 1.0001y = 1$$
$$x + 0.9999y = 2$$

have the solution $x = 5001.5, y = -5000$. If we change only a couple of digits in the fourth decimal place of the coefficient of y and consider the equations

$$x + 0.9999y = 1$$
$$x + 1.0001y = 2$$

the solution is $x = -4998.5, y = 5000$. Here, a change in the fourth decimal place of two coefficients both almost equal to 1 has made a change of nearly 10000 in the values of both x and y. The source of the problem becomes clear if we round the coefficient of y in each set of equations to the nearest integer. i.e., 1. Then the matrix of the coefficients of x and y becomes singular, and the equations have no solution since they represent a pair of parallel lines. Equations where the matrix of coefficients is almost, but not exactly singular, are often said to be *ill-conditioned*. Ill conditioning gives rise to numerical solutions that are highly susceptible to rounding. It is important to distinguish between such numerical instability and chaos.

22

Self-similarity and Fractals

22.1 The Road to Chaos

In Sect. 21.3 we met *self-similarity* as a characteristic of bifurcations. We associated with it a limiting constant ratio epitomized in the Feigenbaum number 4.669201609. This may be looked upon as a scaling or magnifying factor that, after many bifurcations, enlarges each self-similar repeat of the bifurcation to the size of the previous one. For the logistic mapping in Sect. 21.3 when $k > 3$, we saw that as k increased we moved ever more rapidly from 2 to 4 phase and then to 8, 16, 32 phase periodic cycles until, eventually, the system became chaotic. In this sense chaos arose as a limiting situation in a process exhibiting self-similarity.

Self-similarity predates chaos

However, self-similarity is an older idea than chaos. One of the best known, and one of the most straightforward examples of a geometric construction based on self-similarity, is a figure first considered in 1904 by Helge von Koch (1870–1924). It is called the Koch snowflake. It has the remarkable property that it is a closed figure with a finite volume, but an infinitely long boundary, or perimeter. Fig. 22.1 indicates the basic self-similarity concept for the Koch snowflake, and Fig. 22.2 indicates how self-similarity applies to the perimeter at each stage.

Fig. 22.1(**a**) shows an equilateral triangle. This is the basis for self-similarity in the Koch snowflake. In Fig. 22.1(**b**) we have added another equilateral triangle of the same size, but rotated through 180°. It is positioned so that it, and the original triangle, form a symmetric six-pointed star.

By elementary geometry, we see that all of the non-overlapping portions of the superimposed triangles are themselves equilateral triangles. Each is of one-third the side length, and consequently of one-ninth the area, of the original triangle. Fig. 22.1(**c**) is generated by applying the same superimposition

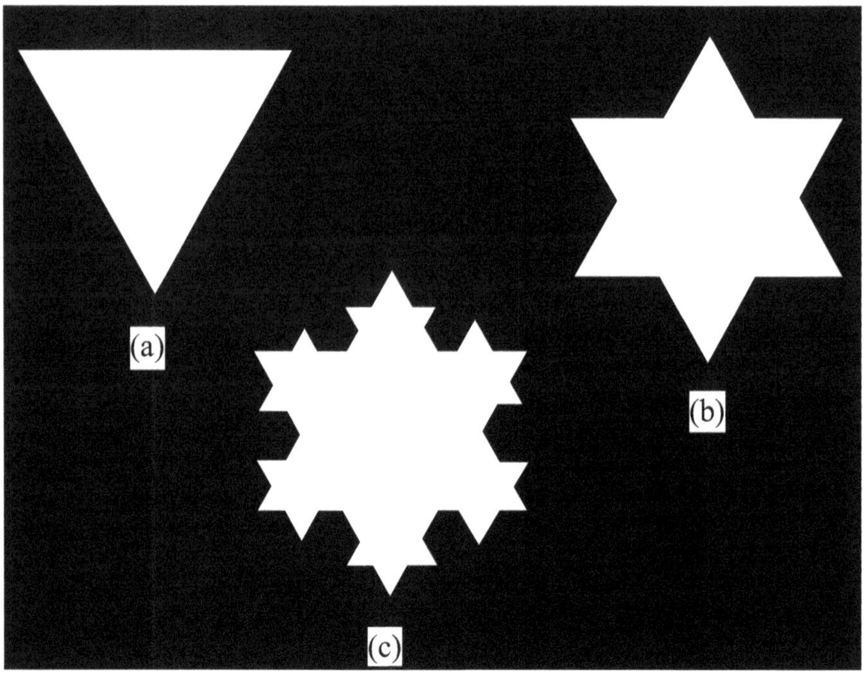

Fig. 22.1. First steps in generating a Koch snowflake. Figure (**a**) is an equilateral triangle and figure (**b**) is a superimposition of two such triangles (see text) while figure (**c**) is obtained by repeating the same superimposition to the smaller non-overlapping portions of the triangles in figure (**b**)

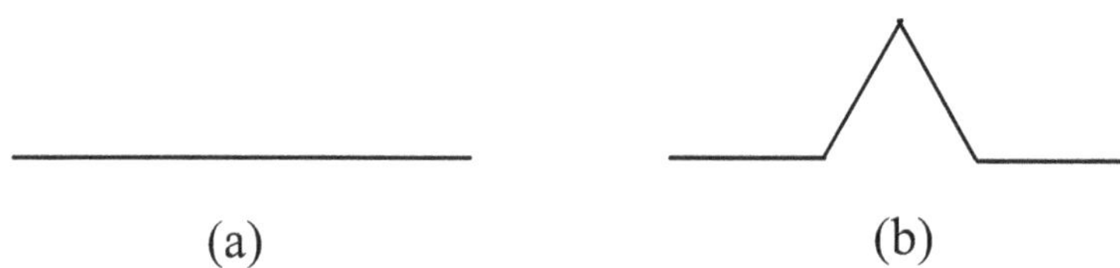

Fig. 22.2. The perimeter segment self-similarity transformation for a Koch Snowflake

process as that for generating Figure 21.1(**b**) to each triangle that is not part of the area common to both of the earlier triangles. The resulting figure is beginning to look like a snowflake.

Self-similarity manifests itself in another way in the Koch snowflake. We see this most easily by looking at the boundary.

In Fig. 22.1(**a**) there are three straight line boundaries of equal length. These are trivially self-similar. Fig. 22.2(**a**) shows one such boundary. What has happened to each boundary when we move from Fig. 22.1(**a**) to Fig.

22.1(**b**)? The middle third of each is replaced by two pieces each the same length as the piece it replaces, and these form two sides of any of the non-overlapping triangles in Fig. 22.1(**b**). This is shown in Fig. 22.2(**b**), which represents the change in any one boundary at the first step in forming the snowflake. The same thing happens for any straight line boundary when we proceed from Fig. 22.1(**a**) to (**b**). This means that each straight line boundary segment in (**a**) is replaced by a four-segment boundary in (**b**), with each segment one-third the length of boundary it replaces in (**a**). Thus, the perimeter of the Fig. 22.1(**b**) is four-thirds the length of the perimeter in Fig. 22.1(**a**).

The same process is repeated on each of the 4 segments in Fig. 22.2(**b**), and this may be continued indefinitely. This self-similarity is the key to the final Koch snowflake. Each straight boundary segment in Fig. 22.1(**b**) — there are 12 in all — is one third the length of each side in Fig. 22.1(**a**). If we now scale down the perimeter transformation in Fig. 22.2(**b**) by one third, and apply this to each line segment in Fig. 22.1(**b**), we arrive at Fig. 22.1(**c**). The total perimeter in the latter is 4/3 that of of the former. We repeat the transformation from three to four segments demonstrated in Fig. 22.2 for all line segments in Fig. 22.1(**c**), with the appropriate self-similarity scaling, and continue this process indefinitely to form the Koch snowflake. In practice the degree of resolution in the resulting figure makes it almost impossible to see further refinements by eye in a small diagram. Even after some 100 such self-similarity scaling steps the final figure will not look very different from that in Fig. 22.3, where we stopped the computer generation after only a few more 'iterations'. The flake is converging rapidly to its final form.

We have seen that the first and second self-similarity transformations, where we replaced a straight line by four segments each one third the length of the original, increased the perimeter of the newly formed enclosed figure by a ratio 4/3 each time we carried out the process. If the original triangle has a perimeter of length 1 unit, then the star shaped figure in Fig. 22.1(**b**) has a perimeter of 4/3. That of the emerging snowflake in Fig. 22.1(**c**) has a perimeter $(4/3)^2$, and so on. The successive perimeters form a geometric progression. Thus, after n such applications the perimeter is $(4/3)^n$. Since $4/3 > 1$ the perimeter becomes infinite as $n \to \infty$.

It is apparent that the ideal Koch snowflake has a finite area that will differ little from that of the approximate flake in Fig. 22.3. Indeed, it is not hard to see that the snowflake lies entirely inside the circumscribing circle of the original equilateral triangle in Fig. 22.1(**a**). The rate at which the area increases at each step is harder to compute than is the corresponding value for the perimeter. However, the final area is approximately 1.6 times the area of the original triangle.

We have an interesting phenomenon — a surface with a finite area but an infinite perimeter. More remarkable perhaps is the highly patterned nature. There is self-similarity in the perimeter pattern, in the repetitions of equilateral triangles associated with the area, and indeed in the various portions of the area. One example is highlighted in Fig. 22.4.

Fig. 22.3. A computer-generated Koch snowflake that closely approaches the limiting form

22.2 New Dimensions

In Chap. 7, and elsewhere, we spoke about a straight line having one dimension, a plane having two dimensions, a solid having three dimensions, without giving those notions anything more than an intuitive interpretation. There are several different ways we can look at dimensions.

A conventional approach outlined briefly in Sect. 7.8 is to say a point has zero dimension, a curve has one dimension, a surface has two dimensions, a solid has three dimensions and to link those ideas with the possible directions

Fig. 22.4. The portion of the snowflake in the small dashed rectangle is self-similar to the portion in the solid rectangle, and to numerous other portions of the figure

of motion a particle may take in each. A particle that is restricted to a point cannot move, so in this sense it is reasonable to say a point has zero dimension.

A particle restricted to lie on a curve is free to move along that curve; it has one direction of motion — along the curve. It may move from any initial point either 'backward' or 'forward' on the curve, but in our concept of one-dimensionality we distinguish between these directions only by a difference in sign. The basic rule is that motions must stick to the curve. The sign determines the direction of motion on the curve. A real-world situation modelled by this mathematical idealization is that of a train travelling along the track between London and Edinburgh, or between Paris and Marseilles.

For a point on a surface, which need not be a plane, but could, for example, be part of the surface of a sphere, we need at least two 'directions of measurement' to determine a direction of motion. Commonly, these are components measured along two different curves on the surface. A real-world example is that of a ship sailing the Atlantic. Its direction of motion, or position at any time, might be measured by an east-west component (longitude) and a north-south component (latitude).

In three dimensions we measure direction of motion, or position, by three components. For example, an aeroplane can change position not only in both latitude and longitude, but also in altitude. The third dimension is above or below a two dimensional surface.

On this basis, it seems that our Koch snowflake is a surface in two dimensions, and that its perimeter — often referred to as the Koch curve — is a curve with one dimension. Here we run into snags. Try and picture what happens to the self-similarities illustrated by Fig. 22.2(**b**) as we apply them to the repeatedly decreasing length segments in the Koch curve. In the limit, as each segment tends to zero length, the curve representing the perimeter changes direction at every point. Any notion of moving along a curve in a specific direction now becomes fuzzy. No matter how small a distance we move along the curve, the direction required to keep on the curve changes. We appear to be in neither one, nor in two, dimensions.

Further, it is plain that since the whole perimeter is infinite in length, so also is any finite non-zero part of that perimeter. For example, the part of the perimeter enclosed in the larger rectangle in Fig. 22.4 is replicated on the same scale five more times in the complete figure. Since the whole perimeter is of infinite length. it follows that each of these six self-similar parts must also have perimeters of infinite length.

The snowflake is confined to the boundaries imposed by the self-similarity condition. Starting from an equilateral triangle of given area, as we have indicated, the snowflake never occupies an area greater than 1.6 times that area. This is a restricted region of a two-dimensional surface.

The 'dimensional' peculiarities of the Koch snowflake can be overcome by a new definition of dimension. We base this on self-similarity. The definition allows non-integer dimensions.

In the conventional concept of dimension, we divide a figure in d dimensions into exactly n self-similar parts. All of these are identical to each other, and similar to the whole. This is a special case of self-similarity. We denote by r a linear scaling factor, the factor by which the whole exceeds the part in size. Fig. 22.5 shows a segment of a straight line divided into n equal parts. Each of these is similar, and of length $1/n$, so the scaling factor is $r = n$, since the whole has a length n times that of each part.

Now consider an equilateral triangle like that in Fig. 22.6(**a**). We may divide each side into three equal parts, and construct another equilateral triangle with each side of length equal to one of these parts (i.e. one third the length of a side of the original triangle). The linear scaling factor is $r = 3$.

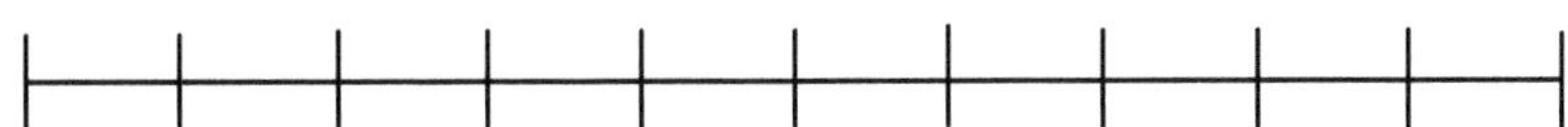

Fig. 22.5. Division of a line segment into n equal self-similar parts, Here $n = 10$

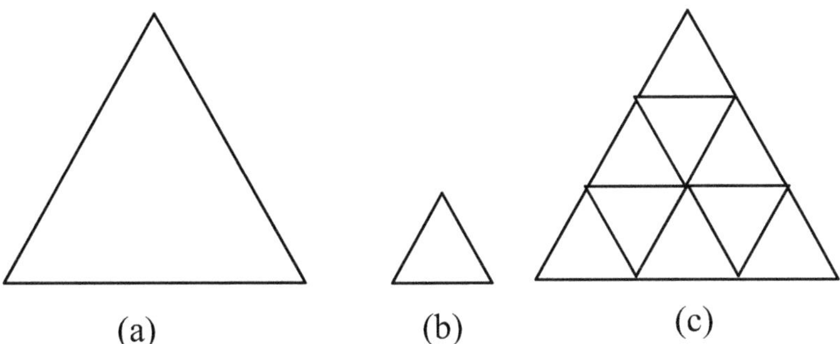

Fig. 22.6. Division of an equilateral triangle into 9 self-similar figures with scaling factor $r = 3$ between (**b**) and (**a**)

One such triangle is shown in Fig. 22.6(**b**) and it is easily seen, as is illustrated in Fig. 22.6(**c**), that the original triangle can be divided into $n = 9$ of these smaller self-similar triangles. Here $n = r^2$. This is simply a reflection of the well-known result that a linear (one dimensional) scaling factor r, has the effect of scaling an area by r^2. For example, if in Fig. 22.6(**a**), we had scaled by taking triangles each with side length one quarter of that of the original ($r = 4$), we would need $n = 16$ of the smaller triangles to cover the area of the original triangle. This is shown in Fig. 22.7, and again $n = r^2$. Similar arguments apply to any two-dimensional surface.

By applying a scaling factor r to a cube, it is not hard to see that we need $n = r^3$ self-similar smaller cubes to fill the original cube. Thus, for the number of dimensions $d = 1, 2, 3$ we have the relationship $n = r^d$, or

$$r = \sqrt[d]{n} \tag{22.1}$$

If, for a given n, r we take (22.1) as a relationship that defines a dimension d, this gives the usual value 1, 2, 3 for the familiar dimensions of classic geometry. In that situation the relationship is little more than an algebraic formalization of the familiar notions of dimension.

We return to the scaling illustrated in Fig. 22.2, demonstrating the self-similarity scaling for the Koch curve. Here $r = 3$, since we divided the initial line into 3 parts to form the similarity figure with $n = 4$ parts. It is repeated applications of this relationship that lead in the limit to the Koch curve. Thus, if we take (22.1) as the relationship to define the dimension of the Koch curve, we have

$$3 = \sqrt[d]{4} \, .$$

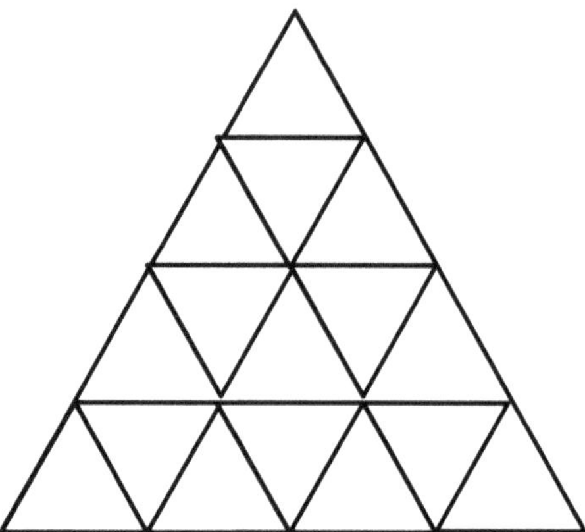

Fig. 22.7. Self-similarity for an equilateral triangle with $r = 4, n = 16$

Taking logarithms, we get $\log 3 = (1/d) \log 4$, whence

$$d = \frac{\log 4}{\log 3} \approx 1.262.$$

So, taking (22.1) as the relationship that defines the dimension of the Koch curve, we conclude it has an approximate dimension 1.262. The idea of using a relationship like (22.1)) as a defining relationship for *dimension* arose in a more general context to determine what is called the *Hausdorff dimension*, being named after Felix Hausdorff (1868–1942), who proposed it in 1919. The importance of non-integer dimensions as a practical reality, rather than a mathematical curiosity, was recognised in 1977 by Benoît Mandelbrot, who gave them the name *fractal dimensions*.

That the Koch curve has a fractal, or nonintegral dimension, is a consequence of its property of having infinite length while enclosing only a finite area. There are several other well-known geometric shapes that have fractal dimensions, yet like the Koch curve when taken in isolation, they seem at first sight only to be a mathematical curiosity rather than having anything to do with real-world applications. Or have they?

22.3 How Long is a Coastline

In 1967 Benoît Mandelbrot posed the apparently innocent question 'How long is the coastline of Britain?' There are fairly obvious technical problems in finding the answer. The length would be greater if measured at the low-water mark than if it were measured at the high-water mark. Even if we select one of these, or some other tide level, as an acceptable definition of a coastline,

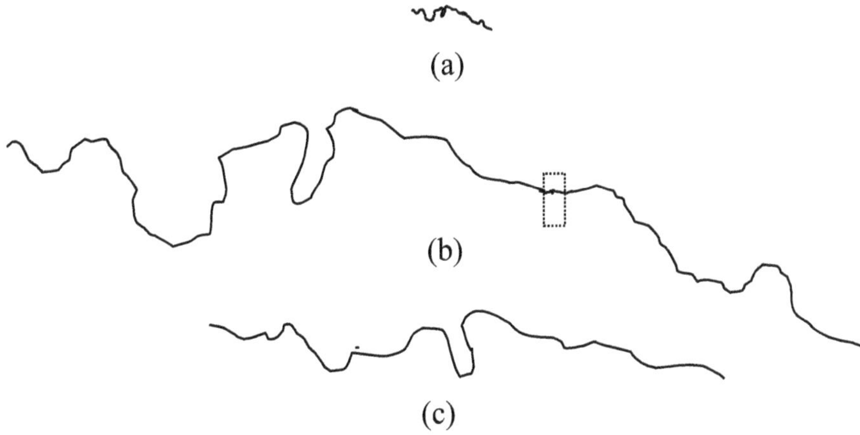

Fig. 22.8. The coastline in (**a**) is magnified 20 times in ((**b**). The section in (**b**) contained in the rectangle is magnified 20 times in (**c**)

there are secondary difficulties of defining a standard such as the high-water mark. The exact maximum tide level differs from day to-day.

There are also problems associated with river estuaries. How far up an estuary does the coastline extend? One might regard coastline as extending to the limit reached by tidal waters, or use some other criterion, such as extending inland only to the point where the width of the estuary first becomes less than half a mile, or perhaps less than 1 km.

Although some arbitrary decisions must be made about these matters, the problems are not insurmountable if conventions are agreed about what constitutes high-water mark, and what we do about estuaries, etc. We might hope to get quite a reasonable approximate answer if we used a set of reliable maps, such as those for the United Kingdom produced by the Ordnance Survey. These use a specified linear scale.

We might carefully measure off on such maps sections of coastline with dividers set at an interval of, say, 5 mm. Knowing the scale of the map — which has the property of self-similarity to the true horizontal lengths it portrays — we could use the appropriate scaling factor to get a good estimate of the coast length. Or could we? It would be a better estimate if we used standard 1 : 50 000 scale Ordnance Survey maps than if we carried out the same exercise on a map of the UK covering one page in a school atlas. The scale of this would be of the order 1 : 10 000 000. The former shows more detail of local bays, headlands, etc.

Fig. 22.8(**a**) is a small scale map of a section of coastline. Fig. 22.8(**b**) shows the same section magnified 20 times. The coastline map is only a fictional one that was computer generated, initially as shown in (**a**), then (**b**) was obtained by scaling up (**a**) by a factor of 20. Fig. 22.8(**c**) shows the section of coast enclosed in the rectangle in Fig. 22.8(**b**), again magnified 20 times.

In both Fig. 22.8(**b**) and (**c**) there are typical features like bays, smaller inlets and headlands, as well as lesser abutments that we expect on a coastline. But remember (**c**) represents a small section of the coast in (**b**), magnified 20 times. In the equivalent bit of coast in the rectangle in (**b**) it is impossible to detect these details.

If we had not known about the magnification and the scale differences, we might have regarded (**b**) and (**c**) simply as maps of rather typical, but different, bits of coastline. We have here a real-world phenomenon that is best described as *approximate self-similarity.* If we measured on that map the length of the coastline displayed in Fig. 22.8(**b**) using dividers set to 5mm and scaled appropriately, we would assign a much shorter total length to the segment of coast included in the rectangle, than we would if we applied the same process to the magnified version in (**c**). This is because on the latter we could then follow inlets and headland shapes not visible in (**b**). Had we used only the original map (**a**), we would have seriously underestimated the length of the coast.

There is no reason why we should stop at the largest possible maps. If we had access to the coastline, and some measuring device such as a metre stick or a tape measure, we could get a more accurate measurement by walking and measuring as near to the shore as possible. Using a metre stick is equivalent to the process we used on the map, but with dividers set to 1 metre. We will get a more accurate and greater estimate of the coast length than we would get from the largest available map.

But why stop with steps of 1 metre? Use a half-metre rule, or better still a flexible tape measure, and we will be able to follow the contours of objects like large rocks to get a new and greater estimate. In practice, some technique like this may give as good an estimate as we can get for any appreciable length of coast. But no section of coastline is exactly straight, even if a very small portion looks nearly straight in our ultimate unit of measurement. If this measurement unit were reduced to 1 mm we would be measuring across particles about the size of large grains of sand.

Why stop there? With modern scientific instruments we can magnify items like a grain of sand of cross section 1mm, and measure round it. Why stop at grains of sand? Why not continue measuring down to atomic particles, even perhaps to sub-atomic levels? At all these levels, with suitable magnification we still find evidence of approximate self-similarity. Shapewise, a magnified grain of sand looks like a rock.

The gist of the paper Mandelbrot wrote on this topic in the journal *Science* was that all real-world coastlines are fractal. Thus, the length of a coastline is infinite. Look again at the Koch snowflake in Fig. 22.3. It is easy to visualize it as a picture of an idealized island. Indeed some writers, e.g., Devlin [14], call it the *Koch Island* rather than the Koch snowflake. Typical coastlines have a fractal dimension not very different from that for the Koch curve.

Fig. 22.9. The first three stages of elimination in forming the Cantor set

22.4 Cantor Sets

An abstract example of a fractal dimension is provided by the Cantor set created in the following way. Consider a unit segment, which we may arbitrarily choose as the segment (0, 1) on a line representing real numbers. This is illustrated by the top line in Fig. 22.9. As a first step, we eliminate the middle third interval $1/3 < x < 2/3$. We then remove the middle third from each of the two remaining segments, i.e., all points x such that $1/9 < x < 2/9$ and $7/9 < x < 8/9$. These steps, and the one following the latter, are illustrated in Fig. 22.9. The process continues indefinitely, where at each step we remove the middle third of any remaining segments.

We have here a self-similarity process where each original segment of a given length is replaced by two, each of length one-third that of the original, so that in the notation of (22.1), $r = 3$ and $n = 2$ giving

$$3 = \sqrt[d]{2}$$

whence $d = \log 2/\log 3 = 0.631$.

It is not difficult to show that the total length of the segments removed is unity. You might to try to do this as an exercise. The only hint I give is that the lengths of the sectors removed at each step form a geometric progression. It is the remaining isolated points, i.e., $1/3, 2/3, 7/9, 8/9, \ldots$, that form the Cantor set, that has the fractal dimension of 0.631.

This result was established before the concept of self-similarity was applied to determine fractal dimensions. It was based on the concept of the more general Hausdorff dimension. Where they occur, the recognition of fractal, or noninteger, dimensions may be compared in importance to the step from integers to rational fractions.

22.5 Mandelbrot's World of Fractals

Mandelbrot's work on fractals established connections between various phenomena in the physical and life sciences and fractals. Details of these are given in nearly all books on, and in many papers about, chaos. The books include the very readable Stewart [43], [44], and at a slightly more general introductory level, giving many historical sidelights, that by Gleick [25], as well as detailed monographs.

We introduced self-similarity in Sect. 21.3, when discussing the logistic difference equation. In this chapter we have demonstrated the link between self-similarity and fractal dimensions. The mathematics involved in firmly establishing the exact nature of that link cannot be adequately covered at the level of this book, but there are close links between the strange or chaotic attractors met when bifurcation breaks down for recursions with the logistic difference equation, and the fractal Koch curve with its infinite length and infinitely frequent changes of direction as we proceed along it.

A mathematically simple function explored by Mandelbrot was the mapping

$$z_{t+1} = z_t^2 + c \tag{22.2}$$

given some initial value z_0, where now both the z_t and c are complex numbers.

In exploring the behaviour of this mapping, two approaches are possible when we are interested in a range of values of c. We may explore what happens when we choose various z_0 (initial values) with some pre-chosen value of c, or we may fix z_0 at, say $z_0 = 0$, and study the behaviour of the mapping over a range of values of c.

For any chosen value of c, and a range of values for z_0, it only requires a few lines of computer code to set up the iterative process to see what happens for large values of t.

In broad terms one of two things usually happens. Either z_t becomes larger and larger in modulus and, in the complex plane of the Argand diagram, diverges or heads off towards infinity, or else it tends to some attractor. This may be a single point, a collection of points visited periodically, or perhaps some strange attractor consisting of some defined region in the complex plane. Generally speaking, there will be sets of initial values for which the iterations diverge, others for which it will go to an attractor, and a further set of values on some boundary region where the point will stay on that boundary region.

The simplest case is where $c = 0$. Any $z = x + \mathrm{i}y$ may be written in the form $z = r\text{cis}\ \theta$, where $r = \sqrt{(x^2 + y^2)}$ and $\text{cis}\ \theta = \cos\theta + \mathrm{i}\sin\theta$. De Moivre's theorem, Sects. 6.1 and 6.2, tells us that $z^n = r^n\text{cis}\ n\theta$, so repeated applications of the mapping $z_{t+1} = z_t^2$, with initial condition $z_0 = r\text{cis}\ \theta$, implies that when $t + 1 = n$

$$z_n = r^{2n}\text{cis}\ 2n\theta.$$

This means that z_n lies on a circle of radius r^{2n} in the complex plane. This implies that $z_t \to 0$ if $r < 1$ and $z_t \to \infty$ if $r > 1$. If $r = 1$, then z_n is somewhere on the unit circle for all n. Just where is determined by the angle $2n\theta$. Thus the attractor is 0 if $r < 1$, and the recursions diverge towards infinity if $r > 1$. The circle with radius 1 is the boundary between initial values leading to a fixed point attractor, and initial values that, in the sense of a dynamic motion, give mappings that wander off to infinity.

In general, if c is set at any value other than zero, the behaviour is more complicated. There will often be more than one attractor. The boundary will be either a fractal curve, a series of fractal curves, or perhaps a dusting of particles over the plane. There is no easy analytical way to decide what it may be for a particular case, but as indicated above a few lines of computer programming enables one, in theory at any rate, to determine the nature of that boundary for a grid of values of z_0 for any chosen fixed c. How long it will take to do this is another matter. A coarse grid of points will only give a rough idea of the boundary pattern. A fine grid is needed to get a clear graphical picture, because in general the boundary will be fractal.

If one chooses a value of c other than zero, say $c = 0.31 + 0.04\mathrm{i}$, and performs the mapping (22.2) with a fine grid of starting values, z_0, one finds a continuous fractal boundary between iterations that converge to a single point and those that diverge after a couple of hundred iterations. The boundary exhibits self-similarity, and has an intricate pattern. These boundaries, as mentioned above, can be extremely complicated, and many are very beautiful. The boundaries are called *Julia sets* and the mapping giving rise to them was first described by Gaston Maurice Julia (1893–1978) in 1918.

An alternative approach to exploring the mapping (22.2) was developed by Mandelbrot. He set $z_0 = 0$ and then examined the mapping for different c. With a modern computer it is feasible to determine all values of c for which z_t converged to a finite attractor, which could be a point attractor, a periodic attractor or a chaotic attractor. Such points form a set in the complex plane. For all other c, the mapping diverges.

The set of points going to an attractor is the Mandelbrot set. Points on the boundary of the set give rise to Julia sets. Fig. 22.10 shows the Mandelbrot set at a magnification that clearly indicates the self similarity.

The small dots in space just visible to the naked eye in Fig. 22.10 are not faulty print marks. They are small clusters of values of c that are sub-sets of the entire Mandelbrot set, and they are self-similar to the large main set.

Mandelbrot sets may be formed for other simple mappings. Indeed that for the mapping $z_{t+1} = z_t^4 + c$ is not much more complicated than that in Fig. 22.10. The computer lends itself well to the generation of both Julia sets and Mandelbrot sets. If you have access to the internet it is worth setting your search engine to work to find sites with literally hundreds of pictures of these sets. Many of the sites explain how the sets were generated. A site I have found useful because it contains a facility for producing and exploring Mandelbrot sets, as well as much explanatory material is located at

www.math.utah.edu/ alfeld/math/mandelbrot/mandelbrot.html

The site is developed and maintained by Peter Alfeld,University of Utah.

Fig. 22.10. The Mandelbrot set for the mapping (22.2) with $z_0 = 0$. Note the fractal self-similarities and the isolated 'dust particles' in white space. These are self-similar to the main body

22.6 Loose Ends

The simple mathematical models of dynamic systems we have used to present some of the main features of chaos, and concepts such as fractal dimensions, give no idea of the impact these developments have had both upon our understanding of major physical dynamic systems and upon the way we view the real world. Newtonian dynamics served us well, and still does, for the description of truly deterministic systems. Einstein's work on relativity showed up limitations of Newtonian dynamics when applied on the cosmic scale (e.g., when velocities approach that of light). At the opposite end of the scale, wave or quantum mechanics are needed to describe the dynamics of particles whose energies can change only in discrete steps. The behaviour of large numbers of such particles is studied in the subject of *statistical mechanics.*

A more detailed study of the links between real dynamic systems and chaos that requires only an understanding of elementary mathematics is given by Stewart [43], [44]. A nonmathematical treatment of the impact of chaos and other modern ideas on the philosophy of science is given by Cohen and Stewart [11] .

Mandelbrot's work on fractals has had an impact well beyond mathematics. This has been made possible only by the advent of fast computers. Indeed many of the animated computer graphics familiar on our television screens are fractal diagrams. Mandelbrot [35] gives a basic account of fractal geometry and its many applications that were still in their infancy at the time he wrote that book.

23

Mathematical Tripstones

23.1 Traps for the Unwary

No intuition – no chaos?

Chapters 21 and 22 sketched some recent developments. We return to more familiar ground with a few examples, some light-hearted, that provide easier reading. One or two drive home that message that intuition may lead us astray. Nevertheless, mathematics would be lost without it. If Benoît Mandelbrot had not used intuition, there might still be no Mandelbrot sets. My other aim is to show by examples that 'teasers' may arise in many different fields. Several of the examples are classics.

23.2 Crossing One's Bridges

The Königsberg bridges problem was put to, and solved, by Leonhard Euler in 1735. There were seven bridges in the town of Königsberg crossing the different branches of the river Pregel in a pattern sketched in Fig. 23.1. The challenge was to start at any point in the town and cross all seven bridges once, and once only, on a round trip back to the starting point.

We may choose the starting point in any of the land areas marked A, B, C, D connected to one another only by the bridges, and follow any of the routes indicated by the dotted curves. The challenge is to cross all the bridges exactly once on a round trip back to the starting point. That is, all the dotted routes are to be traversed once and once only.

We use a trick given in Sects. 9.2 and 9.3 for the four-colour problem. There we dubbed the points A, B, C, D as vertices, and the paths joining them as edges. These terms are widely used, as explained there, in graph theory. The essentials of the bridge problem in terms of vertices and edges are

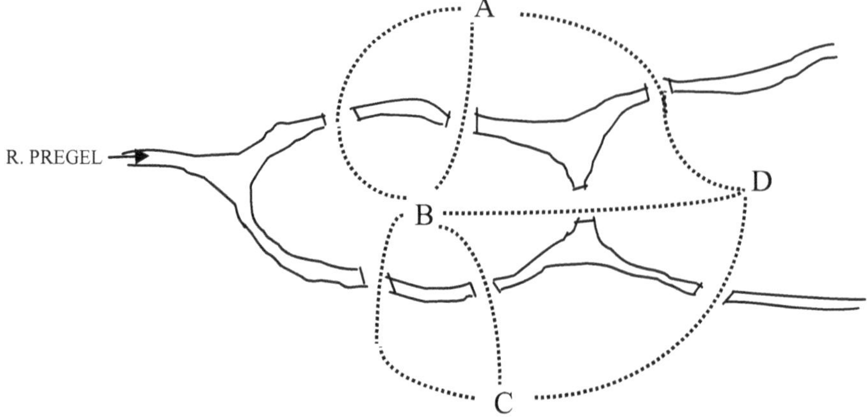

Fig. 23.1. The seven bridges of Königsberg

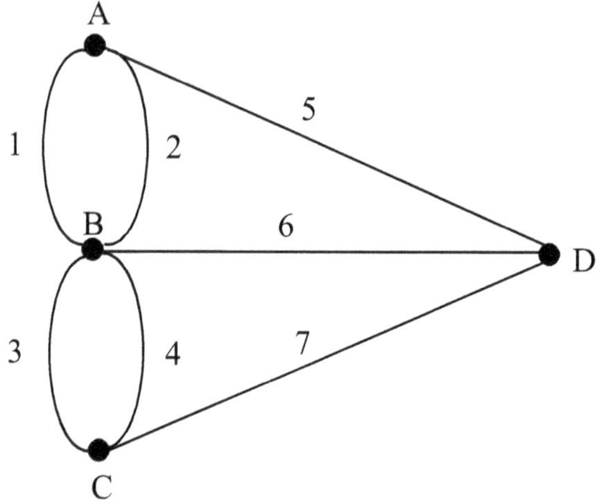

Fig. 23.2. The essentials of the Königsberg bridge problem

epitomized in Fig. 23.2 . It is easy to see that the number of vertices is 4, one corresponding to each land area. There are 7 paths, or edges, (numbered 1 to 7). Each path is associated with one, and only one, bridge. No matter from which vertex we start, it is impossible to cross all bridges just once and get back to that starting vertex.

To cross all bridges (i.e. traverse all edges in Fig. 23.2) and return to the starting point, we must visit every vertex at least once. Further, each time we visit a vertex there must be an edge leading out that we have not yet traversed. A little reflection shows this is only possible if there is an even number of edges meeting at each vertex. Figure 23.2 shows there are an odd

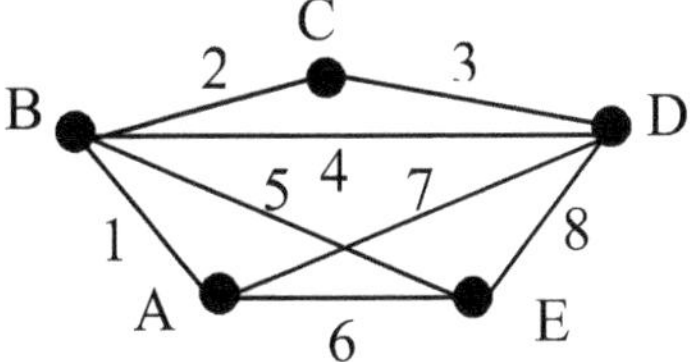

Fig. 23.3. A semi-Eulerian graph; three edges meet at vertices A and E

number of edges meeting at each vertex. There are 5 at B and 3 at each of A, C and D.

Euler solved the problem when he was in the process of developing his ideas about graph theory.

Graphs that have an even number of edges meeting at each vertex, and so can be traversed with a return to the starting point using each edge once and once only, are said to be *Eulerian.* An Eulerian graph may be drawn without lifting our pencil from the paper, and without retracing any curve representing an edge.

If we relax the restriction that we must return to the starting point, it may be possible to traverse each edge of the graph once, and once only, and thus draw the graph without lifting pencil from paper, or retracing any edge already drawn. Fig. 23.3 is an example. For this graph, starting from A we may draw the edges in the order 1, 2, 3, 4, 5, 6, 7 ,8 indicated on that diagram, but it is impossible to return to the starting vertex A without using one of the edges already traversed. The vertices A and E each have three edges meeting at them. Such graphs are sometimes called *semi-Eulerian.* It is not difficult to establish that a graph will be semi-Eulerian if there are only two vertices with an odd number of edges meeting at them. A little careful thought should soon convince you this is so.

Graphs like that for the Königsberg bridge problem are sometimes referred to as *non-Eulerian.*

Meaning may depend on context

As in other branches of mathematics there are several everyday words used in graph theory with a special technical meaning. The word *graph* itself, as we have seen, is used in mathematics with meanings ranging from a diagram showing the relationship between two variables to the patterns of vertices and edges we have considered here and in other places, e.g., in Sect. 9.3 (in topology) and in Sect. 20.2 (in CPA). The basic notion in the latter cases is that a graph is a set of points called vertices, or nodes, connected by a set of curves called edges. If we associate a direction with each edge, as in critical path analysis, the graph is a *directed graph* or *network*. A closely related idea is that of a *weighted graph*, where a weight or penalty or reward is associated with each edge. We gave some practical examples in Sect. 20.4 (heights of mountain passes, etc).

Two classic problems involving networks are that of the *Chinese postman* and that of the *travelling salesman*. The adjective 'Chinese' is used because the problem is believed to have been first posed by a Chinese mathematician, but the circumstances are relevant to any postman who has to visit all streets in an area to deliver mail, then return to his starting point. A postman or postwoman has to visit every edge at least once (one edge corresponding to each street), and return to the starting point. One seeks a route that minimizes the total distance to be travelled. The travelling salesman problem is about a salesman who has to visit a number of cities, and return to his starting point, in such a way that he covers the least possible distance. In terms of a weighted graph, this is a matter of determining a route that visits all vertices at least once. Again we seek a solution that minimizes the distance covered. The emphasis in the Chinese postman problem is on edges ,and in the travelling salesman problem on vertices. For more than a few vertices or edges, the solution to either problem is not trivial.

23.3 A Cost Cutting Tool

The link between what at first appears an abstract mathematical notion, and its role in providing a simple solution to a practical problem, may not be self-evident at first meeting.

To illustrate one such link, we first pose a practical problem. We then define what might seem a somewhat abstract idea. We use this to solve our problem.

Here is the problem. Farmer Brown wants to apply fertilizer with a phosphate content at least 35 per cent. He can use a basic fertilizer, or a mixture of two or more from 5 basic fertilizers. The phosphate content, and the price per bag, all bags being the same size, for each basic fertilizer are:

Brand A 20 per cent phosphate £10
Brand B 30 per cent phosphate £9.50
Brand C 50 per cent phosphate £11
Brand D 80 per cent phosphate £13
Brand E 40 per cent phosphate £10

What is the cheapest way of attaining the required phosphate level?

It is easy to see that Farmer Brown will meet the 35 per cent minimum phosphate requirement if he take an equal mix of brands A and C. The cost per bag will then be the mean of the prices for each, i.e., £10.50. He could do better by taking some mix of other brands. For example, an equal mix of brands B and C would more than suffice, giving 40 per cent phosphate. It would cost

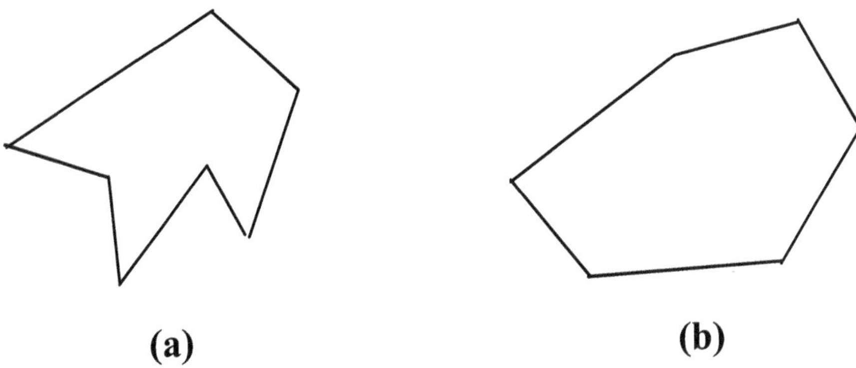

Fig. 23.4. A concave polygon (**a**) and a convex polygon (**b**)

only £10.25 per bag. A further straightforward calculation establishes that a mix of 2 bags of brand B with one of brand C, would reduce the cost per bag to £10, while still meeting the requirement. The requirement will also be met by a variety of mixes of two, three, four or all five brands.

To explore all possible mixes leading to feasible solutions is tedious, and as we shall see, unnecessary.

To develop the abstract idea that gives us a simple solution, we need some definitions. Fig. 23.4(**a**) and (**b**) each represent polygons. The polygon in (**a**) has some of its interior angles less than 180°, and some greater than 180°. The polygon in (**b**) has all its interior angles less than 180°. A polygon with all interior angles not greater than 180° is called a *convex polygon.* Polygons that are not convex, are sometimes described as *concave.*

A distinguishing characteristic of a convex polygon is that if we start to walk around the perimeter from any vertex in a clockwise direction, at each succeeding vertex we must bear right to get to the next vertex. We must continue to bear right until our return to the starting vertex. If we proceed anticlockwise, we must bear left at each vertex. For a concave polygon, no matter whether we proceed clockwise or anticlockwise, at some vertices we must bear left, while at others we must bear right. You should check these assertions hold for the polygons in Fig. 23.4(**a**) and (**b**).

Any concave polygon can always be enclosed in a convex polygon. The enclosing polygon is not unique. Fig. 23.5(**a**) and (**b**), both feature the same concave polygon ABCDEF. Each is enclosed within a convex polygon. This is the polygon PQRST in (**a**) and the polygon ACEF in (**b**). In the first case, the enclosing convex polygon has no vertices in common with the concave polygon ABCDEF. In the second, case all vertices of the convex polygon are vertices of the original polygon, but not all vertices of the polygon ABCDEF are included.

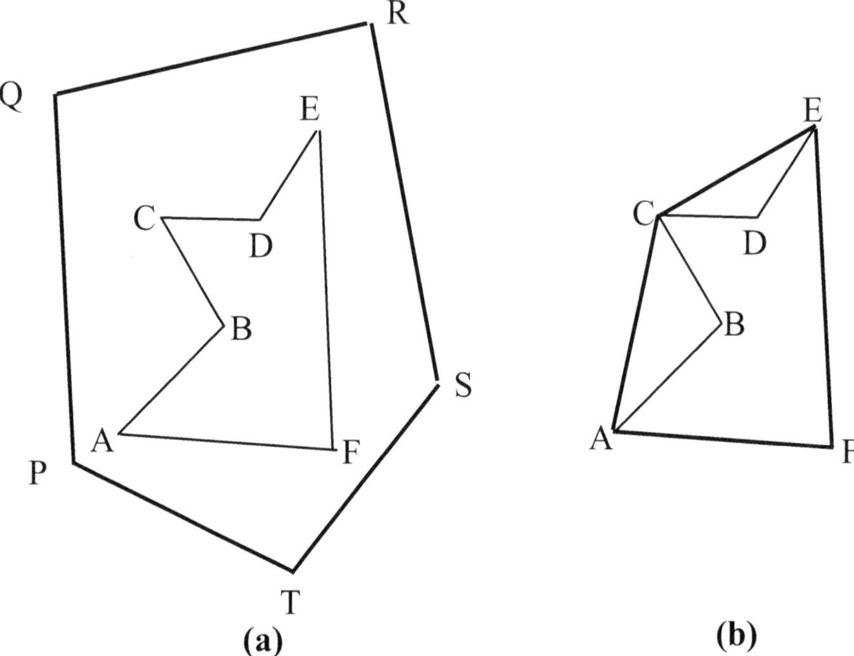

Fig. 23.5. An arbitrary convex polygon that encloses a concave polygon ABCDEF in (**a**) and the corresponding convex hull in (**b**)

For these particular examples, it is easy to see by inspecting the figures, that ACEF is the convex polygon of minimum area that encloses the concave polygon ABCDEF. If we walk clockwise from the vertex A around the perimeter of the polygon ABCDEF we must bear left at B and D, and bear right at A, C, E, and F. To form this minimum area enclosing convex polygon, we omit the vertices corresponding to interior angles greater than 180°. This minimum area convex polygon is called the *convex hull* of the original polygon.

Convex hulls also feature in graph theory. If we have a set of n points, and join every pair of points by a line segment, the graph is said to be *complete*. In graph theory terminology, the points are vertices and the joins are edges. An example is given in Fig. 23.6. For any such set of points it is posible, by a suitable choice of edges, to form a convex polygon that includes all points, the chosen edges being the sides of that polygon. We do not prove it formally, but this convex polygon has a lesser area that any other convex polygon enclosing all the points in the set. It is called the *convex hull* of that set. In Fig. 23.6 the convex hull is depicted by thicker lines. In Fig. 23.6 the angle BCD is 180°, since the vertex C lies on the join of BD. This is often expressed by saying the

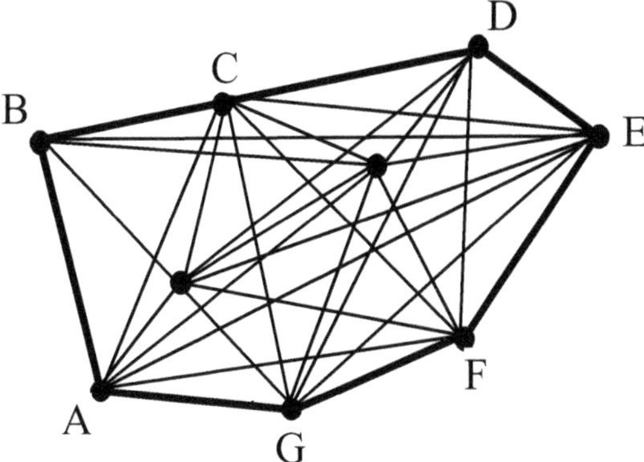

Fig. 23.6. A complete graph where convex hull for the set of all vertices is the polygon ABCDEFG

vertices, or points, B, C, D are *collinear*. It is to cope with situations like this that we include angles of exactly 180° in our definition of a convex polygon.

The convex hull is the abstract idea we need to help solve the fertilizer problem posed at the start of this section. Fig. 23.7 is a plot of cost per bag (y) against percentage phosphate (x) for each of the five brands. On this graph the dotted line at $x = 35$ is relevant to the constraint that $x \geq 35$.

Any mixture that corresponds to a point on or to the right of this line, satisfies the constraint on phosphate content. Other constraints are the physical ones of precisely which phosphate levels can be attained by possible mixtures of brands.

The straight lines joining each pair of the points A, B, C, D, E indicates the cost and phosphate level of possible mixtures of that pair of brands. It is also straightforward, as we show below, to calculate the proportion of each brand in the mixture corresponding to all possible price/phosphate proportions.

Here are the calculations for the join AD in Fig. 23.8. From the original data the coordinates of the points A, D are respectively (20, 10) and (80, 13) whence, from (7.1), the equation to the line AD is

$$\frac{y - 10}{13 - 10} = \frac{x - 20}{80 - 20} \,.$$

This reduces to

$$20y - x = 180.$$

Thus, when $x = 35$ we find $y = 10.75$, i.e., the cost is £10.75 per bag. Any combination of only A and D with a higher phosphate content will cost more. A glance at Fig. 23.8 confirms this. It is also easy to work out the proportions of A and D that give this optimum mix. If we denote by x_{A}, x_{D} and x_{M} the x-coordinates of A, D and M where M is any point on the segment AD, it

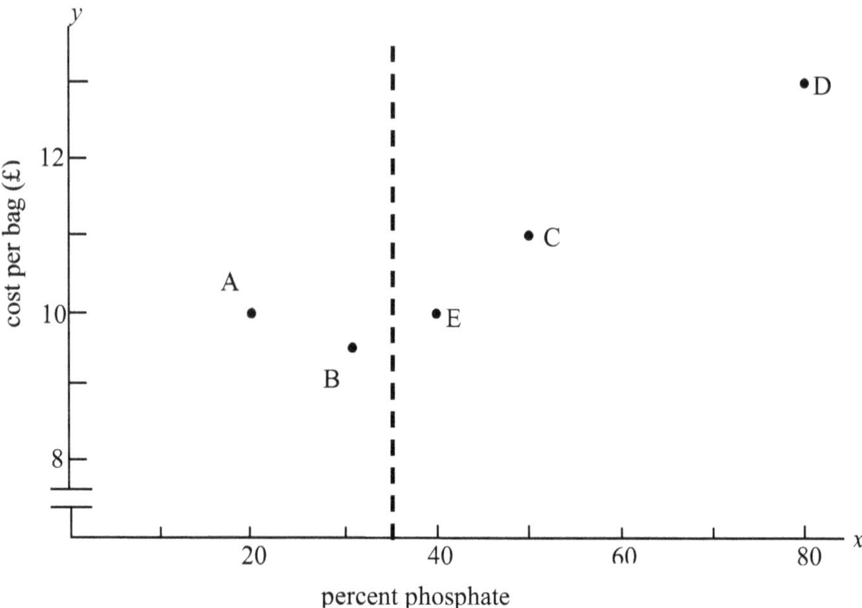

Fig. 23.7. Cost per bag and phosphate content of 5 fertilizer brands A, B, C, D, E. The broken line corresponds to the required minimum phosphate content of 35 per cent

is not difficult to show that the proportion of D in a mixture that gives the phosphate content x_{M} is given by

$$p = \frac{x_{\mathrm{M}} - x_{\mathrm{A}}}{x_{\mathrm{D}} - x_{\mathrm{A}}} . \tag{23.1}$$

If M coincides with A, (23.1) gives the proportion of D to be zero, as it should. If M coincides with D, this gives a proportion 1 (i.e. 100 per cent of D), again as it should. If M corresponds to P in Fig. 23.8 with coordinate $x = 35$, then, since $x_{\mathrm{A}} = 20$ and $x_{\mathrm{D}} = 80$, (23.1) gives $p = 1/4$. This indicates that the mixture consists of 3 parts of A with one part of D. We may confirm this by noting that, for such a mixture, the percentage phosphate is $(1/4) \times 80 + (3/4) \times 20 = 35$.

Possible combinations and costs for pairs of fertilizers may be obtained by joining all pairs of points A, B, C, D, E in Fig. 23.7 and obtaining the relevant equations for each. This forms a complete graph.

In this particular problem, we shall soon see that mixtures of three, four or or even five fertilizers will not lead to an optimum solution. If one required the cost and phosphate levels of these for other purpose, they may be worked out. Fig. 23.9 illustrates the way to determine cost of combinations of fertilizers A, C, D.

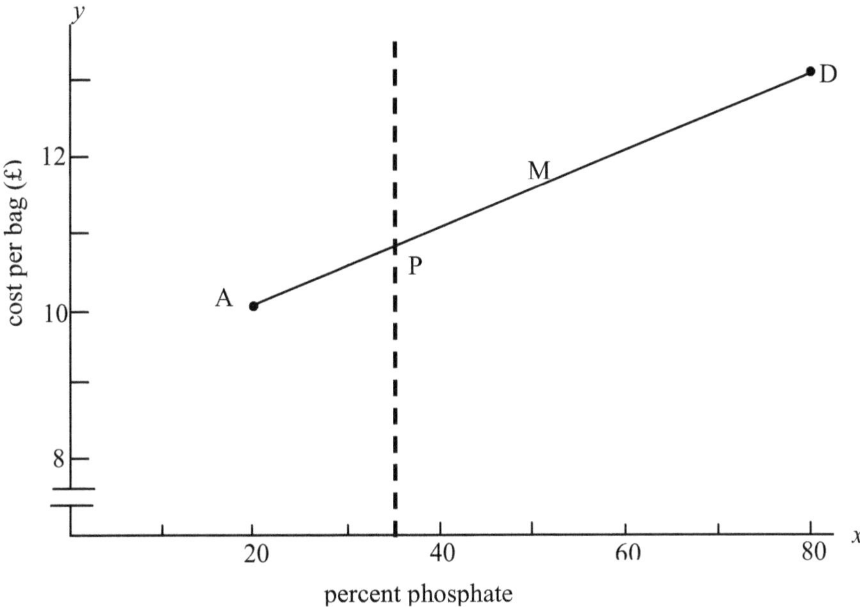

Fig. 23.8. The line segment AD represents cost and phosphate level of all possible combinations of brands A and D. The point P represents the minimum requirement of 35 per cent phosphate content for a mixture

We can get relevant mixtures in two steps. The point K on AD corresponds to some mixture of A and D. We may regard this as a new basic fertilizer. Possible mixtures of this with brand C give cost/proportions on the line KC. As K moves from A to D, the line KC sweeps out the triangle ACD. Thus, all possible cost/phosphate mixtures obtainable are represented by points in the triangle ACD. The extension to mixtures of four is straightforward. Since any point in the triangle ACD represents a mixture of the basic brands A, C, D, by joining any such points to B, we form a line that sweeps out the triangle ABD, which is easily added to Fig. 23.9. We leave it as an exercise to work out the possible cost/proportions structure for all other possible combinations of two, three, four or all five fertilisers. After some careful thought you should see that all points within or on the convex hull of the set of points A, B, C, D, E lead to possible cost/phosphate proportions and that no points outside this hull are possible combinations.

The constraint that the phosphate content must be at least 35 per cent, means that the basic feasible solutions are contained in that part of the convex hull to the right of the broken line $x = 35$ in Fig. 23.10. This is the hatched area in that figure. It is immediately obvious that the minimum cost mixture consists of equal parts of B and E giving exactly 35 per cent phosphate at a cost of £9.75 per bag.

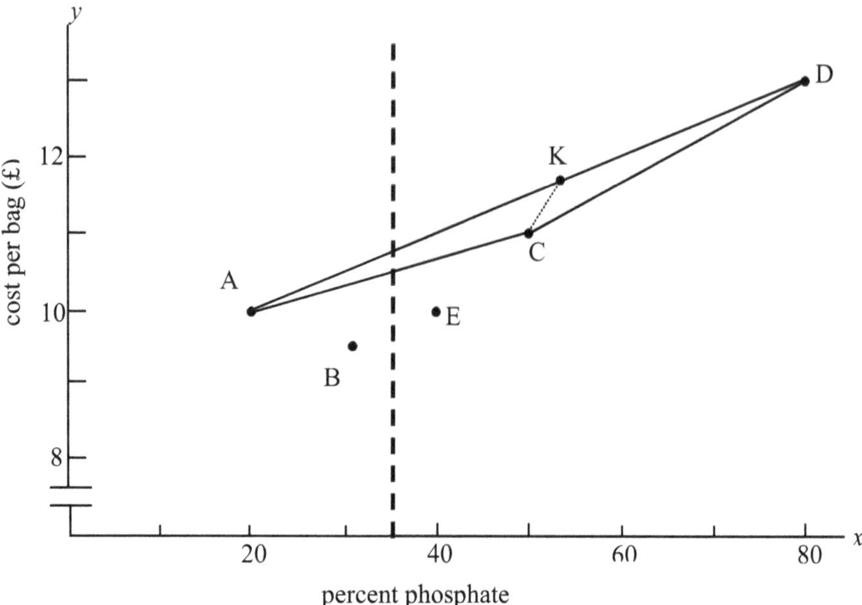

Fig. 23.9. Available mixtures of three fertilizers A, C, D give cost/phosphate proportions in triangle ACD

This is an example of a particularly simple linear programming problem. All we need to do to solve it is determine the convex hull of the data points corresponding to the cost/phosphate proportions for single fertilizers. The solution will lie at a vertex of that part of the convex hull that forms the feasible region, or at a point where that hull is intersected by the constraint line $x = 35$. In some cases where there is not a unique solution there may be a set of points on one of the boundaries of the feasible region that are all optimum solutions.

23.4 Improving your Average Without Really Trying

Cricket was initially an English game, although the French sometimes claim to have dreamt it up. It is now often played more effectively, and with more enthusiasm, by people in other parts of the world. It is a mystery to those not addicted to the game.

However, for this exercise all we need to know is that a *bowling average* is calculated by dividing the runs scored by all batsmen when facing that bowler by the number of batsmen that that bowler dismisses — this latter referred to in cricket jargon as 'the number of wickets taken'. A bowler likes

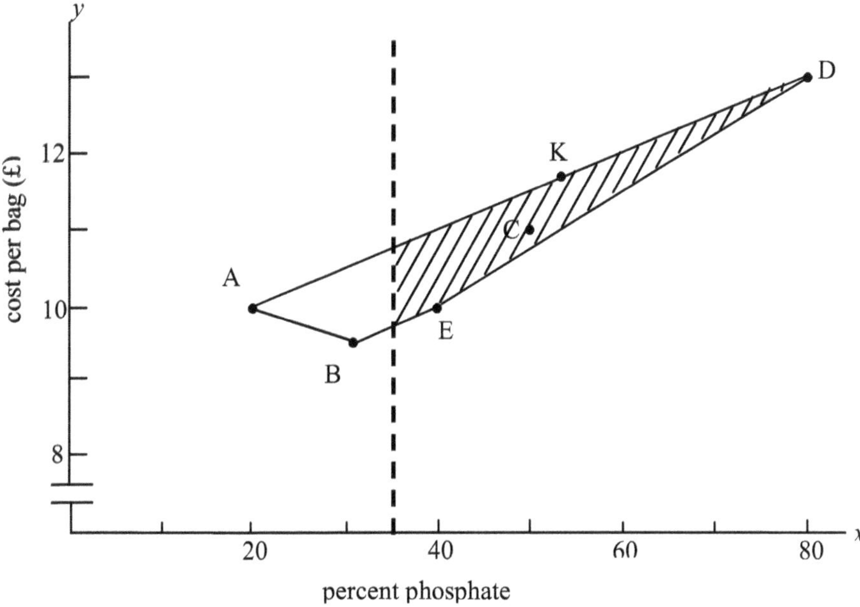

Fig. 23.10. The optimum mixture is equal parts of B and E giving 35 per cent phosphate at a cost of £9.75 per bag

Table 23.1. A bowling analysis for 15 games

	Brown	*Smith*
Runs scored	210	214
No of wickets	30	30
Average	7.00	7.13

a low average because his aim is to dismiss batsmen for as few runs scored as possible. The average calculated here is the arithmetic mean.

A cricket club gives a prize to the bowler with the best (i.e. lowest) average. Last year there was a close contest for the prize between Bill Smith and Joe Brown. The club secretary used information from the scorebook for all 15 matches on the fixture list to compile Table 23.1. In this table the averages are obtained, as explained above, by dividing runs scored by number of wickets taken. The secretary concludes Brown has won the prize by a whisker and informs the club president accordingly.

The president is a careful person who likes things to be just right. He points out that the secretary has not included the results for an extra match

Table 23.2. Bowling analysis for extra game

	Brown	*Smith*
Runs scored	70	39
No of wickets	5	2
Average	14.00	19.50

Table 23.3. Bowling analysis, all games

	Brown	*Smith*
Runs scored	280	253
No of wickets	35	32
Average	8.00	7.91

that was not on the original fixture list. He insists that these be included. The secretary looks out the results for the extra match, and these are shown in Table 23.2.

Again, in this game Brown has the better average, so the secretary decides that the results only enhance Brown's lead. He reports this to the president.

Fusspot president

Being a stickler for detail, the president insists that to keep the record straight the secretary should calculate the final average by combining the runs scored and numbers of wickets in Tables 23.1 and 23.2. He should then recalculate the averages in the way the rules require for all 16 matches. The result, together with the recalculated averages are given in Table 23.3 where, for example, for Brown the total runs scored is now 210 + 70 = 280, and so on. So Smith is the winner.

Check the calculations to make sure I am not cheating. Before reading on, try to puzzle out why we have this apparent anomaly.

The key is that in the match that was not included in the fixture list, both bowlers have performed relatively poorly. their averages in this match are both more than twice what they were over the other 15 matches. In the one extra match Smith has done worse than Brown in the sense that he has taken fewer wickets, and has a higher average. But in one respect he has done better than Brown. Fewer runs have been scored off him — 39 compared with 70. Combining these results with those from the other games, and calculating averages afresh, penalizes Brown more heavily for those extra runs, despite his having taken more wickets.

It is clearer to look at things this way. In the extra match Brown has 5 relatively poor dismissing performances, while Smith has only 2. In the final analysis these 2 get less weight because they are 2 outr of 32 dismissals, or only one sixteenth of all Smith's dismissals. On the other hand, Brown takes 5 from 35 dismissals in this match, and they represent one seventh of his total take. The arithmetic mean takes account of these proportions, using them as *weights* in calculating the overall average.

Paradoxes like that just described manifest themselves in more complicated situations. Here is an example.

An experiment is planned to compare the efficacy of two drugs for curing a respiratory disease. It is believed that the efficacy of each may differ between patients living in city, and those living in rural, areas. The reasons for this are that in cities the levels of atmospheric pollution are often higher, and this may affect severity of the disease which, in turn, may reduce the efficacy of a drug. Also, in cities people are likely to be engaged largely in indoor activities, which may aggravate respiratory conditions. In a rural environment many people work out of doors, and this may be beneficial. The opposite is also possible, working out of doors in a cool or damp climate might aggravate the condition.

Because of the uncertainty raised by such factors, it is good practice to design an experiment to make it possible to look at results for urban and rural areas separately. If one of the drugs were a standard treatment, and the other a new treatment, health officers might decide that 2000 urban, and 1000 rural patients, be used in the experiment. These numbers might be chosen for practical reasons, because records show that these are the number of cases to be expected over, say, a period of approximately 3 months.

Ideally, they would like to give about half the patients in each group the standard treatment and half the new treatment. However, because of factors such as the amount of the new drug available, or difficulty in its administration that require more visits from the doctor, and country visits are more time consuming, it is decided to give the new treatment to 70 per cent of the urban cases and 30 per cent of the rural cases. The remainder are to receive the standard drug. A statistician may accept, perhaps reluctantly, that this is reasonable. He or she is confident that the different proportions can be taken into account when analysing the results. The statistician would also advise a random choice of patients to receive each drug — subject only to the constraints on total numbers to receive each treatment. The random choice avoids possible bias. Without it, some doctors might give the new drug only to severe cases because they know the other is then almost useless, but they hope the new one might do better. If patients were selected in this way it would be impossible to interpret what might happen in more general use.

There may be ethical considerations that come into play concerning a doctor's duty to prescribe what he believes to be in the best interest of patients. These can, and should be taken into account in conducting the experiment, but we do not look at that issue here. We shall assume it is not relevant in the circumstances here. Results for an experiment of this type are summarized in

Table 23.4. Numbers responding to two drugs

	Urban		Rural	
	Standard	*New*	*Standard*	*New*
No effect	500	1050	350	120
Cure	100	350	350	180
Total	600	1400	700	300

Table 23.5. Total numbers responding to two drugs in both urban and rural areas

	Standard	*New*
No effect	850	1170
Cure	450	530
Total	1300	1700

Table 23.4. The numbers are chosen to keep the arithmetic simple, but this does not invalidate the argument now given.

It is not difficult to see from this table that, in both urban and rural areas, the proportion of cures with the new drug is higher than it is with the old. For instance, in urban areas 100 cures among 600 patients with the standard drug is a percentage cure rate of 16.67, while 350 cures in 1400 patients with the new drug is a cure rate of 25 per cent. Similarly, for rural areas there is a 50 per cent cure rate with the standard drug, and a 60 per cent rate with the new drug. These higher cure rates in rural areas would be consistent with a pattern where the infection is less severe in rural areas, and both drugs are more effective when symptoms are less severe.

Experimenters faced with a situation like this are tempted to combine results for urban and rural areas to get an overall impression of the efficacy of the drugs. Table 23.5 combines the corresponding totals for urban and rural areas given in Table 23.4.

The cure rates now look much the same for each drug. Indeed, the percentage cure rates are 34.62 per cent for the standard drug, and 31.18 per cent for the new drug, so the standard drug does slightly better. We have a paradox. According to the information in Table 23.4, the new drug does better in both rural and urban areas, but according to Table 23.5, among all patients treated the standard drug performs better. This paradox is called *Simpson's paradox* after Edward H. Simpson, who first explored it in Simpson [40]. It is likely to arise if we combine totals where different numbers and proportions of the population receive each treatment in two or more environments, especially when the response rates differ greatly between environments. In these circumstances direct combination of numbers in each of the four possible response categories may hide, or as in this example, even reverse, the apparent effect. The moral is that differences between numbers treated, and in the response rates in the separate tables, must be allowed for in assessing the overall effect.

In the bowling average example I suggested we should combine the tables for the main set of matches and the one special fixture. By doing so we exhibited what is really a case of Simpson's paradox. We allowed this combination because the club rules said that was how the prizewinner was to be chosen. In the case of the drug trial in urban and rural areas we reasoned that it was inappropriate to combine results for each. Do you think the cricket club should change the rules for determing the better bowler? If so, how do you think they might do it?

23.5 Random Selections

An area of mathematics in everyday life where there is a lot of confusion, is that associated with chance and randomness. We may toss a coin ten times and note the outcomes for each toss and record them in the order in which they occur. If H denotes a head, and T a tail, which of the following ordered outcomes do you consider the most likely to occur, and which do you consider the least likely?

(i) HHHHTTTTTT
(ii) THTHTHTTHH
(iii) TTTTTTTTTT

Surely we can eliminate (iii) as unlikely straight away. Everyone knows we are unlikely to get 10 head in 10 tosses. The outcome in (i) also seems unlikely. Four heads one after the other do appear from time to time in coin tossing. When they do, getting 6 tails in the next 6 tosses would again seem rather freakish. On the other hand outcome (ii) looks much the sort of thing one expects in 10 tosses, a good mix of heads or tails in a somewhat higgledy-piggledy order. So (ii) seems the most likely outcome.

Here, intuition, or should we call it commonsense, looks to serve us well. There is one problem. It gives the wrong answer.

The correct answer is that all these *ordered* sequences are equally likely. If the coin is a fair coin, each has a probability 1/1024 of occurring. This follows from simple applications of the multiplication rules for probabilities given in Sects. 17.3 and 17.4. The outcome at each toss is independent of that at any other, and at each toss $\Pr(\mathrm{H}) = \Pr(\mathrm{T}) = 1/2$. For 10 tosses it immediately follows that the probability associated with each of the outcomes above is $(1/2)^{10} = 1/1024$.

Where intuition lets us down is that, in applying 'commonsense', we ignored the fact that the outcomes were recorded in the order in which each 'head' or 'tail' appeared. If the question had been which of the following outcomes is most likely

- (i) 4 heads and 6 tails,
- (ii) 5 heads and 5 tails,
- (iii) 10 tails

the intuitive answer given above would have been correct. The binomial distribution given in Sect. 17.6 is relevant here, with $n = 10$, $p = q = 1/2$ and r = the number of heads in each of the cases listed in (i) to (iii). We leave it as an exercise to check that the relevant probabilities, P, are (i) $P = 210/1024$, (ii) $P = 252/1024$, (iii) $P = 1/1024$. Have another look at Sects. 17.5 and 17.6 if you are not sure how to go about this.

The reason people believe that an ordered sequence like HTTHTTHTHT is more likely than one such as TTTTTTTTTT, is that there are so many different ordered sequences or 'outcomes' – in fact 210 – that contain 4 heads and 6 tails, but only one that consist of 10 tails. There is, however, only *one* ordered sequence HTTHTTHTHT.

We leave this example with something for you to think about. Suppose you toss a coin ten times, and observe 10 tails, and are asked to bet on the outcome at the next toss. To maximize your chances of winning should you bet on it being a *head*, or on it being a *tail*?

A more sophisticated gambling situation where chance plays a role that is often misunderstand is that of lotteries. The United Kingdom national lottery exhibits features that, apart from minor variations, are common to major lotteries world-wide. Each player selects 6 numbers which, in the case of the UK lottery, must be different integers between 1 and 49. Care is taken by the organisers to see that in the draw of six numbers to decide the winner, each possible set of 6 different numbers from the integers 1 to 49 is equally likely to be drawn. The top, or jackpot, prize money is shared between those players (there may be more than one) who have selected the six numbers that are drawn. Do you think you have a better chance of winning if you select the numbers 3, 17, 25, 31, 35, 42 or if you select the numbers 1, 2, 3, 4, 5, 6? Thinking back to our remarks about sequences of heads and tails in coin tossing, what conclusion do you reach? The situation is not quite the same here. The random sampling system for selecting the six winning numbers at each draw does not quite follow the binomial distribution. However, the broad principle that all possible sequences of six numbers are equally likely to be drawn still holds.

So far as I know the sequence 1, 2, 3, 4, 5, 6 has never won a major lottery that is similar to the UK lottery. There was a draw in the UK lottery the night before I wrote this, and the winning numbers were 11, 18, 24, 34, 41, 49. As in the coin tossing situation, most people are less surprised about a set of numbers like this than they would be if the winning numbers were 1, 2, 3, 4, 5, 6. But, nevertheless, both sets of numbers have the same probability of being a winning sequence. Any one wide scattering of numbers between 1 and 49 seems more likely than a close and highly patterned sequence only because of the misconception that randomness implies scattering as distinct from pattern. This is not so. What is true is that there are many more outcomes that show scattering, than there are outcomes that are highly patterned.

Shortly after the UK national lottery started, a number of eminent statisticians made predictions about how many jackpot winners could be expected at each draw. They made slightly different, but usually quite reasonable, assumptions about how many players there would be for each game, and came up with fairly consistent estimates which varied only slightly according to assumptions. All agreed that more than about half a dozen jackpot winners at any one draw would be unlikely.

Many of these statisticians had red faces when one draw produced 133 jackpot winners. The statisticians had not allowed for the fact that although the drawing mechanism was tuned to give a random selection, many players do not select their numbers at random. A lot of people choose dates of birth, or other important anniversaries, in their selection of numbers. Since there are 7 days in a week, 12 months in a year and at most 31 days in a month, this means that numbers below 12 tend to be more often selected than higher numbers. Numbers above about 30 would have, in general, a lower probability of selection than do smaller numbers if a majority of participants select numbers this way. The effect of this would be that if a draw results in a preponderance of low numbers, we would expect more jackpot winners than if the draw gave predominantly high numbers, or a good mix of low and high numbers.

There are off-setting effects. Some people have worked out that the anniversary effect just described will lead to people ignoring high numbers. They conclude that if one chooses high numbers, and the Gods, i.e. the randomization process of the draw, favours high numbers — and the particular high numbers they have chosen — they are likely to get a bigger share of the jackpot. This is because they believe fewer people are likely to have made the same selection, than would be the case for a set of low winning numbers. Other factors also come into play. For instance, because many people think that scattering and randomness are much the same thing, they choose widely scattered numbers. This introduces a pattern of extreme scattering.

It has also been suggested that people are less likely to select numbers near the edges of the rectangular array in which the numbers are laid out on the entry slip, than they are to select numbers near the centre of the array. A more detailed account of the vagaries associated with lotteries is given by Henze and Riedwyl [29].

23.6 Happy Birthdays

You may have had this question put to you before. It is a a favourite in probability courses. You are at a party. Assuming there are no twins, triplets or other multiple-birth guests present, what is the minimum number of people that must be at the party to make the probability exceed 0.5 that there are at least two people in the room who share a common birthday date? By 'birthday date' we mean born on the same day of the same month, but not necessarily in the same year. To make progress we assume that people are equally likely to be born on any of the 365 days in a year and we ignore leap years. The assumption of births being uniformly distributed throughout the year is not quite accurate but it is a fairly good approximation. There may be slightly fewer births on days such as Christmas day, thanks to artificial means of inducing birth. The effect of this and leap years have only a small influence on the solution given below. In effect, we assume that the probability of any person in the room being born on a specific day of the year, say September 15, for example, is 1/365. We also assume that the birth dates of individuals are independent of one another.

We share a birthday.
Coincidence or not?

If you have not met this poser before, make a guess at the answer. It is one of the following:

$$17, 19, 23, 27, 31, 35, 42, 44, 51, 63, 72, 95, 110, 113, 142, 157, 163, 207$$

This is how we calculate the number. For any two people at the party, the first person's birthday may be any day. The second person has a different birthday if it occurs on any of the remaining 364 days. This means the probability the members of the pair have different birthdays is 364/365. A third person will have a different birthday to each of the first pair, if it occurs on any of the remaining 363 days. That event has a probability 363/365, on the assumption that births on any day are equally likely. Carrying on this way, a fourth person has a different birthday to the other 3 with probability 362/365. Keep going, and it is easy to see that if r people all have different birthdays, then the probability that an $(r + 1)$th person has a different birthday to all these r people, is $(365-r)/365$. Then, by the multiplication rule for probabilities, the probability that all $r + 1$ people have different birthdays is

$$\frac{364}{365} \times \frac{363}{365} \times \frac{362}{365} \times \ldots \times \frac{365-r}{365}. \tag{23.2}$$

If we set $r = 21$, this gives the probability that $r + 1 = 22$ people all have different birthdays. Multiplying out you should find that the probability that all 22 have different birthdays is, to 3 decimal places, 0.524. Be careful if you are using a pocket calculator, because if you multiply all numerator terms together and try to divide by 365^{21} you are likely to meet severe rounding, or even overflow, problems. You should instead divide 364 by 365, then multiply the answer by 363, divide the result by 365 again, and continue in that way. This keeps all numbers less than 1 after each division.

The probability $p = 0.524$ just calculated is that for all 22 having different birthdays. The opposite event is that at least two people have the same birthday. This has probability $1 - p = 1 - 0.524 = 0.476$.

If we set $r = 22$ so that $r + 1 = 23$ the last relevant term in (23.2) becomes 343/365. We find the probability that all 23 people have different birthdays is now 0.493. If you are using a pocket calculator you don't have to start the computation afresh. It is easy to see that having got the answer $p = 0.524$ for 22 people, you need only multiply this by 343/365 to get the probability that 23 people all have different birthdays, i.e.,

$$0.524 \times \frac{343}{365} = 0.492.$$

There is slight rounding here, since 0.524 was rounded. Keeping more decimal places in the result for 22 gives the slightly more accurate 0.493. The event we are interested in is the opposite event, i.e., that at least two people have the same birthday. This has probability $1 - 0.493 = 0.507$. Thus, if there are 23 or more people in the room, the probability of at least one shared birthday date exceeds one half. What was your guess? Most people unfamiliar with the solution guess a larger number. You may like to amuse yourself working out the probability of at least one paired birthday if there are 50 people at the party. It is approximately 0.970. Taking the relative frequency view of probability, this implies that if you consider lots of parties attended by 50 people, and with different individuals present at each of them, in about 97 out of every 100 such parties there is likely to be at least one set of two people sharing a birthday.

24

Some Sins of Omission

24.1 A Vast Subject

Many interesting topics — both classic and modern, and some of considerable importance — have not yet been mentioned in this book.

In this chapter we look briefly at a few of these — some fascinating, others important even if less intriguing. To give a full introductory treatment to each would double, even treble, the length of what is already a long book. Nevertheless, they are concepts that any budding mathematician should know about. I make many assertions either without proof, or backed only by indications of the line of proof. Section 24.7 has an historical flavour. It shows how even the application of relatively simple mathematics may only be possible if the technology to apply it is available, and how later technologies may, in turn, sometimes remove the need for an application.

24.2 The Golden Ratio and Fibonacci Numbers

Buried treasure

At first sight the number $(\sqrt{5}+1)/2$, called the *golden ratio*, or its decimal equivalent to 4 place, 1.6180, looks no more interesting than, say, $7+\sqrt{2}$. Its form suggests that $(\sqrt{5}+1)/2$ is one root of some quadratic equation. Indeed, it is a root of the equation $x^2 - x - 1 = 0$, again something that does not appear to be of special interest. Greek geometers stumbled upon this number when they found that:

> If, in a line segment AB a point P is chosen such that AP/PB = AB/AP then the ratio AP/PB = $(\sqrt{5}+1)/2$.

This is easy to verify. There is no loss of generality in assuming AB is of unit length, since we need not specify any particular unit of measurement. If, measured in the same unit, AP $= x$, then we have to find P such that

$$\frac{x}{1-x} = \frac{1}{x}\,.$$

It needs only elementary algebra to show that this implies

$$x = \frac{1}{2}(-1 \pm \sqrt{5}).$$

Since x must be positive and less than 1, the solution is $x = (\sqrt{5}-1)/2$. Whence

$$\mathrm{AP/PB} = \mathrm{AB/AP} = \frac{1}{x} = \frac{2}{\sqrt{5}-1} = \frac{2(\sqrt{5}+1)}{(\sqrt{5}-1)(\sqrt{5}+1)} = \frac{1}{2}(\sqrt{5}+1)$$

Things now look more interesting, especially if the choice P that gives this ratio arises naturally.

This number does indeed turn out to have interesting geometric properties that were discovered by the Greeks. In Fig. 24.1 ABCDE is a regular pentagon. A regular pentagon has all 5 sides of equal length. If the 5 diagonals AC, AD, BD, BE and CE are drawn and they intersect at $\mathrm{A'.B', C', D', E'}$ as shown in that figure, then each of these points divides the diagonal on which it lies in the golden ratio.

The Greeks also had a ruler and compass construction to find a point P that divided any given line segment AB in the golden ratio. In Fig. 24.2 the square ABCD is constructed, and the mid-point E of the side AD is determined. The point F is located so that EF = EB. The square AFGP is now constructed. On completion of these standard Euclidean constructions we easily establish, by appropriate application of Pythagoras' theorem, that P divides AB in the golden ratio.

The name *golden ratio* was introduced by artists during the Renaissance. Rectangles with sides in this ratio came to be regarded as an ideal or divine proportion in art, architecture and sculpture. Although Greek architects did not use the name, some rectangles (e.g., windows) with approximate dimensions in this ratio are found in classic Greek buildings.

The ratio is also associated with a famous sequence of numbers called the *Fibonacci numbers.* This sequence commences with the numbers 1, 1, and each subsequent number is the sum of its two predecessors, i.e., the first few terms of the sequence are

$$1, 1, 2, 3, 5, 8, 13, 21, 34, 55, 89, 144, 233, 377, 610, 987, 1597, 2584, \ldots.$$

Writing the nth term u_n, the ratios u_{n+1}/u_n for successive values of n in the above sequence are (rounded to 4 decimal places)

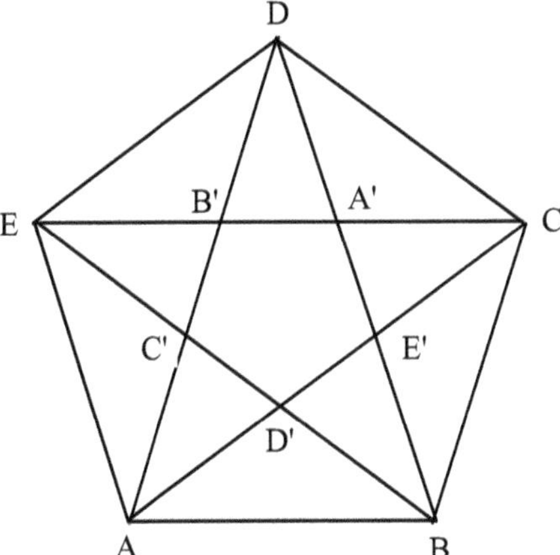

Fig. 24.1. The diagonals of a regular pentagon are divided in the golden ratio at the points A′, B′, C′, D′, E′

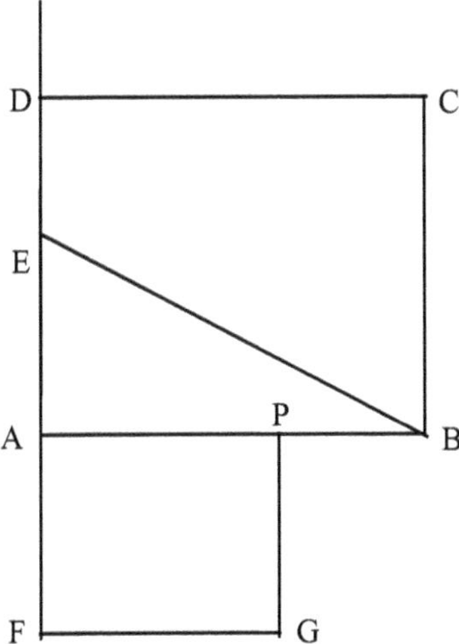

Fig. 24.2. A Euclidean construction of a golden ratio as described in the text

$$1, 2, 1.5, 1.6667, 1.6, 1.625, 1.6154, 1.6190, 1.6176, 1.6182, 1.6180, 1.6181,$$
$$1.6180, 1.6180, 1.6180, 1.6180, 1.6180, \ldots$$

Intuition suggests that the ratio is tending to a limit that is the golden ratio. Here intuition is correct. A rigorous proof requires the solution of the difference equation defining the Fibonacci sequence, but we omit details. The relevant difference equation is obviously

$$u_{n+2} = u_{n+1} + u_n \ ,$$

with initial conditions $u_1 = u_2 = 1$.

Has the Fibonacci sequence a practical interpretation as a model of some real-world situation? Or is it just a mathematical, and perhaps an artistic, curiosity?

Rabbits and their habits

One of the earliest and best known problems giving rise to Fibonacci numbers is the so-called *rabbit problem.* A pair of rabbits of opposite sex is placed in an enclosure. This pair, and every pair of their progeny produce one new pair each month, starting when they are two months old. After one month there is still only one pair of rabbits. After two months there will be two pairs, after 3 months $1 + 2 = 3$ pairs, after 4 months there will be $2 + 3 = 5$ pairs, after 5 months there will be $3 + 5 = 8$ pairs and so on, giving rise to a Fibonacci sequence.

Since this simple breeding model was formulated, examples of Fibonacci sequences have been given in fields as diverse as botany and music. The golden ratio is also linked to a type of spiral known as the logarithmic spiral — a shape associated in nature with certain sea-shells such as that of the Nautilus.

A more detailed account of Fibonacci numbers and the golden ratio, together with many further references is given by Banks [6]

24.3 Monte Carlo Sampling

Experiments to evaluate, at least approximately, a constant such as π have a long history. Once we know the area of a circle is πr^2, a simple experiment yields an approximation for π. We illustrate this using Fig. 24.3.

The figure shows a circle of radius r, and thus of area πr^2, inscribed in a square ABCD, with each side of length $2r$, and hence of area $4r^2$. Thus, the ratio of the area of the circle to that of the square is $(\pi r^2)/(4r^2) = \pi/4$. We have imposed a $15 \times 15 = 225$ grid of smaller squares in ABCD. This enables us to find a crude estimate of π by counting the numbers of small grid squares inside the circle. Suppose this is n, then, since the total number of grid squares in ABCD is 225, this implies that $\pi/4 = n/225$.

A difficulty lies in counting, or estimating, the total number of smaller grid squares in the circle. It is easy to count the number of complete grid squares, but one can only estimate a total for partial squares intersected by the circle.

For example, consider the second row of grid squares in Fig. 24.3. In that row there are 7 complete grid squares, but four squares that are neither completely in, nor completely outwith, the circle. We might estimate that about half of these four squares lie within the circle, making a total of $7 + 2 = 9$ squares within the circle. This is a subjective approximation that could only

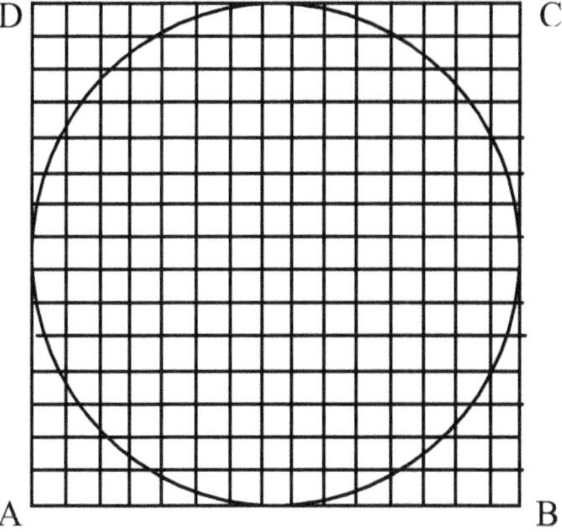

Fig. 24.3. A circle of radius r and area πr^2 inscribed in a square of area $4r^2$

be improved by using a closer grid or alternatively (and equivalently) maintaining the same grid size, but increasing the size of the original square and its inscribed circle. Because of the subjective element in dealing with squares that lie partly within and partly outwith the circle, different observers are likely to get different total counts for the number of grid squares within the circle. Treating partially included squares in the way outlined above, I estimated the total number of squares within the circle to be $n = 175$, implying that $\pi/4 = 175/225 = 7/9$. This yields an estimate $\pi = 28/9 \approx 3.111$, an underestimate of the true value.

Modern computing facilities enable us to get a better estimate of π using the basic principle of a circle inscribed in a square, yet avoiding the rigmarole of drawing the relevant figure, inserting a grid, then counting grid squares within the circle. Once a few lines of program code are written a good approximation to π can be obtained in a second or two. The vital tool is a random number generator, a feature of any statistical software package worthy of the name.

Before we show how this works, we look at a physical experiment to estimate π. This indicates the logic behind the way a random number generator is used in situations like this.

Imagine a large circle inscribed in a square. Suppose we scatter pebbles so that they fall at random within the square. By 'at random' we mean that each pebble is equally likely to fall within any small non-overlapping area, each of the same size, and within the square. Then, with a large number of pebbles, we expect the number falling within the circle, n, say, to be roughly proportional to the area of the circle, and the total number N falling within the square, to be proportional to its area. The ratio of the two areas is $\pi/4$, and n/N is an estimate of this. For large N, this approximation effectively uses the limiting relative frequency concept of probability introduced in Sect. 17.1. Like the counting of grid squares, this experiment requires a lot of physical effort to perform.

A few lines of computer code does the equivalent in less than 1 second. Such computer imitations of real experiments are referred to as *simulations.* More specifically, when they involve random number generators to produce samples with properties equivalent to some real sampling situation, as *Monte Carlo sampling.* The illusion here is to the games of chance played in the casinos of that well-known resort.

This is how it works. Random number generators have a range of uses. A basic one is to produce an independent string of numbers that follow the uniform distribution over any given interval. The uniform distribution was described in Sect. 17.7. As pointed out there, a sample from that distribution has the property that any member of the sample is equally likely to lie in any small sub-interval of length δx that lies within the range (a, b) that specifies the distribution.

Figure 24.4 is a slight modification of Fig. 24.3 where we have drawn a circle of radius 1 with centre at the origin, and which is inscribed in a square with vertices at the points $(1, 1), (-1, 1), (-1, -1)$ and $(1, -1)$. Again, the ratio of the area of the circle to that of the square is $\pi/4$, and since r does not appear in that ratio there is no loss of generality in taking a circle of unit radius.

We now use a random number generator to select N points at random in the square. We may regard each as equivalent to a point where a pebble falls in the random distribution of pebbles. Any point within the square has coordinates (x, y), where both x and y may take values between -1 and 1. We might set $N = 10\ 000$ and use the random number generator to select 10 000 values for x, and a further independent set of 10 000 values for y. We then pair these off, i.e., pair the first x selected with the first y selected, and so on.

Each point with co-ordinates (x, y) is equally likely to lie anywhere within the square. We don't need to plot the points on a graph to find out how many lie within the circle. The computer can work this out. For all points within the circle, the coordinates (x, y) are such that $x^2 + y^2 < 1$. For any point outside the circle $x^2 + y^2 > 1$. In practice, using random number generators it is unlikely that we shall find cases where $x^2 + y^2 = 1$ exactly. If we do, half of these might be regarded and counted as lying within the circle, and half as lying outside. A few lines of code will program a computer to calculate the number of points, n, for which $x^2 + y^2 < 1$.

I set $N = 10\ 000$ to generate that number of points randomly located in the square. I found, or at least the computer found, that for 7882 of these $x^2 + y^2 < 1$. This gives an estimate $7882/10000$ for $\pi/4$, implying $\pi \approx 3.153$. For practical Monte Carlo simulations $N = 10\ 000$ is a relatively small number. In this example setting $N = 100\ 000$ or even $500\ 000$ would be more appropriate, and should give an improvement in the estimate of π with computing time probably still of the order of 1 second.

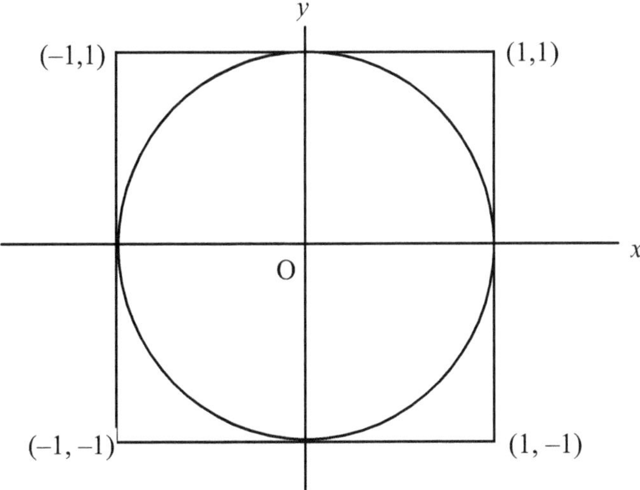

Fig. 24.4. If N points are selected at random in the above square and n of these lie within the circle then for large N, $\pi \approx 4n/N$

The above example illustrates the basic principles of Monte Carlo sampling. In practice, it is applied to more sophisticated problems, where exact analytic solutions may not exist, or may be difficult to obtain.

We pointed out in Sect. 17.7 that the integral

$$F(x) = \frac{1}{\sqrt{2\pi}} \int_{-\infty}^{x} \mathrm{e}^{-u^2/2} \mathrm{d}u$$

could not be obtained analytically. Indeed even the integral, I, over (0, 1) i.e.,

$$I = \frac{1}{\sqrt{2\pi}} \int_{0}^{1} \mathrm{e}^{-u^2/2} \mathrm{d}u$$

cannot be obtained analytically. Numerical evaluation is possible, as we showed in Sect. 15.8. Alternatively, it is easy to calculate the value of the function

$$f(x) = \frac{1}{\sqrt{2\pi}} \mathrm{e}^{-x^2/2} \tag{24.1}$$

for any x. Figure 24.5 is a sketch of $f(x)$ for x over the interval (0, 1). The function decreases from the value 0.3989 when $x = 0$, to 0.2420 when $x = 1$.

We use a similar scheme to that in the previous example. We select at random N points in some larger known area such as a unit square with vertices at (0, 0), (1, 0), (1, 1), (0, 1). We then compute the number, n, of such points that lie between the x-axis and the curve $f(x)$ over the interval (0, 1). Since

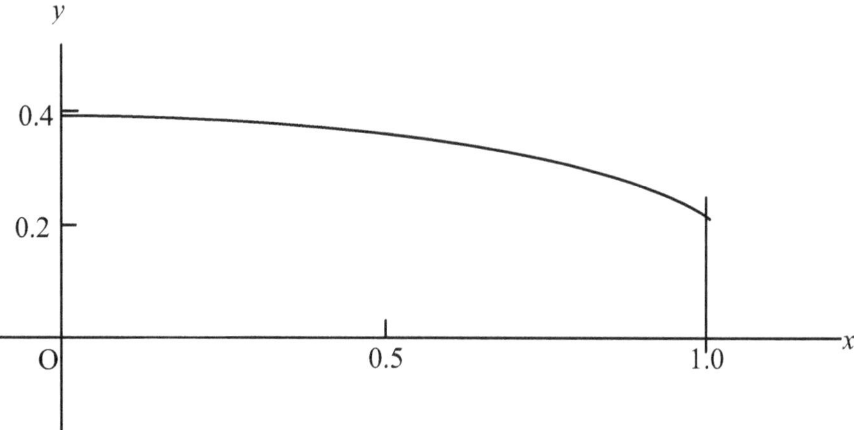

Fig. 24.5. Graph of the function (24.1) over the interval (0,1)

we know that the square has an associated area 1, it follows that n/N gives an estimate of the required integral (area under the curve), i.e., of

$$I = \frac{1}{\sqrt{2\pi}} \int_0^1 \mathrm{e}^{-u^2/2} du \, .$$

To determine n, we need to calculate for each random point (x, y) the difference $y - f(x)$ where

$$f(x) = \frac{1}{\sqrt{2\pi}} \mathrm{e}^{-x^2/2} \, .$$

If this difference is negative, then (x, y) lies beneath the curve. If it is positive, (x, y) is a point in the unit area square above the curve. By similar arguments to those used in the previous example, we see that n/N gives an approximation to the required integral. Only a few lines of computer program are needed to obtain this information for large N. With $N = 50\ 000$ I found $n = 17123$ for this Monte Carlo sampling, giving an estimate 0.3425 for the integral. We pointed out in Sect. 17.7 that this integral is widely tabulated. Tables indicate the true value is 0.3413, a value we also obtained using Simpson's rule in Sect. 15.8. If you have access to a PC and know a little about programming, try to write a program which allows you to choose various values of N and use this to estimate the area for, say, $N = 10\,000$, $20\,000$, $50\,000$, $100\,000$, $200\,000$. You will find, in general, that the approximation improves as N increases. It may not do so uniformly, e.g., you may happen to find for some samples that you get a better approximation for $N = 10\,000$ than you do for $N = 50\,000$. However, if you repeat the experiment, say 10 times, using a different set of samples each time, you should find in general that you get more accurate estimates for larger values of N.

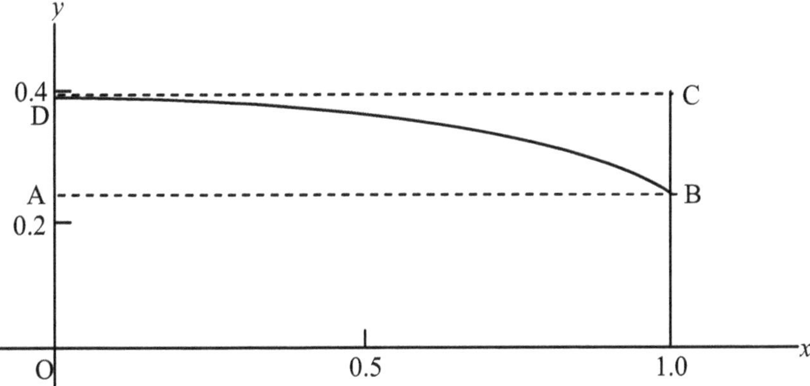

Fig. 24.6. A random choice of N points in the rectangle ABCD may be used to estimate the integral of $f(x)$ over (0,1)

The sampling scheme used above was not very efficient, because the points selected at random were distributed over a unit square. If we can get the same number of points concentrated in a smaller known area, we can increase the accuracy of the estimate. Fig. 24.6 illustrates one way to do this. When $0 < x < 1$ (24.1) only takes values in the interval (0.2420, 0.3989). We know for values of y outside that interval, with certainty, whether or not they lie above or below $f(x)$ irrespective of the value x may take. This is clear from Fig. 24.6 where the dotted lines have the equations $y = 0.2420$ and $y = 0.3989$ respectively. It is not hard to see that the area between the x-axis and the lower of these lines associated with the interval (0, 1) for x is 0.2420. To get the total area under the curve, we only have to add the size of the portion of the area of the rectangle ABCD that lies below $f(x)$. The area of the rectangle ABCD is easily seen to be

$$(0.3989 - 0.2420) \times 1 = 0.1569.$$

Random points within ABCD are obtained by using a uniform distribution random number generator to get x coordinates in the interval (0, 1). We get y coordinates independently by using a uniform random number generator with the interval (0.2420, 0.3989). We then pair each x with the corresponding y, i.e., first x with first y, etc. For each of the N points so generated we then compute $y - f(x)$. As before, the number of negative values gives an estimate of n, the number of points in ABCD that are below $f(x)$. The area under the curve is then estimated by $(n/N) \times$ (area of rectangle ABCD), that is by $n/N \times 0.1569$. To get the total required area, we add the area 0.2430 of the rectangle between the x-axis and AB. I set $N = 50\,000$ for Monte Carlo sampling and estimated the area to be 0.3414, close to the correct value 0.3413.

Monte Carlo sampling has many variants. It is particularly useful for studying the effect of changing conditions in, for example, queuing systems. These are often intractable, or impossible, to study mathematically. For example, if it is anticipated that more people will be making use of a facility such as a bank or post office, one may use Monte Carlo sampling to study the effect on service times and get some indication of whether it is desirable to increase the number of servers, and if so by how many, to avoid undesirable delays to customers. At the same time one will not want to leave servers idle for long periods. The study may be done making different assumptions about changing patterns of arrival and server times. We do not describe such studies in detail, as the mathematical or statistical models involved are often complicated. To increase accuracy we often need to refine sampling procedures appropriately, as we did above for estimating the area under a curve.

24.4 Gambler's Ruin and Random Walks

The classic gambler's ruin problem is about a game between two players A and B. Initially player A has a units and player B has b units of capital, a, b each being positive integers. At each play there is a non-zero probability p that A will pay 1 unit to B (i.e., B wins), so that B now has $b + 1$ units and A has $a - 1$ units. There is a probability $q = 1 - p$ that B will pay 1 unit to A (i.e., A wins), so that A then has $a + 1$ units and B has $b - 1$ units. The next play starts with this new allocation of capital and the same probabilities p, q associated with each player winning. It continues in this way until either A has acquired all $a + b$ units of capital, in which case B has zero capital (i.e., B is ruined), or B has acquired all $a + b$ units, in which case A is ruined.

A gambler's walk

The gambler's ruin problem is best studied using what statisticians call a *simple random walk with two absorbing barriers.* We mimic the game by considering the behaviour of a particle initially at the origin, or zero point, on a horizontal straight line. At the first play the particle moves either one step to the right (i.e., to $+1$) with probability p, or one step to the left (i.e. to -1) with probability $q = 1 - p$. At the next play there is a move one step left, or right, of the new position with the same probabilities p, q. This process is repeated until the particle first reaches either the point a or the point $-b$. For the gamblers' ruin problem, in the former case A is ruined, and in the latter case B is ruined. In either case the game then ceases and the walk is said to be absorbed at a or at $-b$. The points $a, -b$ are the *absorbing barriers.*

We are interested in the respective probabilities of absorption at a or of absorption at $-b$. At this stage it is clear that these probabilities are mutually exclusive, but not that they are in general exhaustive, for it seems possible, especially if a or b are both large, that the game might continue indefinitely

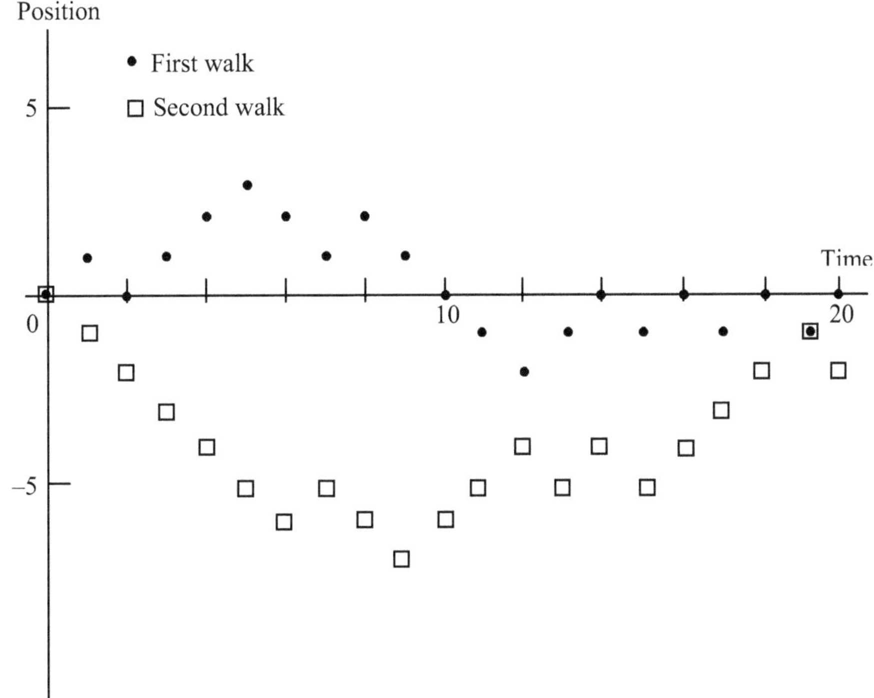

Fig. 24.7. Two realisations of first 20 steps of a simple random walk when $p = \frac{1}{2}$

without reaching either a or b. Although we do not establish this, in fact one of the barriers will always be reached.

The position of the particle at any step defines the *state of the system*. Mathematically, a simple random walk can be expressed in terms of a difference equation containing a term which is a random variable X. We denote by u_t the position of the particle after t plays. Here t takes positive integral values between 0 and n, the total number of games that have been played, and u_t takes integer values between $-n$ and n. At each play there is a step of $+1$ or -1. What that step is to be is determined by a random variable X_{t+1} that for all t is identically distributed, and independently for each t takes the value $+1$ with probability p, or -1 with probability $q = 1 - p$.

The difference equation that determines the simple random walk may be written

$$u_{t+1} = u_t + X_{t+1} \; ,$$

with the initial condition $u_0 = 0$. Thus $u_1 = X_1$, $u_2 = u_1 + X_2 = X_1 + X_2$, and continuing these step-by-step solutions we see that after n plays the position of the particle is

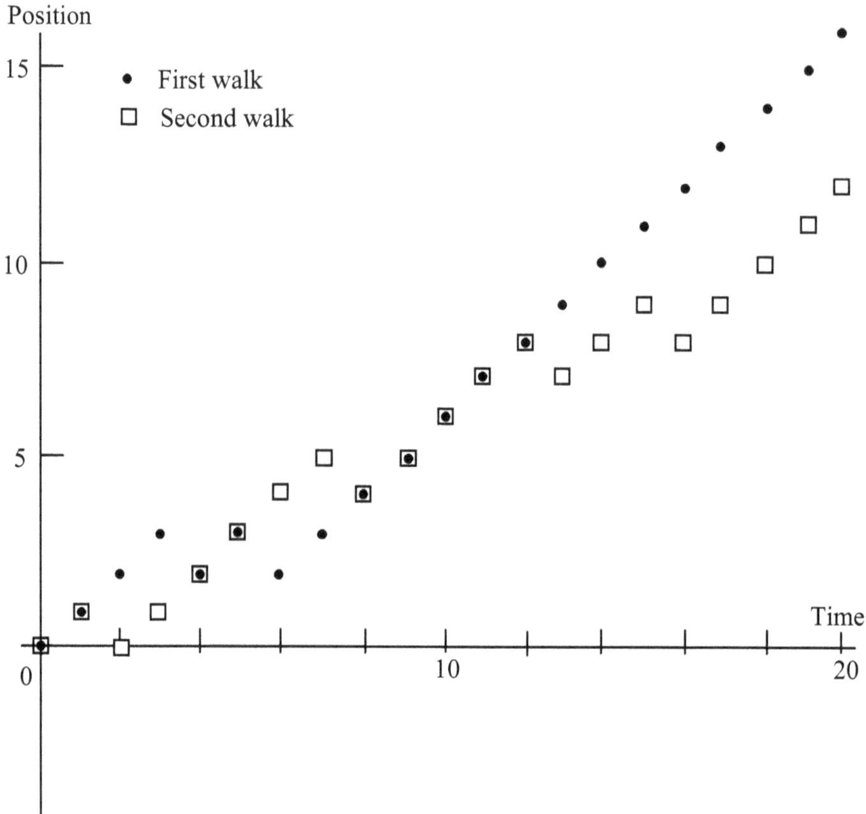

Fig. 24.8. Two realisations of first 20 steps of a simple random walk when $p = 0.8$

$$u_n = X_1 + X_2 + X_3 + \ldots + X_{n-1} + X_n \, .$$

The precise value of u_n depends upon the numbers of X_i that take each of the values $+1$ and -1. The total number of positive X_i (each having value 1) has a binomial distribution with parameters n and p. If there are r positive X_i, the remaining $n - r$ of the X_i must all take the value -1.

Fig. 24.7 shows the position of a particle initially at zero that performs a simple random walk with $p = 1/2$, for the first 20 steps for each of two realizations. The positions in each realization are markedly different, but each realization tends to centre not too far from zero despite, in the second case, being always below zero for these first 20 steps.

Fig. 24.8 shows 20 steps for each of two realizations when $p = 0.8$. Not surprisingly, there is evidence in each case of a drift towards increasing positive values.

To obtain a general formula for the probability of absorption at each barrier the simplest procedure is to make use of what are called *probability generating functions*. We do not use these here, but a brief introduction to generating functions is given in Sect. 25.2. Using these, it is a reasonably straightforward, but not trivial, mathematical exercise to show that when $p \neq q$ the probabilities of absorption at $-b$ and a for a simple walk starting in state 0 are respectively

$$\Pr(\text{absorption at } -b) = q^b \frac{p^a - q^a}{p^{a+b} - q^{a+b}}, \qquad (24.2)$$

and

$$\Pr(\text{absorption at } a) = q^a \frac{p^b - q^b}{p^{a+b} - q^{a+b}}.$$

It is a relatively straightforward exercise in elementary algebra to confirm that these probabilities sum to 1, confirming that absorption is certain.

A minor problem is that these probabilities are not defined if $p = q = 1/2$. This is unfortunate since this case corresponds to the simplest form of a gambling game, repeated tosses of a fair coin. However, it is not difficult to show, using a little ingenuity, that the limiting values of these probabilities as p, q tend to $1/2$ are respectively $a/(a+b)$ and $b/(a+b)$. We leave this as an exercise, with the hint that the result follows reasonably easily by setting $p = 1/2 + \delta$ and taking the limit as $\delta \to 0$ in the relevant expansions given above for the probabilities with given p and q.

The importance of the concept of random walks lies in the possible generalizations of the simple random walk. Some of the possibilities may be illustrated in a gambling context. Taking this approach, here are a few.

Playing an infinitely rich adversary. If player A has infinite capital and player B only finite capital b, the simple random walk becomes one with a single absorbing barrier at $-b$. In this case it can be shown that absorption at $-b$ is certain if $q \geq p$, but if $q < p$ then the probability of absorption at $-b$ is $(q/p)^b$. These results may be deduced from (24.2) by considering the limit as $a \to \infty$ when b is fixed.

Reflecting barriers. These are exemplified by a situation where one player, say A, shows generosity by advancing 1 unit of capital to B to allow the walk to continue each time the barrier at $-b$, now called a *reflecting barrier*, is reached. This situation might arise if a generous opponent was keen to allow the game to continue perhaps just for its entertainment value. A more mercenary benefactor may find it worthwhile if he or she charged a high interest rate on the 'overdraft' generated by the advance. If player B is not prepared to reciprocate an opponent's generosity if the walk arrives at barrier a, this barrier remains an absorbing barrier. The walk then has one absorbing and one reflecting barrier.

Elastic barriers. If we modify the situation in the above example, and player A is less generous and only advances a unit of capital with probability p^* if the barrier at $-b$ is reached, otherwise the game ends, then the barrier at $-b$ is called an *elastic barrier.*

Mixed barriers. It is possible to have a mixture of absorbing, reflecting or elastic barriers with any simple random walk.

Modification to steps. A straightforward modification to the simple random walk is to allow an additional zero step. Possible steps then are $X = 1$ with probability p, $X = -1$ with probability q and $X = 0$ with probability $1-p-q$. Here $p+q < 1$. In a gambling context this is similar to the one we considered for the simple random walk, but it allows for a third outcome in which neither player wins. The analysis of possible outcomes differs little from that for a simple random walk, and indeed some writers generalize the simple random walk to include this possibility of 'drawn' games.

More elaborate modifications of steps include the situation where not all steps are unit steps. For example, we may consider a walk where possible steps are $X = +5$ with probability p, $X = -5$ with probability q and $X = -1$ with probability $1 - p - q$.

We may also generalize to situations where steps are not integer steps, but at each step its size is specified by a random variable Y which may be discrete or continuous. The steps at each stage are independently and identically distributed.

In studies of various physical systems, such as modelling the diffusion of gases, the model may be one of a random walk for individual molecules that involve many very small steps. We are then often interested in the limiting behaviour of the system as a whole as the steps become smaller and smaller for each of a large number of molecules.

Extension of random walks to more than one dimension has been studied in some detail. The simplest example is the two dimensional analogue of the simple random walk, using a square grid. Here a particle may move either one step east, one step west, one step north or one step south, with fixed probabilities p_1, p_2, p_3, p_4, such that $p_1 + p_2 + p_3 + p_4 = 1$. Many of the modifications that are possible in one dimension may be extended to this case. Barriers may now well take the form of lines confining the walk to certain geometric configurations in two dimensional space. All these ideas may be extended to walks in three or higher dimensions.

Computer simulations of random walks, especially those of a complicated nature, often give a valuable indication of their likely behaviour. If you have access to a computer program (and many statistical software packages have this facility) that generates random samples of a variable that takes the value $-1, 1$, you might like to amuse yourself by writing a macro to do 1000 simulations of a simple random walk starting at the origin with $p = q = 1/2$ and absorbing barriers at -1 and 2. According to the result we got above the probability of absorption at -1 is $2/3$. In your simulation you should find absorption occurs at -1 in approximately two thirds of your simulations.

24.5 Flow Diagrams for Computer Programs

Most of us have at some time cursed, or branded as stupid, a computer for its instant refusal to even try and deliver an e-mail addressed to *joebloggs@aol,com* because we think it should have been obvious that we meant *joebloggs@aol.com*.

Computers only obey instructions which are precise and unambiguous in the context of their operating rules. These rules vary from system to system, but they are consistent in any one system. For example, some internet service providers allow you to use a full stop within your name in the email address, e.g., writing it *joe.bloggs*, but others do not allow such a full stop. There may be some flexibility. In many systems it does not matter whether an email address is written in upper case (capitals) or lower case, but this is not invariably so.

Catering for every emergency

When it comes to programs to carry out mathematical calculations, the program must be designed with logical procedures to cover all eventualities. In the early days of computing, many programs came to a grinding halt with simple operations like division when suddenly confronted with a requirement to divide by zero. This was usually because the programmer had failed to appreciate that that situation might arise in a real problem, and so had made no provision for the program to give a message along the lines of 'division by zero not defined; programme terminated'. If one was lucky some of the earlier machines would give a message such as 'error 5729' in such circumstances. After much rummaging around in the manufacturer's documentation. one might find that this implied what was called 'floating point overflow' or sometimes a more enigmatic 'impossible operation attempted'. Even with more recent operating systems, most of us are familiar with unhelpful messages along the lines 'this program has attempted to perform an illegal operation and will now close'.

There is a wide range of commercial software for mathematical and statistical and other logical calculations from the simple to the sophisticated. Few users have to do much programming, although many packages have a facility for writing what are called 'macros'. In essence, these are little bits of program that can be slotted into the main software package to perform special tasks.

To ensure one proceeds logically in program or macro writing a useful device is to set up a flow diagram. Fig. 24.9 is a flow diagram that illustrates the logical structure a programmer might follow to write a macro to calculate the mean of n numbers $x_i, i = 1, 2, \ldots, n$. The advantage of flow diagrams for sorting out the logic of a problem is that they are not tied to any one programming language.

We gave the formula for calculating the arithmetic mean in (14.1). A flow diagram indicates a possible set of steps that a computer program might use to perform this calculation given values for n and the x_i. The diagram has many similarities to some of the network diagrams developed earlier in this book. Nearly all flow diagrams feature at least three different kinds of nodes. It is convenient to distinguish between these types by giving them different shapes. A common, but not universal, convention is to use, *ellipses*, *rectangles* and *diamonds*. The main node types and shapes are

- *Start and finish nodes* - circles or ellipses,
- *Action nodes* - rectangles,
- *Decision nodes* - diamonds.

Paths through the diagram are indicated by directed lines, often called routes, the direction being indicated by an arrow. There is at least one route into any node other than the start node. The start node and all action nodes have one route out. A decision node must have two routes out. A finish node must have one, or more, inward routes, but no outward routes.

Fig. 24.9 tells us that to calculate a mean, we must at the start set two variables to zero. We have called these COUNT and SUM to indicate their role, but their precise names in a program will depend upon the conventions of the programming language or software being used.

Next, the node labelled READ N indicates that the program must be made aware of the number of datum values that are to follow. This may be entered by keyboard, or most modern software will allow entry from another program or a spreadsheet, or from a CD or other storage device.

READ X is an instruction to enter the first datum value. After it has done this, the program is instructed to increase the count variable by 1, and to change the sum variable (initially zero) by adding the datum value that was read in. A decision is now required. The decision box labelled COUNT = N? means the computer is told to compare the current value of the count (at this stage having the value 1) with N, the number of observations. If, as will usually be the case in practice, N is greater than 1 the path marked NO is followed. This leads back to the instruction READ X. At this stage the second datum value is entered, the count is increased from 1 to 2, and the new X is added to the SUM. The decision node COUNT = N? is again reached. If N > 2 the NO path is again followed.

This procedure forms what is called a LOOP. The loop is enclosed in a dotted rectangle in Fig. 24.9. It is repeated until a stage where the decision node COUNT = N? gives a positive (YES) response. Following the YES response, the flow diagram indicates the obvious procedure given by the remaining action nodes. DIVIDE SUM BY N gives the value of the mean, and the program should then PRINT ANSWER. Here the word PRINT should not be taken literally. It may include such actions as displaying the answer on the screen, or saving it to a named file. The program then terminates as indicated by the finish node END.

If we write a program and follow the rules of the programming language in which it is written, we then have a program that should always find the mean of a specified data set of n observations.

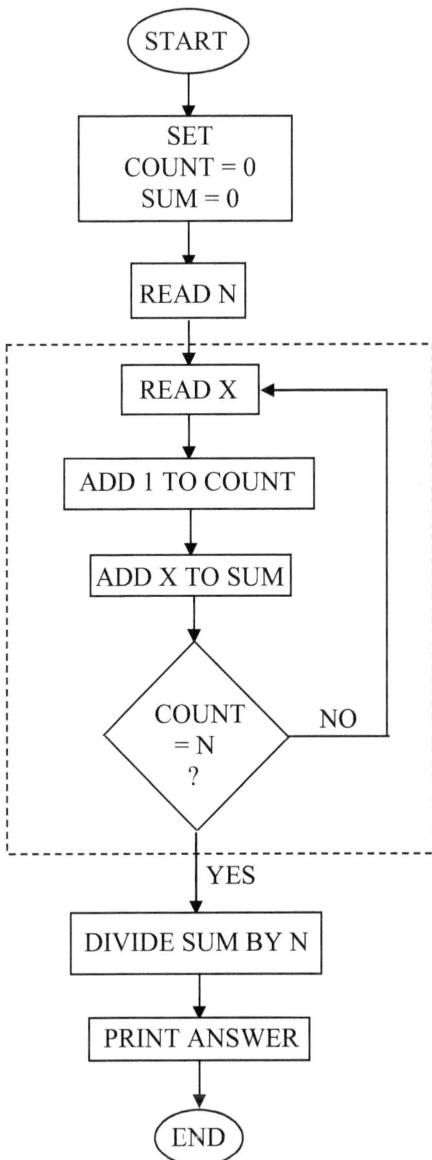

Fig. 24.9. A flow diagram for a program to calculate the mean of n numbers

An experienced computer programmer would be quick to tell us we have not written a good program. The main objection he or she would raise is that we have made no allowance for certain common data errors. Mistakes in entering data are common. We might forget to enter n. The program will then take the first datum value x_1 as the number of observations. Another common mistake is to enter n correctly, but to miss one of the data values (or perhaps to include some value twice). A good computer program should be able to detect such errors. For example, if we have omitted n and as a consequence the first data value is then mistakenly assumed by the program to be the number of datum, we need a trap built into the flow diagram to detect this. For instance, if that first datum value is not a positive integer there is clearly a mistake. This could be detected by adding extra steps between READ N and READ X in Fig. 24.9. For example, the path from READ N might lead to a decision node 'N A POSITIVE INTEGER?'. If the answer is 'YES' the path should continue to the READ X node. The 'NO' path should lead to an action node saying something like 'PRINT:IMPERMISSIBLE VALUE OF N', with an out path leading to the END node.

If the real N has been omitted, and the first datum value is a positive integer this will not be detected by the above addition to the program. In that case, however, unless there is some peculiar fluke where the first data value is equal to one less than the correct N, the program will appear to have either too many, or too few, data values. If you are interested, think about what additional logical steps may be added to the flow diagram to cover this situation, or that where N is given correctly, but too few or too many data follow in the input.

The action to be taken at decision nodes are basically of the form of testing whether some assertion e.g., COUNT = N is true or false. In computer programming languages the decisions are often expressed in forms such as

$$\text{IF COUNT} = \text{N THEN GO TO 23 ELSE GO TO END} \,,$$

but terminology varies between programming languages.

In more complicated situations decision nodes are often nested. For example in the case of a decision node COUNT = N?, if the answer is NO it may be important to know if 'COUNT > N?'. This may then lead to further decisions being required depending on whether the answer is 'YES' or 'NO'. For instance, if the answer is 'YES', one might want to know if the count is even, requiring a further decision node 'COUNT EVEN?'.

In such situations the logic of the problem can often be expressed using the notion of *truth functions*, *truth tables* and *truth values*, all of which are important in a branch of logic known as *propositional calculus*. The nature of truth functions, tables and values is described briefly in Sect. 25.2.

24.6 Dynamic Systems

In painting a broad canvas of what mathematics is about and what it does we have paid less attention than many writers to applications in physics. The consequences of Newton's laws of motion have only been touched upon in passing. Relativity, quantum theory and statistical mechanics have only been mentioned. Relativity and quantum theory owe their importance in part, at least, to the fact that Newtonian mechanics breaks down when we deal with systems either at the atomic level, or with some large scale (e.g., cosmic) or high energy systems. Between these extremes classical mechanics based on Newton's laws works well. The law of conservation of mass, for example, holds as an adequate approximation until we consider particles moving at or near the velocity of light. Here the theory of relativity tells us that mass may be converted to energy.

The theory of relativity is crucial for the explanation of mass-energy relations when we consider systems involving velocities approaching that of light. At the other extreme quantum theory is needed to explain dynamic behaviour at the atomic level. A feature of quantum theory is the assumption that at the atomic level certain basic physical quantities, in particular energy or momentum, are quantized — another way of saying they can take only certain discrete values. Quantum theory also gave rise to the notion of wave-particle duality, a recognition that very small particles can act or behave like waves, or vice versa.

In relativity there is a well-known and basically simple relationship between energy, E and mass m, namely $E = mc^2$, where c is the velocity of light. Any in-depth study of either quantum theory or relativity soon leads to highly specialized mathematics.

An interesting development in the twentieth century was that of the subject *statistical mechanics.* This is relevant to a study of large scale systems where there are millions of particles that behave more or less in a random manner, but where we are more interested in the behaviour of the system as a whole, than in that of individual particles. Gases are typical of such systems. In these, despite the fact that individual molecules may be moving around at random, the effect of this when averaged over many millions of molecules is predictable. This predictable behaviour is described in terms of familiar gas laws such as *Boyles's law* or *Charles's law.* These laws are applicable only to gases as a whole, and not to the individual molecules of which a gas is composed.

In all forms of mechanics — classical, quantum, statistical or relativistic — differential or difference equations are nearly always basic to any mathematical model set up to describe the system. Many of these are (mathematically) formidable.

In Sect. 12.8, when describing the motion of a pendulum that swung through a very small arc, we dealt briefly with the equation for *simple harmonic motion.* There, simple harmonic motion was used only as an approximation to indicate the motion of the pendulum.

The differential equation associated with simple harmonic motion, and its physical implication, is perhaps the simplest and most widely studied topic in Newtonian dynamics, apart from the constant acceleration equation (12.12).

We look a little more closely at simple harmonic motion, not just as an approximation to describe the motion of a certain pendulum, but rather to illustrate how we set up a mathematical model for a simple dynamic system and study some of the implications of that model.

Both the equation (12.12), namely

$$\frac{\mathrm{d}^2y}{\mathrm{d}t^2} = a \ ,$$

and (12.17)

$$\frac{\mathrm{d}^2y}{\mathrm{d}t^2} = -\omega^2 y \ ,$$

are based on Newton's second law of motion. They are relevant to a particle moving in a straight line. If you are not familiar with the second law, one of its implications is that the acceleration of a particle is directly proportional to the applied force. Equation (12.12) is appropriate if it is assumed that the acceleration, or equivalently the applied force, is constant. In equation (12.17), y represents the displacement from some initial or equilibrium position $y = 0$ and $-\omega^2$ is, for real ω, a necessarily negative constant, which implies there is a restoring force acting towards the equilibrium position. Further, This force is proportional to the particle's displacement y, from this equilibrium position.

We stated in Sect. 12.8 that the equation (12.17) had a solution

$$y = d\cos(\omega t + a) \ , \tag{24.3}$$

where d and a are constants. We have not shown how to solve (12.17), but it is easily verified by differentiating y with respect to t twice, and using the rules given in Sect. 11.8 that (24.3) satisfies (12.17); i.e., on differentiating

$$\frac{\mathrm{d}y}{\mathrm{d}t} = -\omega d\sin(\omega t + a) \ ,$$

and differentiating again

$$\frac{\mathrm{d}^2y}{\mathrm{d}t^2} = -\omega^2 d\cos(\omega t + a) = -\omega^2 y \ .$$

We see from (24.3) that d is the maximum displacement (called the amplitude), and that at time $t = 0$ the displacement is $d\cos a$. In physical terminology, ω is called the *angular frequency* and $\omega t + a$ is the phase. In particular, a is the initial phase, i.e., the phase at $t = 0$. The motion repeats itself at time intervals $2\pi/\omega$, since substitution of this value in (24.3) gives

$$y = d\cos(2\pi + a) = d\cos a \ .$$

For this reason $2\pi/\omega$ is called the period of the motion. For a given d and a, (24.3) is said to describe a pure sinusoidal displacement in time. This is

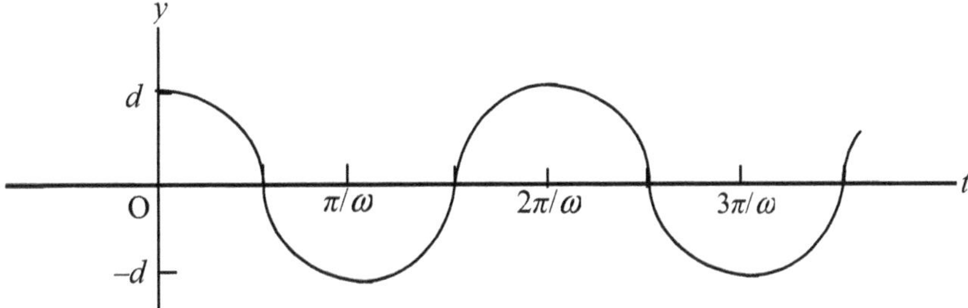

Fig. 24.10. Part of sinusoidal curve of displacement y at time t illustrating simple harmonic motion with amplitude d and period $2\pi/\omega$ where ω is the angular frequency. The initial phase angle a is zero

associated with a specific amplitude and frequency. Fig. 24.10 is a plot of y against t that clarifies the meaning of the terms described above.

Key characteristics of simple harmonic motion are that it is a repetitive back and forth motion about a central or equilibrium position. The maximum displacements d on either side of this central position are equal in magnitude, and the period, or time interval between each complete cycle or vibration, is always the same. The force that governs the motion is, in accordance with equation (12.17), always directed towards the equilibrium position, and is directly proportional to the displacement from it.

Under idealized assumptions of no energy loss, such as that due to air resistance or friction, many physical systems exhibit simple harmonic motion. These include a vibrating mass attached to a completely elastic spring, the vibrating particles in a medium such as air that carry a sound wave of constant frequency, and the electrons in a wire carrying an alternating current.

Harmony or discord?

Musical sounds in their purest form are simple harmonic motions. More often they are a combination of motions of different amplitudes and frequencies. The effect may or may not be harmonious. If the equations for two simple harmonic motions are of the form

$$y_1 = d_1 \cos(\omega t + a_1)$$

and

$$y_2 = d_2 \cos(\omega t + a_2)$$

they are said to be out of phase unless $a_1 = a_2$.

24.7 The Longitude Problem

We have given several examples in this book where mathematical theories or concepts have been developed before it was possible to apply them in practice, except perhaps in very restricted cases. Two such cases were that of matrix inversion, and that of getting good approximations to integrals by Monte Carlo sampling. Widespread use of these techniques only became possible with the advent of electronic computers.

Global positioning systems (GPS) which will, at the press of a button tell us our present latitude, longitude and altitude accurate to within a few metres at virtually any point on earth are commonplace. So it is perhaps surprising that a major problem less than 300 years ago was accurate determination of the longitude of a ship at sea. The mathematics needed to do this was simple and had been long understood. The problem was that the technology that would enable sailors to get one piece of information needed to apply that mathematics, was not available.

It comes as something of a surprise to those who study the history of navigation to learn that while the accurate determination of latitude based on astronomical observations had been developed long before, by the early 1700s there was still no practical way of determining longitude while at sea. It was realised, in theory, that it was possible to do so by means of astronomical observations. That approach was hampered first because the vast amount of astronomical information needed as a data base for determining longitude had not yet been collected. Second because, even as and when the data became available, laborious calculations would be needed to use it in conjunction with, not-easy-to-make, observations. In contrast to an astronomical approach, a simple mathematical formula existed for measuring longitude.

This used the relationship between longitude and time. Since the earth rotates through 360° every 24 hours, it follows that it rotates through 15° each hour, or 1° every 4 minutes. If the true local time at any point on the earth's surface is defined by designating the time the sun reaches its highest point in the sky as midday, or 12 noon, then an accurate fixed-position clock set at noon on one day will, after a lapse of 24 hours, again indicate the time as 12 noon.

If it is 12 noon exact local time, as measured in the way just described, at a port from which a ship sails, then at a point 15 degrees longitude east, the time will be 1.00 p.m. Similarly, at a point, say, 45 degrees of longitude west, it will be only 9.00 a.m., and so on.

Using the basic relationship between longitude and time — i.e., a one hour difference between true local times at each of two points corresponds to a 15° longitude difference — a ship's captain can determine how far east or west he has sailed from the point of departure. To do so he must know the true local time both at the point of departure, and at the ship's current position. The difference between the two is used to work out how far east or west the ship has sailed from its home port. One only has to apply the rule that each hour of time difference represents a longitude difference of 15 degrees. Equivalently, each minute represents a longitude difference of one-quarter degree. Remembering that, at or near the equator, one degree of longitude is approximately 70 miles, then a one minute error in determining the time difference, introduces an error of some 18 miles in determining the east-west component of one's current position. Such an error has potentially disastrous consequences for a ship that is navigating a narrow channel in fog, or one passing close to a reef or a dangerous rock. Even though the difference in distance between degrees of longitude decreases as one moves away from the equator, any errors in longitude determination are still of practical importance.

In the early eighteenth century there was little difficulty in determining true local time accurately by astronomical observations, the simplest being that the sun was at its highest point in the sky at noon local time (due south in the northern hemisphere and due north in the southern hemisphere). On land accurate clocks existed to record local time. The simple solution to the latitude problem might appear to be to carry such a clock on board ship, this being set to true local time at the port of departure. Then one only had to take the necessary observations (or to have access to a clock) that indicated true local time at the ship's current position to determine current longitude. The snag was that the pendulum clocks of that era that recorded time on land were virtually useless when taken to sea. Not only was the swing of the pendulum seriously disturbed by the ship's motion, but changes in temperature, pressure and humidity also affected the accuracy.

There were additional complications such as that caused by the clock stopping temporarily during the rewinding process, and needing resetting thereafter. Thus, it was almost impossible to keep an accurate record of home port true local time while at sea for even a few hours, let alone for days. Remember, that in the 1700s there were no 'instant' communication methods such as radio for determining departure-port time.

In the early eighteenth century the world's leading trading nations were becoming increasingly dependent on merchant shipping. All realized that if ships were able to obtain accurate estimates of longitude this would stem the often huge commercial losses (not to mention loss of human lives) associated with shipwrecks. Many of these were solely attributable to inaccurate estimates, often little more than guesstimates, of longitude.

In 1714 the British parliament passed the Longitude Act. This set up the *Board of Longitude* to adjudicate the award of prizes for solutions to the longitude problem. In today's monies, the prizes offered would amount to millions of pounds or dollars, being in the sterling currency of 1714

- £20 000 for a method to determine longitude to an accuracy of half a degree;
- £15 000 for a method to determine longitude to an accuracy of two-thirds of a degree;

- £10 000 for a method to determine longitude with an error of no more than one degree.

We omit the story of the controversies and counter-claims and intrigues generated during the eighteenth century by the Longitude Act, and the machinations of the Longitude Board. A nontechnical account of these is given by Sobel [42]. That account is recommended reading for anyone unfamiliar with the story of the longitude problem, and the attempts to solve it.

It suffices here to reiterate that this was a problem where there was both a simple mathematical solution based on time differences, and a more complicated mathematical solution based on collection of, and on calculations using, astronomical data. Both were in 1714 impossible solutions in practice. The first because the technology for producing clocks that would be accurate at sea over extended periods did not exist. The second because some of the basic astronomical data was not available, and some of what was available was not of the required accuracy.

The Longitude Act, with its tempting prizes, stimulated work based on the time and on the astronomical approaches to solving the basic problem. There were also many spin-offs. These included the opening of the Royal Observatory at Greenwich, primarily to explore and develop the astronomical approach and to collect the data needed for its use. There were also technological developments coming from the work of an English clockmaker, John Harrison(1693–1776). He perfected the chronometer as an accurate seagoing clock.

The outstanding work at Greenwich led not only to a worldwide acceptance of the designation of the meridian passing through that observatory as the zero meridian of longitude, but to the acceptance of Greenwich time as an international time reference.

A spin off from Harrison's chronometer was his invention of the bimetallic strip to counteract the effect of temperature changes. Without it we would not have the modern thermostat. He also developed a number of components that made possible the move from pendulum clocks to the much more compact spring-based watches. The latter have only recently given away to battery powered or similar timekeepers.

In this short account, I have used the term true local time to refer to the true time relevant to longitude determinations at any given position. This must not be confused with local time associated with *zones of longitude* such as Greenwich mean time, Central European time, Australian Eastern standard time, and so on.

For example, Great Britain and Eire observe Greenwich mean time (with a one-hour adjustment in the summer). This is the true local time based on the Greenwich or zero meridian, but it applies to a range of longitudes covering the United Kingdom and Ireland. Observing such a standard time over a range of longitudes, rather than using the precise true local time for each longitude within a zone, avoids chaos in, for example, railway timetables. In practice the world zones are, with a few exceptions, so arranged that there are one-hour time differences between neighbouring zones. The standard times within each zone are locally those at longitudes that pass through that zone, and differ from that of Greenwich by integer multiples of 15°.

The longitude problem is a good example of a situation where simple mathematics could not be applied until the technology to use it (in this case the chronometer) became a reality. Today the chronometer is virtually superfluous, thanks mainly to other technical developments that were not thought of originally as potential navigational aids. Many of these were themselves only made possible by the application of, usually quite complicated, mathematics such as that associated with spatial dynamics, electromagnetism and wave motion. These developments include radio, and later the space technology that gave us satellites, and the satellite transmitters that made possible global positioning systems.

25

Thumbing Through the Dictionaries

25.1 Concepts Worth a Mention

We have not mentioned, or made only passing references to many important mathematical concepts, including some that most mathematicians meet at some stage in their careers. The ones each of us meets and when we meet them depends on our fields of interest, on whether one is a professional mathematician or concerned mainly with applications in another discipline. To pick up some themes that have fallen by the wayside I thumbed through several well-known mathematics dictionaries and selected topics that I felt deserve a mention.

I have confined the selection to terms, or topics, that can be explained, at least in a elementary way, using concepts already met in this book. This means excluding references to some items that feature prominently in certain undergraduate mathematics courses. Attempts to describe these outside the wider context in which they are important become at best formalistic, and sometimes virtually meaningless without an understanding of other terms specific to the context where they are used. It is more appropiate to learn about these topics either in coursework or from relevant textbooks. A few examples of the many specialist terms that have been excluded because they fall into this category are *bijection, contour integral, eigenvalue, elliptic integral, Galois field, homomorphism, lattices, Lebesgue integral, manifold, Markov processes, sufficient statistics.*

Rather than giving dictionary-style definitions of the terms that are included, I only indicate here what they are about and why and where they are relevant. Topics are in alphabetical order. An asterisk before a term introduced in an explanation indicates a separate entry under that headword in this chapter.

25.2 Some Important Concepts

Bessel functions. We have pointed out that it is almost the exception for differential equations arising in real world situations to have solutions that are simple analytic functions. Many problems in physics and engineering lead to differential equations of the form:

$$x^2\frac{\mathrm{d}^2y}{\mathrm{d}x^2}+x\frac{\mathrm{d}y}{\mathrm{d}x}+(x^2-n^2)y=0.$$

Here n is a constant. The equation is called a Bessel equation of order n. For integral n, one solution is a Bessel function of the *first kind* of order n. This takes the form of an infinite series

$$J_n(x)=\sum_{r=0}^{\infty}\frac{(-1)^r}{r!(n+r)!}\left(\frac{x}{2}\right)^{n+2r}.$$

There are also Bessel functions of the *second kind.* After simple functions, Bessel functions are among the most widely occurring functions in physics, engineering and astronomy. They occur, for example, in studies of electromagnetism, heat conduction, vibrating membranes and communication systems based on frequency modulation.

Beta function. A function denoted by $\mathrm{B}(m,n)$ where

$$\mathrm{B}(m,n)=\int_0^1 x^{m-1}(1-x)^{n-1}\mathrm{d}x.$$

and which is related to the *gamma function, $\Gamma(m)$, since

$$\mathrm{B}(m,n)=\frac{\Gamma(m)\,\Gamma(n)}{\Gamma(m+n)}.$$

The beta function plays an important role in statistics in association with a versatile distribution of random variables called the *beta distribution.* For integral m, n the beta function is expressible in terms of factorials.

Canonical form. A normal, or standard, form of an equation, matrix or other mathematical expression often obtained by transformation. It is usually, in some sense, a simplest form. The canonical equation to a circle of radius r expressed in Cartesian coordinates is

$$x^2+y^2=r^2.$$

The equation of any circle of radius r is reduced to this form by transferring the origin to the centre of the circle. The canonical form of the equation to a sphere of radius r is

$$x^2 + y^2 + z^2 = r^2.$$

A square symmetric matrix can be reduced to the canonical form of an equivalent diagonal matrix (one where the only nonzero elements lie on the diagonal) by applying what are called *similarity transformations.*

Catastrophe theory. A theory that models the dynamic behaviour of systems where slow growth is accompanied by sudden changes in form. These sudden changes or 'catastrophes' represent changes in the topology of a system. An example often cited where the theory is relevant is that of crashes in the stock market, were a small change in economic circumstances may lead to a sudden dramatic fall in share prices. Catastrophe theory was largely developed by René Frédéric Thom (1923–2002).

Catenary. A plane curve that is important because it is the shape that a uniform flexible chain assumes if hung between two points. The equation may be written

$$y = \frac{1}{2}c\left(\mathrm{e}^{x/c} + \mathrm{e}^{-x/c}\right),$$

or in terms of *hyperbolic functions, as

$$y = c\cosh\left(\frac{x}{c}\right).$$

It is symmetric about the y-axis and c is the height of the lowest point on the curve (i.e., the intercept on the y-axis).

Central limit theorem. A theorem in statistics that states that, under very general conditions, the distribution of the mean of n random variables tends to a normal distribution as $n \to \infty$. In practice convergence towards normality is often rapid even for small n. This is the key justification for the widespread use of assumptions of normality in many statistical inference methods, where inferences are based on the mean of a sample of n observations $x_1, x_2, \ldots, x_n$ from a distribution that may itself be non-normal.

Change of variable in integration. An important tool in integration. The method is also known as integration by substitution. Details and many examples are given in any book on the calculus.

An example that illustrates the basic idea is that where we require

$$I = \int_0^{\pi/2} \sin^3 x \cos x \mathrm{d}x .$$

If we change the variable to $t = \sin x$ then $\mathrm{d}t/\mathrm{d}x = \cos x$. It is intuitively reasonable, and indeed can be established formally, though we do not do so here, that I can be expressed in terms of t in the form

$$I = \int_0^1 t^3 \mathrm{d}t \; ,$$

giving $I = 1/4$, since the integral takes the form $t^4/4$ and integration is over the interval (0, 1). This is the appropriate interval for t since, if $t = \sin x$ it follows that when $x = 0$ then $t = 0$, and when $x = \pi/2$ then $t = 1$.

In the above simple example the result can in fact be obtained more directly if one recognises that $\sin^3 x \cos x$ is the derivative of $(\sin^4 x)/4$.

Continued fractions. A fraction in which the denominator is an integer plus another fraction, which in turn may have a denominator of the same form. Such fractions may either be terminating or infinite (non-terminating). A continued fraction that terminates corresponds to a rational number. One that is nonterminating represents a real number that is the limit of the sequence:

$$a_1 \; , \; a_1 + \frac{1}{a_2} \; , \; a_1 + \frac{1}{a_2 + \frac{1}{a_3}} \; , \; \ldots .$$

Setting $a_1 = 2$ and all subsequent $a_i = 4$ the above sequence has as its limit $\sqrt{5}$.

It can be shown that the following non-terminating fraction is equal to the golden ratio described in Sect. 24.2.

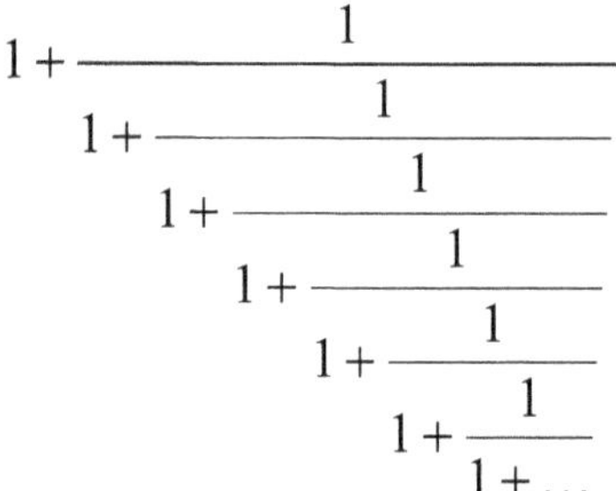

Curvature. The rate of change of direction of a curve at a specified point on that curve. At all points on a circle, the curvature is the reciprocal of the radius. Therefore it is a constant. One requires a formal definition of curvature to establish this result. If, the difference between the direction of the tangent to a curve at a point P and the direction at a nearby point Q is $\delta\psi$, and the length of the arc of the curve between P and Q is δs, then the curvature at the point P is defined as the limit as $\delta s \to 0$ of

$$\frac{\delta\psi}{\delta s}, \text{ i.e., by the derivative } \frac{\mathrm{d}\psi}{\mathrm{d}s}.$$

The result quoted above for a circle follows because, for any arc PQ of a circle with centre O , the angle POQ $= \delta\psi$. If PQ $= \delta s$, it follows that if the radius of the circle is r, then $\delta s = r\delta\psi$. In the limit this means

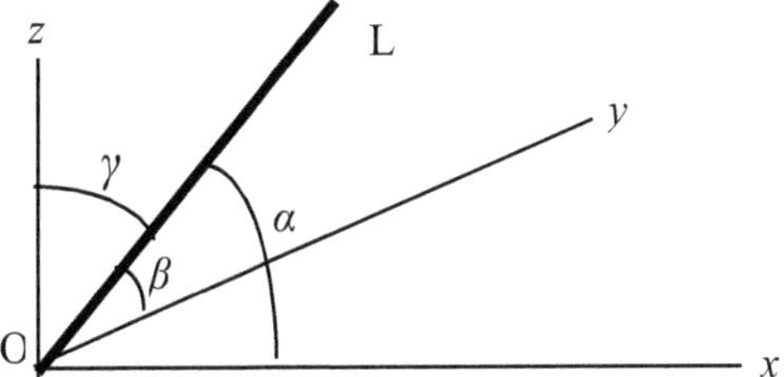

Fig. 25.1. Angles between a line L and the axes

$$\frac{\mathrm{d}\psi}{\mathrm{d}s} = \frac{1}{r} \,.$$

For any curve the reciprocal of the curvature, i.e., $\mathrm{d}s/\mathrm{d}\psi$, is called the *radius of curvature.* For a circle the radius of curvature at any point is a constant, equal to the radius of the circle.

Differential geometry. As the name suggests, this involves the study of geometric properties of curves and surfaces (e.g., the area or curvature of a surface) that are found by the differential calculus. Applications, many involving partial derivatives, extend to any type of space in any number of dimensions.

Direction cosines. If, in a three dimensional coordinate system, the angles between a line and the three positive axial directions are α, β, γ (see Fig. 25.1), then the direction cosines of the line are $l = \cos\alpha, m = \cos\beta$, and $n = \cos\gamma$. The angle, θ, between two lines with direction cosines l_1, m_1, n_1 and l_2, m_2, n_2 satisfies the relationship

$$\cos\theta = l_1 l_2 + m_1 m_2 + n_1 n_2 \,. \tag{25.1}$$

If the lines are coincident the angle between them is zero. This implies that for any line

$$l^2 + m^2 + n^2 = 1 \,.$$

It also follows from (25.1) that if two lines are at right angles then, since the angle between them is 90°,

$$l_1 l_2 + m_1 m_2 + n_1 n_2 = 0 \,.$$

The direction cosines of the x-, y- and z-axes are respectively (1, 0, 0), (0,1, 0) and (0, 0, 1). In *vector analysis the corresponding vectors are denoted by $\mathbf{i}, \mathbf{j}, \mathbf{k}$.

Discriminant. A term used in several contexts. In particular, it is used in polynomial equations in one or more dimensions for functions of the coefficients that give information on the nature of the roots of an equation, or the form of the curves, or surfaces, represented by the polynomial. The quadratic equation $ax^2 + bx + c = 0$ has discriminant $\mathrm{D} = b^2 - 4ac$. If $\mathrm{D} > 0$ the two roots are real and different, if $\mathrm{D} = 0$ the roots are real and equal, and if $\mathrm{D} < 0$ the roots are complex.

If the general equation of a conic section is written

$$ax^2 + 2hxy + by^2 + 2gx + 2fy + c = 0 \tag{25.2}$$

the discriminant may be written as a determinant

$$\Delta = \begin{vmatrix} a & h & g \\ h & b & f \\ g & f & c \end{vmatrix}$$

A determinant is a function of a square matrix, and in the case of the 3×3 matrix involved here

$$\Delta = a(bc - f^2) - h(hc - fg) + g(hf - gb) .$$

If $\Delta \neq 0$, then if

$$\begin{array}{ll} h^2 - ab < 0 & \text{(25.2) is an ellipse,} \\ h^2 - ab = 0 & \text{(25.2) is a parabola,} \\ h^2 - ab > 0 & \text{(25.2) is a hyperbola.} \end{array}$$

If $\Delta = 0$, the conic section is said to be degenerate, and then if

$$\begin{array}{ll} h^2 - ab < 0 & \text{(25.2) is a point,} \\ h^2 - ab = 0 & \text{(25.2) is a pair of parallel} \\ & \text{or coincident lines or the} \\ & \text{locus is imaginary,} \\ h^2 - ab > 0 & \text{(25.2) is a pair of} \\ & \text{intersecting straight lines.} \end{array}$$

Envelopes. An envelope is a curve that is tangential to, or touches, every member of a family of curves. For example, if we consider a family of circles each of which has its centre on the circumference of a fixed circle of radius r, and each member of the family has radius $a < r$, then it is easily seen from Fig. 25.2 that the family has an envelope consisting of a circle of radius $r + a$ and a circle of radius $r - a$. The concept extends to surfaces that are tangential to all members of a given family of surfaces.

Expectation and expected values. The first *moment about the origin for a random variable X is the expected value, or mean, of X. For a discrete random variable that takes values x_i with probabilities p_i it is defined as

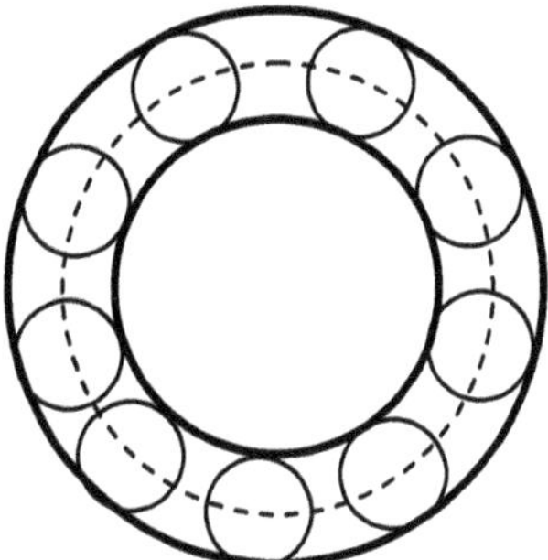

Fig. 25.2. The thicker circles represent the envelope to a family of circles of equal radius with their centres on the dotted circle

$$\mathrm{E}(X) = \sum_i x_i p_i \, .$$

For a continuous random variable with probability density function $f(x)$ it is defined as

$$\mathrm{E}(X) = \int_{-\infty}^{\infty} x f(x) \mathrm{d}x.$$

The expected value of a function $g(X)$ of X is defined as

$$\mathrm{E}[g(X)] = \sum_i g(x_1) p_i \text{ or } \mathrm{E}[g(X)] = \int_{-\infty}^{\infty} g(x) f(x) \mathrm{d}x.$$

See also *moment, *moment generating function, *probability generating function.

Fourier series. An infinite series of the form

$$f(x) = \frac{1}{2} a_0 + \sum_{n=1}^{\infty} (a_n \cos nx + b_n \sin nx).$$

This has period 2π, since both the sine and cosine components have that period. The series is important because by suitable choices of a_n and b_n it can be made to converge to any periodic function $f(x)$ defined over the interval $(-\pi, \pi)$. This is invaluable in the study of wave forms composed of constituent sine waves of different frequencies and amplitudes.

For a given periodic function $f(x)$ the coefficient a_0 is given by

$$a_0 = \frac{1}{2\pi} \int_{-\pi}^{\pi} f(x) \, \mathrm{d}x \, ,$$

and a_n, b_n are given by

$$a_n = \frac{1}{\pi}\int_{-\pi}^{\pi} f(x)\cos nx \, dx \text{ and } b_n = \frac{1}{\pi}\int_{-\pi}^{\pi} f(x)\sin nx \, dx.$$

Fuzzy logic. In classic set theory any object is, or is not, a member of a given set. Membership may be represented by the number 1, and nonmembership by 0. In fuzzy *set theory* – the concept behind fuzzy logic – set membership is represented by a real number between 0 and 1. For instance, whether an animal is large or small, is on its own not a statement that is either *true* or *false*, but more a matter of degree. Systems of fuzzy logic have found application in fields as diverse as control of dishwashers and studies of how the human brain works.

Gamma function. The function

$$\Gamma(n) = \int_0^{\infty} x^{n-1} e^{-x} dx, \; x \geq 0$$

is called the *gamma function.* It satisfies the relationship

$$\Gamma(n+1) = n\Gamma(n).$$

In particular, if n is a positive integer $\Gamma(n+1) = n!$. It can also be shown that $\Gamma(1/2) = \sqrt{\pi}$. In statistics the function is closely related to a versatile family of distributions of random variables. The function also arises in other areas of applied mathematics.

Generating functions. A simple generating function, often referred to as an *ordinary generating function,* is a function $G(a_i, x)$, that has a formal series expansion

$$G(a_i, x) = \sum_{i=0}^{\infty} a_i x^i \; .$$

The word *generating* is used to reflect the property that the coefficient of x^i is a_i. That is, $G(a_i, x)$ generates the sequence $a_0,\ a_1,\ a_2,\ \ldots$. For example, the generating function of the sequence $1, 1, 1, \ldots$ is

$$G(1, x) = \frac{1}{1-x}$$

because

$$\frac{1}{1-x} = 1 + x + x^2 + x^3 + \ldots \; .$$

We already know that this expansion is only convergent if $|x| < 1$.

The generating function for the sequence with $a_i = 2^i$ is

$$G(2^i, x) = \frac{1}{1-2x}$$

since

$$\frac{1}{1-2x} = 1 + 2x + 2^2x^2 + 2^3x^3 + \ \dots \ .$$

The concept is applicable to situations where the a_i may themselves be functions, P_i of some other variable u, say, where $a_i = P_i(u)$.

Generating functions are often met in physics and in statistics. See also *moment generating function, *probability generating function.

Heat equation. An important partial differential equation that models the flow of heat in a body. The equation is often solvable if the temperature on the boundary of the body is known as a function of time. The heat equation for the temperature T at time t at the point (x, y, z) is

$$\frac{\partial T}{\partial t} = \frac{\partial^2 T}{\partial x^2} + \frac{\partial^2 T}{\partial y^2} + \frac{\partial^2 T}{\partial z^2} .$$

Hydrostatics and hydrodynamics. The study of the mechanical properties of fluids, especially liquids. Hydrostatics is relevant to fluids that are not in motion and hydrodynamics to those that are.

Hyperbolic functions. The hyperbolic cosine and sine of x are written $\cosh x$ and $\sinh x$, and

$$\cosh x = \frac{1}{2}\left(\mathrm{e}^x + \mathrm{e}^{-x}\right) \text{ and } \sinh x = \frac{1}{2}\left(\mathrm{e}^x - \mathrm{e}^{-x}\right).$$

In Sect. 12.7 we saw that

$$\cos x = \frac{1}{2}\left(\mathrm{e}^{\mathrm{i}x} + \mathrm{e}^{-\mathrm{i}x}\right) \text{ and } \sin x = \frac{1}{2\mathrm{i}}\left(\mathrm{e}^{\mathrm{i}x} - \mathrm{e}^{-\mathrm{i}x}\right).$$

From these definitions it is easy to show that

$$\cosh \mathrm{i}x = \cos x \text{ and that } \sinh \mathrm{i}x = \mathrm{i}\sin x \ ,$$

and that

$$\mathrm{e}^x = \cosh x + \sinh x \text{ and } \cosh^2 x - \sinh^2 x = 1.$$

We also define

$$\tanh x = \frac{\sinh x}{\cosh x}, \ \operatorname{sech} x = \frac{1}{\cosh x}, \ \operatorname{cosech} x = \frac{1}{\sinh x}, \ \coth x = \frac{1}{\tanh x}.$$

There are other close analogies between hyperbolic functions and the trigonometric, or circular, functions. The name *hyperbolic functions* arises from their use in establishing *parametric equations for a hyperbola.

There are several conventions for getting over pronunciation difficulties, the easiest being to refer to 'hyperbolic sin', 'hyperbolic cosine', etc.; $\tanh x$ is sometimes pronounced as 'than x', or less commonly as 'tank x'.

Kronecker delta. A useful notational shorthand δ_{ij} for a function that takes the values

$$\delta_{ij} = 0 \text{ if } i \neq j \text{ and } \delta_{ii} = 1 \text{ for all } i\ ,$$

where i, j, are positive integers. For example, the $n \times n$ identity matrix with all diagonal elements 1 and all off-diagonal elements zero may be written $[\delta_{ij}]_n$.

Laplace's equation. Physicists studying potential theory meet this equation early in their studies. It takes the form

$$\frac{\partial^2 V}{\partial x^2} + \frac{\partial^2 V}{\partial y^2} + \frac{\partial^2 V}{\partial z^2} = 0.$$

It is sometimes written $\nabla^2 V = 0$, where ∇ (pronounced 'del') is an example of what is called as a *differential operator.* In this example ∇ operates on V.

L'Hôpital's rule. If two functions $f(x)$ and $g(x)$ both tend to zero as $x \to a$ and $f(a) = g(a) = 0$, then the ratio $f(a)/g(a) = 0/0$ which is not defined. L'Hôpital's rule gives a method for determining the limit as $x \to a$ in these circumstances. It states that

$$\lim_{x \to a} \frac{f(x)}{g(x)} = \lim_{x \to a} \frac{f'(x)}{g'(x)}.$$

For example, if $f(x) = \sin x$ and $g(x) = x$, then $f'(x) = \cos x$ and $g'(x) = 1$ and L'Hôpital's rule confirms the well known limit (see Sect. 10.2) that $\sin x/x$ tends to 1 as x tends to zero. If the ratio of the first derivatives is also undefined when $x = a$, one may repeat the process to ratios of higher order derivatives until a defined limit is obtained.

Magic square. A square array of numbers in which the sum of the numbers in any row, column, or full diagonal (i.e., from top left to bottom right or top right to bottom left) is the same. An example that is included in an engraving by Albrecht Dürer (1471–1528) is

16	3	2	13
5	10	11	8
9	6	7	12
4	15	14	1

In this, and some other well known examples of $n \times n$ squares, the numbers used are the integers $1, 2, 3, \ldots, n^2$.

Mean-value theorem. If a function $f(x)$ is continuous in the interval $a < x < b$ and $f(x)$ exists for all x in that interval, then for some value of x between a and b,

$$f'(x) = \frac{f(b) - f(a)}{b - a} .$$

This is the mean value theorem for the differential calculus.

There is also a mean value theorem for integrals. This states that there exists some x between a and b for which

$$\int_a^b f(u)\mathrm{d}u = (b-a)f(x).$$

Metric space. A set of points forms a *metric space* if there is a measure of distance, d, which specifies, for any pair of points x, y, a nonnegative distance $d(x, y)$ which has the properties

- (i) $d(x, y) = 0$ if and only if $x = y$,
- (ii) $d(x, y) = d(y, x)$,
- (iii) for any three points x, y, z the triangle inequality holds, i.e., $d(x, y) + d(y, z) \geq d(x, z)$.

Ordinary. or Euclidean, space is a metric space where d is the Euclidean or straight-line distance between any two points.

Moments. The word *moment* is widely used in a physical context including *moment of a couple*, *moment of a force* and *moment of inertia.*

The term is also used in statistics. If X is a random variable, the rth moment about the origin is defined as the *expectation of X^r. It is written as $\mathrm{E}(X^r)$. The rth moment about the point a is $\mathrm{E}(X-a)^r$. Moments are also defined for bivariate distributions, and for random samples. Moments about the origin and moments about the mean are frequently met. The first moment about the origin defines the mean, often denoted by μ. The rth moment about the mean is often denoted by μ_r. The second moment about the mean is called the *variance. See also *moment generating function.

Moment generating function. For a random variable X the *expectation of e^{tX}, where t is a constant, is called the moment generating function of X It is usually denoted by $\mathrm{M}(t)$, i.e.,

$$\mathrm{M}(t) = \mathrm{E}\left(\mathrm{e}^{tX}\right).$$

If the sum, or integral, associated with the exponential series expansion exists for some $t > 0$ the coefficient of $t^r/r!$ is the rth moment about the origin, $\mathrm{E}(X^r)$, of X. This is also the rth derivative of $\mathrm{M}(t)$ at $t = 0$. That is, $\mathrm{E}(X) = \mathrm{M}'(0)$, and $\mathrm{E}(X^r) = \mathrm{M}^{(r)}(0)$.

An important distribution arising, among other places, in studies of survival times, is called the *exponential distribution.* A random variable X is said to have an exponential distribution with parameter λ if it has a probability density function $f(x) = \lambda \mathrm{e}^{-\lambda x}$ for all non-negative x. By definition

$$\mathrm{M}(t) = \int_0^\infty \mathrm{e}^{tx} \lambda \mathrm{e}^{-\lambda x} \mathrm{d}x.$$

Using the basic rules of integration it is easy to establish that

$$\mathrm{M}(t) = \frac{1}{1 - t/\lambda},$$

providing $t < \lambda$. This gives the series expansion (see Sect. 12.2)

$$\mathrm{M}(t) = 1 + \frac{t}{\lambda} + \left(\frac{t}{\lambda}\right)^2 + \ldots + \left(\frac{t}{\lambda}\right)^r + \ldots$$

where the coefficient of $t^r/r!$ is $(r!)/\lambda^r$, so that $\mathrm{E}(X^r) = (r!)/\lambda^r$. In particular, $\mathrm{E}(X) = 1/\lambda$. The *variance of X (i.e. the second moment about the mean) is given by $\mathrm{Var}(X) = \mathrm{E}(X^2) - [\mathrm{E}(X)]^2$, i.e.,

$$\mathrm{Var}X = \frac{2}{\lambda^2} - \frac{1}{\lambda^2} = \frac{1}{\lambda^2}.$$

See also *probability generating function.

Multinomial theorem. A generalization of the binomial theorem for a positive integer N. It takes the form

$$(x_1 + x_2 + \ldots + x_r)^N = \sum \frac{N!}{(n_1)!(n_2)!\ldots(n_k)!} x_1^{n_1} x_2^{n_2} \ldots x_k^{n_k},$$

where the summation is over all possible integer values of $n_1, n_2, \ldots, n_k$ subject to the constraint

$$\sum_{i=1}^{k} n_i = N.$$

Orthogonality. The concept of orthogonality stems from the notion of being at right angles. For example, two curves are said to be orthogonal to one another at a point of intersection if their respective tangents at that point are at right angles to each other.

In the context of integration, a set of functions $\{f_i(x); i = 1, 2, , n\}$ are said to be orthogonal functions over the interval (a, b) if, for all $i \neq j$,

$$\int_a^b f_i(x) f_j(x) \mathrm{d}x = 0.$$

If, in addition for all i

$$\int_a^b [f_i(x)]^2 \mathrm{d}x = 1,$$

the functions are also said to be normalized. The set of functions is then said to be *orthonormal.*

Parameters. In a general mathematical expression that specifies a family of curves with specific properties, parameters are quantities which, when they are given different values, determine distinct members of that family. For example, the general expression $y = mx + c$ specifies the equation to any straight line. Here the parameters are m and c. By giving these specific values such as $m = 3$ and $c = 0$ we identify the particular member of that family with equation $y = 3x$.

If a random variable, X, has a distribution belonging to the binomial family of distributions, we distinguish between different members of that family by assigning specific values to the two parameters n (the number of trials) and p (the probability of one of two specified events occurring at any trial).

Parametric equations. These are equations that determine the coordinates of a point on a curve in terms of a single variable. For Cartesian coordinate systems, if p is the parameter, then the co-ordinates take the form $x = f(p)$, $y = g(p)$. For example, consider the ellipse with equation

$$\frac{x^2}{a^2} + \frac{y^2}{b^2} = 1\ .$$

This equation is easily seen to be satisfied if we introduce a parameter, θ, such that $x = a\cos\theta$ and $y = b\sin\theta$. For the parabola $y^2 = 4ax$ a suitable parametric specification is in terms of a parameter, t, such that $x = at^2$ and $y = 2at$. The equation

$$\frac{x^2}{a^2} - \frac{y^2}{b^2} = 1\ ,$$

represents a hyperbola. It is easily seen to be satisfied if we introduce a parameter u such that $x = a\cosh u$ and $y = b\sinh u$. This follows from properties of the *hyperbolic functions.

Probability generating function. If X is a random variable that takes non-negative integral values and t is some constant, the probability generating function $\mathrm{P}(t)$ of X is the *expectation, $\mathrm{E}(t^X)$. If $\mathrm{P}(t)$ generates a convergent series for some t, the coefficient of t^r is equal to $\Pr(X = r)$. For the binomial distribution

$$\mathrm{P}(t) = \sum_{r=0}^{n} t^r \binom{n}{r} p^r q^{n-r} = (pt + q)^n\ ,$$

from which the coefficient of t^r is

$$\Pr(X = r) = \binom{n}{r} p^r q^{n-r}.$$

Rings. A concept in modern algebra that is associated with two binary operations on a closed set. It is an extension from the notion of a group (see Sect. 16.5).

Conventionally, the pair of binary operations are denoted by $+$ and $\times$, but these need not be the operations of addition and multiplication, although for convenience they are often referred to by those names. If a, b, c are any elements of a set, R, that is closed with respect to binary operations $+$, $\times$, then R forms a ring if for all $a, b.c$:-

- (i) $a + b = b + a$, i.e., addition is commutative;
- (ii) $a + (b + c) = (a + b) + c$, i.e., addition is associative;
- (iii) $a \times (b \times c) = (a \times b) \times c$, i.e., multiplication is associative;
- (iv) there exists $0 \in \mathrm{R}$ such that $a + 0 = 0 + a = a$;
- (v) there exists for all a, an element $-a$, such that $a + (-a) = 0$;
- (vi) $a \times (b + c) = a \times b + a \times c$ and $(a + b) \times c = a \times c + b \times c$.

Axiom (vi) is the distributive law. If multiplication is also commutative, i.e. $a \times b = b \times a$ the ring is called a *commutative ring.*

If there is also an element 1 such that for all a we have $a \times 1 = 1 \times a = a$ the ring is called a *commutative ring* with identity (or unity).

The set of all 2×2 matrices with the matrix operations of addition and multiplication form a noncommutative ring (multiplication is not commutative). The set of all even integers (positive or negative together with zero) form a commutative ring with ordinary addition and multiplication, but it has no identity element. However, the set of all integers with addition and multiplication form a commutative ring with identity. The set $\{0, 1, 2, 3, 4, 5\}$ modulo 6 forms a ring with respect to addition and multiplication.

Standard deviation. See *variance.

Symmetric functions. A symmetric function of several variables is one that is unaltered if the variables are interchanged pairwise. For example, the function

$$f(x) = x^3 + y^3 + z^3 - 8(xy + yz + zx)$$

is symmetric. If such pairwise interchanges always alter the sign of the function it is said to be *skew-symmetric.* The function $(x - y)(y - z)(z - x)$ is skew symmetric.

Truth functions and truth tables. In logic truth functions are functions where the variables or arguments are truth values. There are two truth values associated with any proposition. These are usually denoted by T for *true*, or F for *false.* The truth values of compound statements can be expressed in terms of the truth values of the constituent parts. We make use of a basic set of truth tables corresponding to the logical concepts *or* (denoted by $\vee$), *and* (denoted by $\wedge$), *not* (denoted by $\neg$), implies (denoted by $\Rightarrow$). The basic tables for two arguments A and B are:

A	B	A ∨ B	A ∧ B	A ⇒ B
T	T	T	T	T
T	F	T	F	F
F	T	T	F	F
F	F	F	F	T

A	¬ A
T	F
F	T

Using the above, we may build up more complicated truth tables. For example, for three components A, B, C, the truth values associated with (A∨ B) ∧ (¬ C) are given in the final column of this table:-

A	B	C	A ∨ B	¬ C	(A∨ B) ∧ (¬ C)
T	T	T	T	F	F
T	T	F	T	T	T
T	F	T	T	F	F
T	F	F	T	T	T
F	T	T	T	F	F
F	T	F	T	T	T
F	F	T	F	F	F
F	F	F	F	T	F

Variance. For a random variable X having mean μ, the variance is defined as the second moment about the mean, $\mathrm{E}(X - \mu)^2$. It is commonly referred to as $\mathrm{Var}(X)$. It is easily shown that $\mathrm{Var}(X) = \mathrm{E}(X^2) - [\mathrm{E}(X)]^2$. The statistical importance of variance is that it indicates the spread of a symmetric distribution. In general terms, the greater the variance the greater the spread. Although it may be calculated for a nonsymmetric or skew distribution, interpreting it as a measure of spread may then be misleading. The square root of the variance is called the *standard deviation.* The variance of a random sample of n observations $x_1, x_2, \dots x_n$ is called the sample variance. This plays an important role in many inference problems in statistics.

Vectors. A vector is an entity in Euclidean space that has both magnitude and direction. It can be represented geometrically by a directed segment of a line. Two vectors are said to be equivalent if they have the same magnitude and the same direction. A vector that is specified by an ordered pair of end points in space is sometimes referred to as a *located vector.*

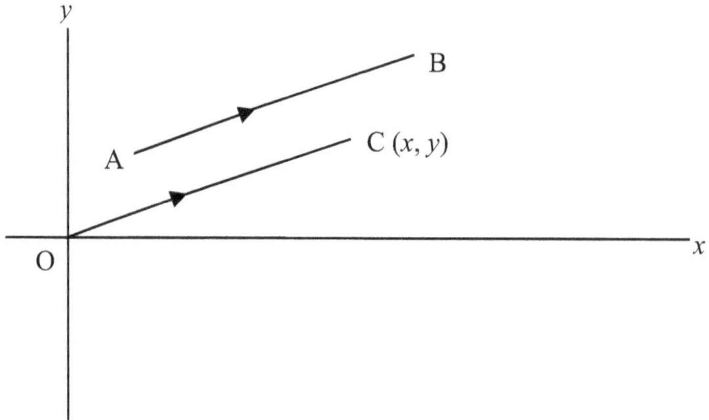

Fig. 25.3. The vector **AB** is a located vector with direction from A to B where A, B have co-ordinates respectively $(x_1, y_1), (x_2, y_2)$. **OC** is an equivalent vector through the origin

Fig. 25.3 illustrates these ideas for a vector in two-dimensional space. The vector **AB** is located by the end points A, with co-ordinates (x_1, y_1), and B with co-ordinates (x_2, y_2). The positive direction is from A to B. The vector **OC**, where O is the origin is an equivalent vector, i.e. it has the same length and direction, but it has a different location.

We use bold **AB** to distinguish between a vector and a line segment AB. Another notation sometimes used for a vector is to write it $\overrightarrow{\mathrm{AB}}$. The fact that for any vector there is an equivalent vector from the origin to some point (x, y) often leads to convenient simplification. For example, if we always take the first ordered co-ordinate pair specifying a vector to be the origin, a vector is uniquely determined if we are given the co-ordinates of the other ordered point determining the vector. Thus, to specify the vector **OC** in Fig. 25.3 we only need to know the co-ordinates (x, y) of C. The length of the vector **OC** is easily seen to be $\sqrt{(x^2 + y^2)}$. This is also the length of the equivalent vector **AB**. These concepts extend to more than two dimensions — indeed to any number, n, of dimensions. In the case of a three-dimensional space, any vector **AB** is equivalent to some vector **OC** where O is the origin, and C has co-ordinates (x, y, z). When we know x, y, z, the direction is uniquely determined and the magnitude is $\sqrt{x^2 + y^2 + z^2}$.

Two or more vectors are added by placing the directed segments end to end. The sum is the vector joining the initial point on the first vector to the final point on the second vector. Fig. 25.4 illustrates vector addition for two vectors **OC**, **CD** in a plane, where for convenience we have located the first vector to start at the origin. The vector **OD** is the sum of **OC** and **CD**.

The rule for addition of vectors is often called the *parallelogram law of addition*, as already mentioned in Sect. 13.5. The reason is clear from Fig. 25.4,

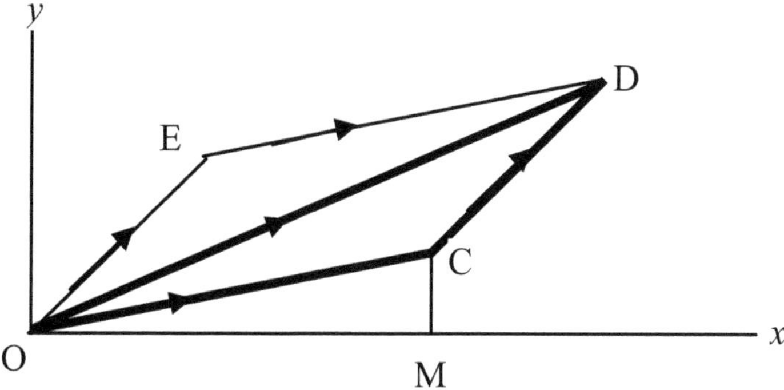

Fig. 25.4. Illustration of the parallelogram law for vector addition, $\mathbf{OC}+\mathbf{CD} = \mathbf{OD}$

where $\mathbf{OE}$ is equivalent to $\mathbf{CD}$ and $\mathbf{ED}$ is equivalent to $\mathbf{OC}$. This equivalence also establishes that vector addition is commutative. It is often convenient to use a single letter to represent a vector, e.g., we might write $\mathbf{AB} = \mathbf{u}$, $\mathbf{BC} = \mathbf{v}$ and so on. In this notation we write $\mathbf{BA}$, which has the opposite direction to $\mathbf{AB}$ as $\mathbf{BA} = -\mathbf{u}$. The rule for addition extends to more than two vectors. It is not difficult to show that it is both commutative and associative.

A vector may also be multiplied by a scalar (i.e. a real number), say λ. The resultant vector, $\lambda\mathbf{u}$ is, if λ is positive, a vector with the same direction as $\mathbf{u}$, but with magnitude λ times that of $\mathbf{u}$. If λ is negative, the magnitude is changed by the same amount, but the direction is the opposite to that of $\mathbf{u}$.

Referring to Fig. 25.4, and recalling the addition rule, it is clear that $\mathbf{OC} = \mathbf{OM} + \mathbf{MC}$. The vectors $\mathbf{OM}$ and $\mathbf{MC}$ are often referred to as the axial components of $\mathbf{OC}$. In many aspects of vector analysis it is convenient to represent a vector in terms of its axial components, and more especially in terms of scalar multiples of unit vectors starting at the origin, and having the positive axial directions. Fig. 25.5 illustrates this for a vector $\mathbf{OC}$ in three dimensions. The unit vectors in the positive axial directions Ox, Oy, Oz are conventionally denoted by $\mathbf{i}, \mathbf{j}$, and $\mathbf{k}$ respectively and are determined by the co-ordinates (1, 0, 0), (0, 1, 0), (0, 0, 1). By the rules of vector addition and multiplication by scalars, if, in Fig. 25.5 the point C has coordinates (x, y, z) then we may write

$$\mathbf{OC} = \mathbf{u} = x\mathbf{i} + y\mathbf{j} + z\mathbf{k}.$$

This follows because the vector from the origin to the point $(x, 0, 0)$ may be written $x\mathbf{i}$, and so on.

When multiplying a vector by a scalar this is equivalent to multiplying each component by the same scalar. Thus if $\mathbf{u} = x\mathbf{i} + y\mathbf{j} + z\mathbf{k}$ then

$$\lambda\mathbf{u} = \lambda x\mathbf{i} + \lambda y\mathbf{j} + \lambda z\mathbf{k}.$$

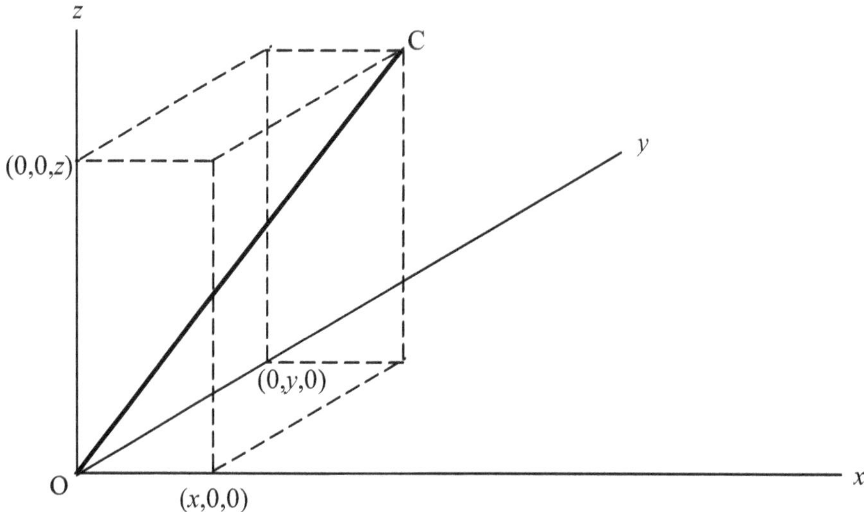

Fig. 25.5. Using the parallelogram rule for addition the vector **OC** may be expressed as the sum of the axial components $x\mathbf{i}, y\mathbf{j}, z\mathbf{k}$ where $\mathbf{i}$ is a unit vector in the positive direction of the x-axis and x is the x co-ordinate of C, etc

It is not hard to see that addition is equivalent to adding components. Thus, if $\mathbf{u} = x_1\mathbf{i} + y_1\mathbf{j} + z_1\mathbf{k}$ and $\mathbf{v} = x_2\mathbf{i} + y_2\mathbf{j} + z_2\mathbf{k}$, then

$$\mathbf{u} + \mathbf{v} = (x_1 + x_2)\mathbf{i} + (y_1 + y_2)\mathbf{j} + (z_1 + z_2)\mathbf{k}.$$

We have not defined the product of vectors. There are indeed two such products, the *scalar* (or dot) product, and the *vector* (or cross) product. Both appear in many applications of mathematics to real-world problems.

A

Appendix

A.1 Inequalities

The rules for manipulating inequalities involving real numbers are the same as those for operating with equalities, apart from two major exceptions. All apply to each of the inequalities

$>$	*is greater than*
$\geq$	*is greater than or equal to*
$<$	*is less than*
$\leq$	*is less than or equal to.*

We illustrate them only for the case *is greater than*. Rules that are analogous to those for operating with equalities are:

(i) if $x > y$ then $x + c > y + c$ for all real c.
(ii) if $x > y$ then $kx > ky$ for all real positive k.

The first exception relates to the equality rule that if $x = y$ then $kx = ky$ no matter whether k is positive or negative. The inequality rule is

(iii) if $x > y$ then $kx > ky$ for all real positive k, but $kx < ky$ if k is negative. There is a similar sign reversal with negative k in the other inequalities.

The second exception is

(iv) if x, y are both of the same sign, taking the reciprocal of an inequality reverses the inequality.

Fig. A.1. The inequality $5x + 11 \geq 2 + 9x$ is satisfied for all $x \leq 2.25$

The above rules also cover subtraction and division implicitly, since subtraction of c is equivalent to addition of $-c$, and division by k is equivalent to multiplication by $1/k$.

Here are some examples:

(a) $12 > -12$, and if we subtract 3 from (i.e. add -3 to) each side we get $12 - 3 > -12 - 3$, i.e. $9 > -15$;
(b) $-17 < -8$, and if we add 20 to each side we get $3 < 12$;
(c) $15 < 100$, and if we divide each side by 10 (equivalent to multiplying by $1/10$) we get $1.5 < 10$;
(d) $25 > -10.2$, and if we multiply each side by -0.5 the left-hand side becomes $25 \times (-0.5) = -12.5$, and the right-hand becomes 5.1 and $-12.5 < 5.1$.
(e) $50 > 25$, and taking reciprocal, $1/50 < 1/25$.

The problem of finding the values of x for which an inequality of the form

$$5x + 11 \geq 2 + 9x$$

is true is referred to as solving a (linear) inequality. The rules for inequalities given above enable us to do this. If we add $-9x$ to each side, rule (i) gives

$$\begin{aligned} 5x + 11 - 9x &\geq 2 + 9x - 9x \\ \text{i.e.,} \quad -4x + 11 &\geq 2 \,. \end{aligned}$$

Adding -11 to each side gives

$$-4x \geq -9 \,,$$

and dividing by -4 (multiplying by $-1/4$) gives, by rule (iii)

$$x \leq 2.25$$

This is the solution to the linear inequality. Unlike a linear equation, there is not just a single value solution, but any x that does not exceed the value 2.25 is a solution.

If we consider the real points as points on a line, the solution in this case is given by the points on the bolder segment of the line in Fig. A.1.

A.2 The Index Laws

If a is a real number we write the products $a \times a = a^2$, $a \times a \times a = a^3$, and more generally,

$$a \times a \times a \times \ldots \times a = a^n \ ,$$

if there are n occurrences of a. In this context n is called an *index* or *exponent*.

It is easy to see that $a^n \times a = a^{n+1}$, $a^n \times a \times a = a^{n+1} \times a = a^{n+2} = a^n \times a^2$. Continuing along these lines, we find that $a^m \times a^n = a^{m+n}$. If $a \neq 0$ and if $m > n$ then $a^m / a^n = a^{m-n}$. Here m and n are positive integers.

We may also confirm easily that $a^m \times a^m \times a^m = (a^m)^3 = a^{3m}$. This generalizes to $(a^m)^n = a^{mn}$. It is also easily verified that $(ab)^n = a^n b^n$. Here again m and n are positive integers.

We can remove the restriction of m, n to integer values, and allow them to be rational numbers by defining $a^0 = 1$, $a^{-n} = 1/a^n$ where $a \neq 0$, and $a^{r/s} = \sqrt[s]{a^r}$ when $a \geq 0$, where r is an integer (positive or negative) and s is a positive integer.

As an example of the last of these, if we set $r = 1$ and $s = 2$ we get $a^{1/2} = \sqrt{a}$. This is consistent with the laws of operation with indices given initially for positive integral indices. For example, $(a^{1/2})^2 = (\sqrt{a})^2 = a$.

If $y = a^x$ the inverse function is $x = \log_a y$, spoken as the logarithm (or log) of y to the base a.

A.3 The Trigonometric Ratios

In introductory courses in trigonometry the ratios sine, cosine, and tangent (abbreviations sin, cos, tan) of an angle θ lying between 0 and 90°, are usually first defined in terms of the ratios of the lengths of the sides of a right-angled triangle that contains an angle θ. In Fig. A.2, if the angle A is of magnitude θ and B is a right angle, then the relevant definitions are

$$\sin\theta = \mathrm{BC}/\mathrm{AC}, \cos\theta = \mathrm{AB}/\mathrm{AC}, \tan\theta = \mathrm{BC}/\mathrm{AB}.$$

It is immediately obvious that $\tan\theta = \sin\theta / \cos\theta$. Also, since the angle C $= 90° - \theta$, that $\sin\theta = \cos(90° - \theta)$ and $\cos\theta = \sin(90° - \theta)$. The reciprocals of the trigonometric ratios also have special names. The reciprocal of the sine is called the cosecant (abbreviation 'cosec'), that of the cosine is called the secant (abbreviation 'sec'), and that of the tangent is called the cotangent (abbreviation 'cotan' or 'cot').

Although the right-angled triangle definitions suffice for many situations arising in numerical computations of heights and distances, in other situations we need definitions of the ratios for angles of any magnitude. These definitions

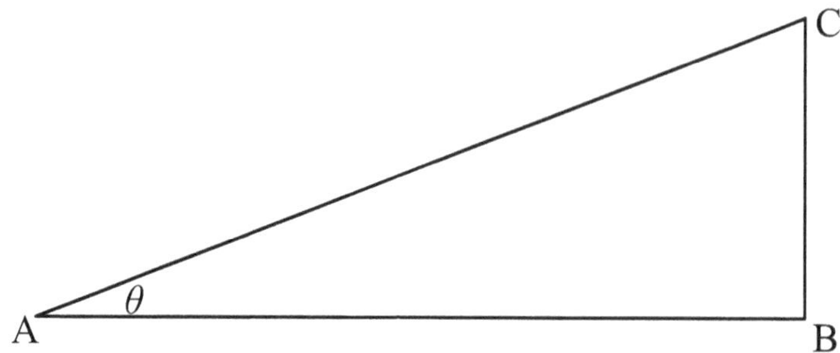

Fig. A.2. A right-angled triangle to illustrate trigonometric ratios for an angle θ

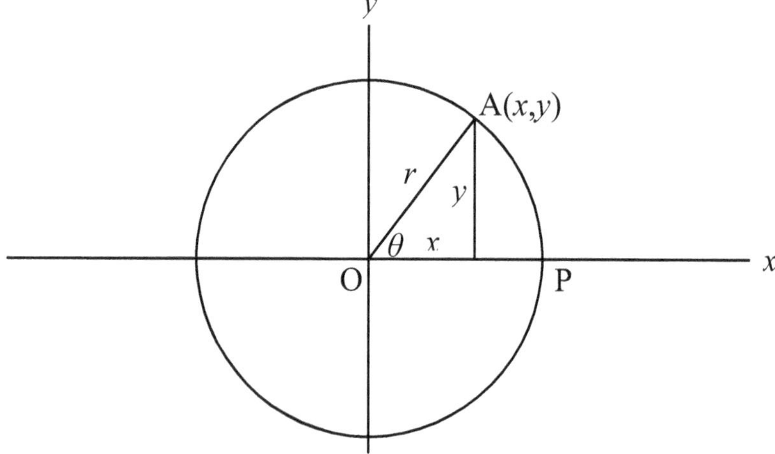

Fig. A.3. Use of a coordinate system for defining trigonometric ratios for angles of any magnitude

are usually expressed in terms of the Cartesian co-ordinates of points on the circumference of a circle of any radius r. In Fig. A.3 the axes Ox and Oy represent a Cartesian co-ordinate system (Chap. 7). OA is a line of fixed length r (positive), which is free to rotate about O. The point A always lies on the circumference of a circle. We consider A to be initially at a point P on the x-axis. We denote the angle POA by θ. Anticlockwise rotation from P specifies a positive angle, and clockwise rotation specifies a negative angle θ. We denote the coordinates of A by (x, y). Then for any θ we define

$$\sin\theta = y/r \text{ and } \cos\theta = x/r \text{ and } \tan\theta = y/x. \tag{A.1}$$

It is easily seen that if θ lies between 0 and 90° these definitions give the same numerical values as those above for right-angled triangles. Since r is always positive, the sign of the ratios depends only on the sign associated

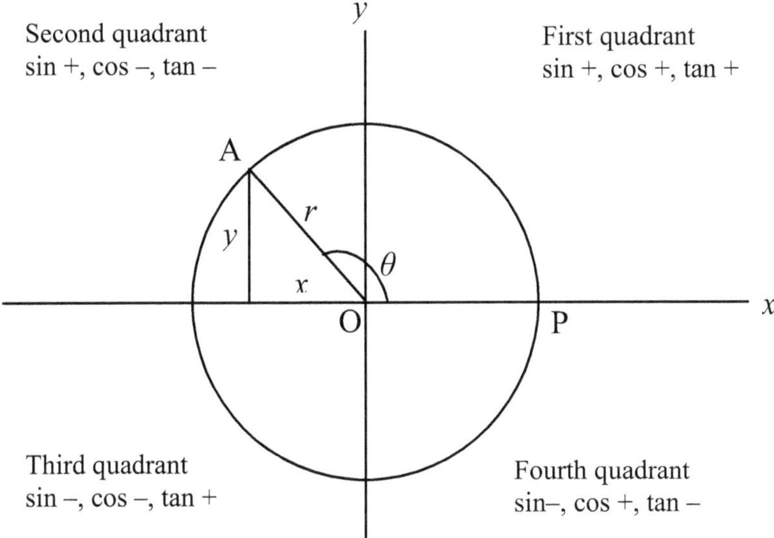

Fig. A.4. Signs of the trigonometric ratios in each quadrant

with x and y. It is easy to verify that in each of the four quadrants the signs associated with each ratio are those indicated in Fig. A.4.

By reference to similar co-ordinate diagrams, a number of relationships between ratios of various angles are easily established. In particular, since θ returns to its original position after each complete rotation of 360°, it follows that for all integers n the ratios are the same for an angle θ and an angle $\phi = \theta + 360n°$. It is easily verified that $\sin(-\theta) = -\sin\theta$, $\cos(-\theta) = \cos\theta$. Also, $\sin(180° - \theta) = \sin\theta$ and $\cos(180° - \theta) = -\cos\theta$ and $\sin(90° + \theta) = \cos\theta$, $\cos(90° + \theta) = -\sin\theta$.

Fig. A.5 is a graph of $y = \sin x°$ over a little more than a 360° cycle. Since the value of $\sin x$ repeats itself after the addition of 360 to x, the pattern repeats through 360° cycles.

So far in this section we have assumed that θ is measured in degrees. We pointed out in Sect. 10.3 that in pure mathematics, unless otherwise stated, angles x are measured in radians. This is particularly important where limiting results are concerned, and in the calculus. Radians and degrees are related by the transformation π radians = 180°.

Unless otherwise stated, all results in this section are valid no matter whether angles are measured in degrees or in radians.

From the definitions (A.1) it follows immediately that

$$\cos^2\theta + \sin^2\theta = \frac{x^2}{r^2} + \frac{y^2}{r^2} = 1 \, ,$$

since $x^2 + y^2 = r^2$.

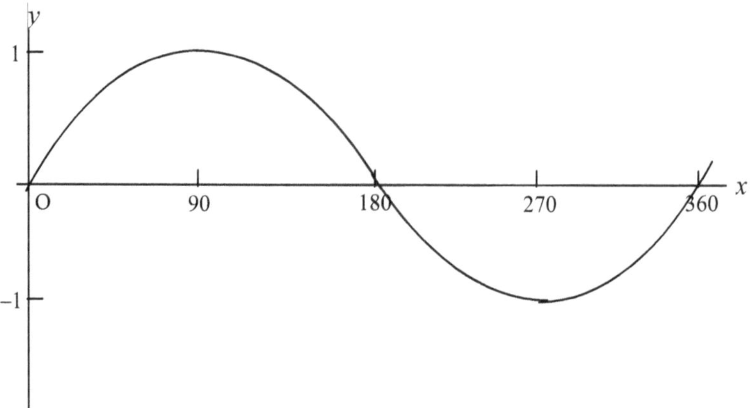

Fig. A.5. The graph of $y = \sin x^\circ$

We state below, without proof, further useful results. Proofs may be found in nearly all standard trigonometry textbooks.

The first group are the basic addition formulae. For any two angles A and B they are:

$$\begin{aligned}\sin(\mathrm{A}+\mathrm{B}) &= \sin\mathrm{A}\cos\mathrm{B} + \cos\mathrm{A}\sin\mathrm{B}\ ,\\ \sin(\mathrm{A}-\mathrm{B}) &= \sin\mathrm{A}\cos\mathrm{B} - \cos\mathrm{A}\sin\mathrm{B}\ ,\\ \cos(\mathrm{A}+\mathrm{B}) &= \cos\mathrm{A}\cos\mathrm{B} - \sin\mathrm{A}\sin\mathrm{B}\ ,\\ \cos(\mathrm{A}-\mathrm{B}) &= \cos\mathrm{A}\cos\mathrm{B} + \sin\mathrm{A}\sin\mathrm{B}\ .\end{aligned}$$

Setting A = B, it immediately follows from the first and third of these formulae that

$$\begin{aligned}\sin 2\mathrm{A} &= 2\sin\mathrm{A}\cos\mathrm{A}\ ,\\ \cos 2\mathrm{A} &= \cos^2\mathrm{A} - \sin^2\mathrm{A}\ .\end{aligned}$$

For any triangle ABC, if we denote the lengths of the sides opposite the angle A, B, C by a, b, c, i.e., BC $= a$, etc., then the area of the triangle ABC is

$$\Delta = \frac{1}{2}bc\sin\mathrm{A} = \frac{1}{2}ca\sin\mathrm{B} = \frac{1}{2}ab\sin\mathrm{C}.$$

The first of these results is evident from Fig. A.6, if we recall that the area of a triangle is (1/2)(base $\times$ height). The other results follow by choosing each of the remaining sides as base.

Another useful formula for the area of a triangle, which we do not derive here, but whose derivation will be found in most standard trigonometry textbooks, is

$$\Delta = \sqrt{s(s-a)(s-b)(s-c)}$$

where $s = (a+b+c)/2$, and s is called the *semi-perimeter*.

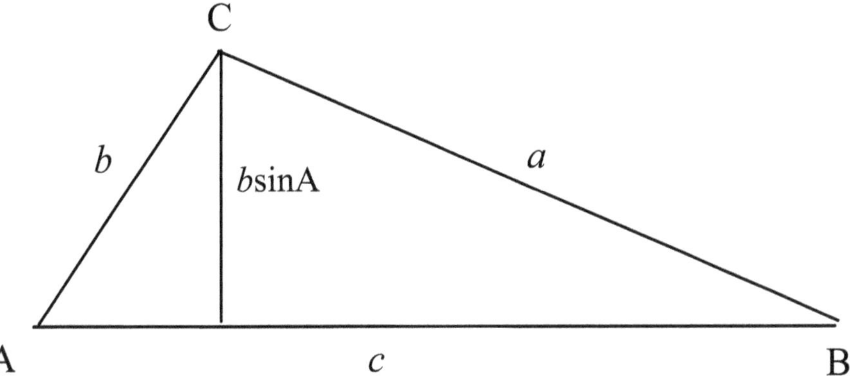

Fig. A.6. Area of the triangle ABC is $\Delta = (1/2)bc \sin \mathrm{A}$

A.4 Permutations and Combinations

The numbers of different ways of selecting 3 objects from 15, more generally of selecting r objects from n, is often required, especially in the calculation of probabilities, or in problems involving the allocation of resources. If there are 15 horses in a race we may want to know how many different possible forecasts can be made of which horses will finish first, second and third. In any one race, any of the 15 horses may come first. In that same race, any of the 14 remaining horses may come second. Thus there are 15×14 possible pairings of first and second, since any one of the 14 second placed horses may be paired with each of the 15 first placed horses. Once particular horses are assigned to first and second places, then each of the remaining 13 horses may come third. Each of these may be associated with one of the 15×14 pairings for first and second, giving $15 \times 14 \times 13 = 2730$ possible 'results' for a race with 15 starters. Here different 'results' correspond to different finishes where first, second and third placed horses only are of interest.

More generally, when selecting r items from n distinguishable ones, where order of selection matters, we may select the first in n ways, i.e., any of the n items may be chosen. We may select the second from any of the remaining $n-1$, giving $n \times (n-1)$ possible selections for the first two places. There are $n \times (n-1) \times (n-2)$ possible orderings for the first 3 places. Generally, for the first r places there are

$$n \times (n-1) \times (n-2) \times \ldots \times (n-r+1)$$

possible orderings. In particular, if $r = n$, it is easy to see that the total number of orderings of all n objects is

$$n \times (n-1) \times (n-2) \times \ldots \times 3 \times 2 \times 1 = n!.$$

If we have four objects A, B, C, D, i.e., $n = 4$, then the number of possible orderings of all 4 is $4! = 24$. It is easy to write them all down for such a small value of n. They are

ABCD	ABDC	ACBD	ACDB	ADBC	ADCB
BACD	BADC	BCAD	BCDA	BDAC	BDCA
CABD	CADB	CBAD	CBDA	CDAB	CDBA
DABC	DACB	DBAC	DBCA	DCAB	DCBA

For allocation of prizes of different values, or for betting on horses, or other competitive sporting events, place order is usually important. The number of ways we may select r objects from n where order is important is given by the product $n \times (n-1) \times (n-2) \times \ldots \times (n-r+1)$. This is called the number of *permutations* of r items from n. It is sometimes written ${}^{n}\mathrm{P}_{r}$.

There are many situations where only the items selected, and not the order in which they are selected, is important. In a bridge or poker game only the actual cards appearing in a hand — and not the order in which they are dealt — are important.

If we have to select 4 item from 5 items A, B, C, D, E the number of permutations of 4 items from 5 is ${}^{5}P_{4} = 5 \times 4 \times 3 \times 2 = 120$. Twenty four of these permutations are the ones given above that involve A, B, C, D. Further, these are the only possible ones involving only A, B, C, D. Of the remaining 96 permutations, 24 are got by replacing A by E in that list, a further 24 by replacing B by E, another 24 by replacing C by E, and the final 24 by replacing D by E. The $4! = 24$ arrangements of ABCD each represent the same selection of items, but in a different order. If we are not interested in order, this is just one selection, which in this case we call a combination of 4 items from 5. The possible combinations of 4 items from 5 are ABCD, ABCE, ABDE, ACDE and BCDE – a total of 5. If we have these combinations, and want the total number of permutations of 4 items from 5, we get this by writing down for each combination the possible 4! permutations of that combination. Thus we have

$$(\text{number of combinations of 4 items from 5}) \times 4! = {}^{5}\mathrm{P}_{4}.$$

A shorthand notation for the number of combinations of r items from n is

$$\binom{n}{r}.$$

In general, we have the relationship

$$\binom{n}{r} r! = {}^{n}\mathrm{P}_{r}\,,$$

or

$$\binom{n}{r} = \frac{{}^n\mathrm{P}_r}{r!} = \frac{n(n-1)(n-2)\dots(n-r+1)}{r!} \,.$$

Multiplying above and below by $(n-r)!$, we get

$$\binom{n}{r} = \frac{n!}{r!(n-r)!} \,.$$

It follows that

$$\binom{n}{r} = \binom{n}{n-r} \,.$$

It is convenient to define $\binom{n}{0} = 1$. In effect, this is equivalent to defining $0! = 1$. The notation for a combination is also referred to as a *binomial coefficient.* This is consistent with the notation introduced in Sect. 12.2. Another notation sometimes used for combinations is ${}^n\mathrm{C}_r$.

A poker hand consists of 5 cards. Since order in which the cards are dealt does not matter, the number of different poker hands that can be dealt from an ordinary pack of 52 cards is

$$\binom{52}{5} = \frac{52 \times 51 \times 50 \times 49 \times 48}{5 \times 4 \times 3 \times 2 \times 1} = 2\,598\,960.$$

Did you realise there were so many?

A.5 The Simplex Method

In Sect, 19.5 we gave a brief introduction to the *simplex method,* using the data for the microwave oven problem from Sect. 19.2. Here all steps in the algorithm are shown. This section should be read in association with Sect. 19.5.

The problem was summarized in equations (19.2) to (19.6)

In this particular example we need one slack variable for each non-zero inequality constraint. There are four such constraints, after excluding the zero inequality constraints $x \geq 0$, $y \geq 0$. We denote the slack variables by s_1, s_2, s_3 and s_4, whence

$$x + s_1 = 140 \tag{A.2}$$

$$y + s_2 = 120 \tag{A.3}$$

$$3x + 4y + s_3 = 640 \tag{A.4}$$

$$x + y + s_4 = 180. \tag{A.5}$$

Here, x, y, s_1, s_2, s_3 and s_4 are all zero or positive.

The equations (A.2) to (A.5) are 4 linear equations in 6 unknowns, so do not have unique solutions, but any solutions of interest have the following features:

- In each the x, y values correspond to a vertex of the feasible region in Fig. 19.4.
- Two from the six variables take the value zero in each solution.

For any vertices of the feasible region in Fig. 19.4 the second property holds. Such solutions form what are known as *basic feasible solutions.*

As explained in Sect. 19.5, the algorithm works like this:

- Step 1. Determine any basic feasible solution.
- Step 2. Examine U to see if it can be increased by moving along any boundary away from the vertex chosen in step 1. If not, U is the maximum. If U can be increased proceed to step 3.
- Step 3. If U can be increased by moving along one or more boundaries, choose one of these and determine the new basic feasible solution corresponding to the next vertex reached. Take this as a new basic feasible solution. Return to step 2 and continue the cycle until a maximum is reached.

Since each slack variable appears in one constraint equation only, if we set $x = y = 0$, equations (A.2)–(A.5) are satisfied by setting the slack variables equal to the numerical values on the right-hand side of the corresponding equation. This gives a basic feasible solution for step 1. We proceed to step 2. We express the non-zero variables, usually called the basic variables, and also U, in terms of the variables that are zero in the basic solution. Thus

$$s_1 = 140 - x \tag{A.6}$$

$$s_2 = 120 - y \tag{A.7}$$

$$s_3 = 640 - 3x - 4y \tag{A.8}$$

$$s_4 = 180 - x - y \tag{A.9}$$

$$U = 10x + 12y - 1200. \tag{A.10}$$

For the basic feasible solution $x = y = 0$, clearly $U = -1200$. Because x and y are never negative, we can increase U by increasing x or y since each has a positive coefficient in (A.10). So we move to step 3.

We increase y. From (A.7), however, we see that we can only increase y to 120. For any greater increase s_2 would become negative. This is not permitted. From (A.6)–(A.9) we find this gives a new basic feasible solution with $y = 120, x = 0, s_1 = 140, s_2 = 0, s_3 = 180, s_4 = 60$.

We now return to step 2 to see whether we can increase U by increasing x. To check, we express the constraints and U in terms of the non-basic variables. These are now x and s_2. Using (A.6)–(A.10) gives

$$s_1 = 140 - x \tag{A.11}$$

$$y = 120 - s_2 \tag{A.12}$$

$$s_3 = 640 - 3x - 4(120 - s_2) = 160 - 3x + 4s_2 \tag{A.13}$$

$$s_4 = 180 - x - (120 - s_2) = 60 - x + s_2 \tag{A.14}$$
$$U = 10x + 12(120 - s_2) - 1200 = 10x - 12s_2 + 240. \tag{A.15}$$

Since the coefficient of x in U is positive, we can increase U by increasing x. An inspection of (A.11)–(A.14) shows the constraints still hold if we increase x by a maximum of $160/3 = 53\frac{1}{3}$. If we increased it any further (A.13) would yield a negative s_3, This is not permissible.

You should check that the return to step 2 with this increase in x, leads to a basic feasible solution with $x = 160/3, s_1 = 260/3, y = 120,\ s_2 = 0,\ s_3 = 0,$ $s_4 = 20/3$. Whence proceeding as above, we obtain

$$y = 120 - s_2 \tag{A.16}$$
$$x = \frac{160}{3} + \frac{4}{3}s_2 - \frac{1}{3}s_3 \tag{A.17}$$
$$s_1 = \frac{260}{3} - \frac{4}{3}s_2 + \frac{1}{3}s_3 \tag{A.18}$$
$$s_4 = \frac{20}{3} - \frac{1}{3}s_2 + \frac{1}{3}s_3 \tag{A.19}$$
$$U = \frac{2320}{3} + \frac{4}{3}s_2 - \frac{10}{3}s_3. \tag{A.20}$$

Since the coefficient of s_2 in U is positive, we may increase U by increasing s_2. We can increase it by a maximum of 20. Otherwise s_4 would become negative in (A.19). Returning to step 2 with this new value we find $y = 100$, $x = 80, s_1 = 60, s_2 = 20, s_3 = 0, s_4 = 0$. Check for yourself that we now have

$$s_2 = 20 - 3s_4 + s_3 \tag{A.21}$$
$$y = 100 + 3s_4 - s_3 \tag{A.22}$$
$$x = 80 - 4s_4 + s_3 \tag{A.23}$$
$$s_1 = 60 + 4s_4 - s_3 \tag{A.24}$$
$$U = 800 - 4s_4 - 2s_3. \tag{A.25}$$

Slack variables must always be nonnegative, so it is clear from (A.25) that we cannot further increase the profit. The implications are discussed in Sect. 19.5.

A.6 Confirming a Requirement

In Sect. 16.5, Table 16.1 was a partial table to check that a group operation defined there was associative. Table A.1 is the complete table required to confirm the associative property.

Table A.1. A check that an arbitrary operation $\circ$ on a set $\mathrm{S} = \{b, p, a\}$ is sssociative

$(bb)b = bb = b$ and $b(bb) = bb = b$
$(bb)p = bp = b$ and $b(bp) = bb = b$
$(bb)a = ba = a$ and $b(ba) = ba = a$
$(bp)b = bb = b$ and $b(pb) = bp = b$
$(bp)p = bp = b$ and $b(pp) = bp = b$
$(bp)a = ba = a$ and $b(pa) = ba = a$
$(ba)b = ab = a$ and $b(ab) = ba = a$
$(ba)p = ap = a$ and $b(ap) = ba = a$
$(ba)a = aa = a$ and $b(aa) = ba = a$
$(pb)b = pp = p$ and $p(bb) = pb = p$
$(pb)p = pp = p$ and $p(bp) = pb = p$
$(pb)a = pa = a$ and $p(ba) = pa = a$
$(pp)b = pb = p$ and $p(pb) = pp = p$
$(pp)p = pp = p$ and $p(pp) = pp = p$
$(pp)a = pa = a$ and $p(pa) = pa = a$
$(pa)b = pb = p$ and $p(pb) = pp = p$
$(pa)p = ap = a$ and $p(ap) = pa = a$
$(pa)a = aa = a$ and $p(aa) = pa = a$
$(ab)b = ab = a$ and $a(bb) = ab = a$
$(ab)p = ap = a$ and $a(bp) = ab = a$
$(ab)a = aa = a$ and $a(ba) = aa = a$
$(ap)b = ab = a$ and $a(pb) = ap = a$
$(ap)p = ap = a$ and $a(pp) = ap = a$
$(ap)a = aa = a$ and $a(pa) = aa = a$
$(aa)b = ab = a$ and $a(ab) = aa = a$
$(aa)p = ap = a$ and $a(ap) = aa = a$
$(aa)a = aa = a$ and $a(aa) = aa = a$

References

[1] Acheson, D. (2002) *1089 and All That.* Oxford: Oxford University Press.

[2] Aczel, A.D. (1996) *Fermat's Last Theorem – Unlocking the Secret of an Ancient Mathematical Problem.* London: Penguin Books.

[3] Appel, K. and Haken, W. (1986) The four-color proof suffices. *The Mathematical Intelligencer*, **8**, 10–20.

[4] Bailey, D.H., Borwein, J.M., Borwein, P.B. and Plouffe, S (1997). The quest for π. *The Mathematical Intelligencer*, **19**, 50–56.

[5] Bailey, D.H. and Borwein, J.M. (2000) Experimental mathematics: recent developments and future outlook. In *Mathematics Unlimited – 2001 and Beyond.* Ed. B. Engquist and W. Schmid. Berlin: Springer-Verlag.

[6] Banks, R. B. (1998) *Towing Icebergs, Falling Dominoes and Other Adventures in Applied Mathematics.* Princeton: Princeton University Press.

[7] Banks, R. B. (1999) *Slicing Pizzas, Racing Turtles and Further Adventures in Applied Mathematics.* Princeton: Princeton University Press

[8] Beck, A., Bleicher, M. N. and Crowe D.W. (2000) *Excursions into Mathematics. The Millennium Edition.* Natick: A. K. Peters, Ltd..

[9] Bell, E.T. (1937, 1986) *Men of Mathematics.* New York: Simon & Schuster.

[10] Clapham, C. (ed) (1996) *The Concise Oxford Dictionary of Mathematics.* 2nd. ed. Oxford: Oxford University Press.

[11] Cohen, J. and Stewart I. (1995) *The Collapse of Chaos - Discovering Simplicity in a Complex World.* London: Penguin Books.

[12] Courant, R., Robbins, H. (1941) *What is Mathematics?* London: Oxford University Press.

[13] Courant, R., Robbins, H. and Stewart, I. (1996) *What is Mathematics?* 2nd.edn. Oxford: Oxford University Press.

[14] Devlin, K. (1988). *Mathematics: The New Golden Age.* London: Penguin Books.

[15] Devlin, K. (1994) *All the Math That's Fit to Print.* Washington: Mathematical Association of America

[16] Devlin, K. (1999) *Life by the Numbers.* New York: Wiley.

[17] Devlin, K. (2000) *The Maths Gene.* London: Weidenfield & Nicolson.

[18] Engquist, B. and Schmid, W. (eds) (2000) *Mathematics Unlimited – 2001 and Beyond.* Berlin: Springer-Verlag

[19] Everitt, B.S. (1999) *Chance Rules: An Informal Guide to Probability, Risk and Statistics.* New York: Springer-Verlag.

[20] Farebrother, R.W, (1999) *Fitting Linear Relationships. A History of the Calculus of Observations, 1750–1900.* New York: Springer Verlag.

[21] Feigenbaum, M. (1979) The universal metric property of nonlinear transformations. *Journal of Statistical Physics*, **21**, 669–706.

[22] Feigenbaum, M. (1981) Universal behaviour in nonlinear systems. *Los Alamos Science*, **1**, 4–27.

[23] Gardner, M. (1981) *Mathematical Circus.* London: Penguin Books. (Most of the material was originally published in Scientific American from 1968 onward.)

[24] Gardner, M. (2001) *A Gardner's Workout. Training the Mind and Entertaining the Spirit.* Natick: A K Peters, Ltd.

[25] Gleick, J. (1988) *Chaos - Making a New Science.* London: Penguin Books.

[26] Gomory, R.E. (1958). Outline of an algorithm for integer solutions to linear programs. *Bulletin of the American Mathematical Society*, **64**, 275–278.

[27] Good, I.J. (1972) What is the most amazing approximate integer in the universe? *Pi Mu Epsilon Journal*, **5**, 314–315.

[28] Gowers, W.T. (2002). *Mathematics: A Very Short Introduction.* Oxford: Oxford University Press.

[29] Henge, N. and Riedwyl H. (1998) *How to Win More: Strategies for Increasing a Lottery Win.* Natick: A K Peters, Ltd.

[30] Higham, C.F.W., Kijngam, A. and Manly, B.F.J. (1980). An Analysis of Prehistoric Canid Remains from Thailand. *Journal of Archaeological Science*, **7**, 149–165.

[31] Koblitz, N. (2000) Cryptography. In *Mathematics Unlimited – 2001 and Beyond.* Ed. B.Engquist and W. Schmid. Berlin: Springer-Verlag.

[32] Lang, R.J. (2003) *Origami Design Secrets: Mathematical Methods for an Ancient Art.* Natick: A.K. Peters, Ltd.

[33] Liebeck, M.W. (2000) *A Concise Introduction to Pure Mathematics.* Boca Raton: Chapman & Hall/CRC Press.

[34] Lorenz, E. (1963) Deterministic nonperiodic flow. *Journal of the Atmospheric Sciences*, **20**, 130–141.

[35] Mandelbrot, B (1982) *The Fractal Geometry of Nature.* San Francisco: W. H. Freeman.

[36] May, R. (1976) Simple mathematical models with very complicated dynamics. *Nature*, **261**, 459–467.

[37] Nelson, R. D. (ed) (2008) *The Penguin Dictionary of Mathematics.* 4th. ed. London: Penguin Books.

[38] Polya, G. (2004) *How to Solve It. A New Aspect of Mathematical Method.* Princeton: Princeton University Press.

[39] Ruelle, D, (1993) *Chance and Chaos* Princeton: Princeton University Press.

[40] Simpson, E.H. (1951) The interpretation of interaction in contingency tables. *Journal of the Royal Statistical Society, B.*, **13**, 238–241.

[41] Singh, S. (2002) *Fermat's Last Theorem.* London: Fourth Estate.

[42] Sobel, D. (1995) *Longitude.* London: Fourth Estate.

[43] Stewart, I. (1989) *Does God Play Dice: The New Mathematics of Chaos.* Oxford: Basil Blackwell.

[44] Stewart, I. (1990) *Does God Play Dice: The New Mathematics of Chaos.* London: Penguin Books.

[45] Stewart, I. (1996) *From Here to Infinity.* Oxford: Oxford University Press.

[46] Wilson, R.A. (2002) *Graphs,Colouring and the Four-Colour Theorem.* Oxford: Oxford University Press.

Index

www.ingramcontent.com/pod-product-compliance
Ingram Content Group UK Ltd.
Pitfield, Milton Keynes, MK11 3LW, UK
UKHW022322190726
13856UKWH00001B/162

9 781409 25670